Whose Backyard, Whose Risk

Whose Backyard, Whose Risk
Fear and Fairness in Toxic and Nuclear
Waste Siting

Michael B. Gerrard

The MIT Press
Cambridge, Massachusetts
London, England

This book was set in Sabon by DEKR Corporation and was printed on recycled paper and bound in the United States of America.

Library of Congress Cataloging-in-Publication Data

Gerrard, Michael.
  Whose backyard, whose risk: fear and fairness in toxic and nuclear
  waste siting/Michael B. Gerrard.
    p.   cm.
  Includes bibliographical references and index.
  ISBN 0-262-07160-6
    1. Hazardous waste sites—Location—Government policy—United States.
    2. Hazardous waste sites—Government policy—United States.
  I. Title.
  HD4483.G47 1994
  363.72′87′0973—dc20   94-18925
                                                              CIP

Table of Contents

Preface

I have spent the past sixteen years as a practicing environmental lawyer with the New York City law firm of Berle, Kass & Case, and as a writer on and (more recently) teacher of environmental law. Much of my practice has involved representing municipalities and community organizations in defending against efforts by higher levels of government—counties, states, and the federal government—and by private developers to build unwanted facilities in their midst: landfills, incinerators, pipelines, highways, airports. I have also represented corporations seeking to build facilities, landowners faced with the responsibility of cleaning up contamination they did not create, and factories trying to cope with the maze of environmental regulations.

One question has come up scores of times. It is some variation of this: How are we (they) supposed to get rid of this (garbage, stuff, glop, gunk)? The question of how to dispose of society's waste engenders a tremendous amount of political conflict, emotional anguish, and wasted transaction costs on all sides. Though the modern era in environmental law is almost twenty-five years old, we have nothing approaching a coherent national system of waste disposal. Instead, there are at least two dozen completely different kinds of waste streams, each the subject of its own scheme of regulation (or chaos). Waste is being trucked from every state to every other state, or sitting in often leaky containers until some other state will take it.

The objective of this book is to try to impose some conceptual order on this chaos and to suggest a coherent scheme for disposing of these many kinds of hazardous and radioactive waste. (Ordinary municipal solid waste is not my prime topic, but it also comes up repeatedly.) The introduction describes in more detail the plan and theses of the book.

Though I have attempted to find and read just about the whole body of literature on the siting of hazardous and radioactive waste disposal facilities (requiring a new pair of glasses in the process), I have also drawn heavily on my own experience. I should disclose here that I have litigated on behalf of clients against several companies named in the book—Waste Management, Inc. and its subsidiaries Chemical Waste Management, Inc., CWM Chemical Services, Inc., Chem Nuclear Services, and Wheelabrator Technologies, Inc.; Occidental Chemical Corp.; and Browning Ferris Industries and its subsidiary CECOS International, Inc. I have also represented several parties whose names come up in the book: Niagara and Cortland counties, New York; the city of Niagara Falls, New York; the towns of Porter and Lewiston, New York; the Natural Resources Defense Council; and the Environmental Defense Fund.

I would like to acknowledge gratefully the insightful comments of the following people on earlier drafts: Vicki Been, Albert K. Butzel, Peter Fox-Penner, Steven B. Gerrard, Rachel D. Godsil, Alice Kaswan, Charles Komanoff, Michael C. Naughton, R. Nils Olsen, Jr., Marvin Resnikoff, Richard Revesz, Tina Santopadre, Ronald Slye, Joseph P. Tomain, Marcia Widder, Philip Weinberg, all of my colleagues at Berle, Kass & Case, and four anonymous reviewers. These people do not necessarily agree with the book's recommendations; needless to say, neither do my firm's clients. William Schaeffer, Carter Strickland, Jacob Levich, and Anthone Damianakis provided invaluable research and cite-checking assistance, and Lance Liebman graciously made available the resources for some of this assistance. My secretary Nancy Provenzano did a wonderful job of processing the manuscript. Madeline Sunley, Melissa Vaughn, and Ann Sochi of The MIT Press provided great encouragement and assistance.

An earlier, considerably shorter version of this book appeared as an article in Volume 68 of the *Tulane Law Review* under the title "Fear and Loathing in the Siting of Hazardous and Radioactive Waste Disposal Facilities." Those portions that appear again here are reprinted with permission of the Tulane Law Review Association.

Most important, I would like to acknowledge the unflagging love, support, and patience of my family, Barbara, David, and William. My greatest hope in writing this book is that it will help point a way for my own and my generation's profligacy not to impose a burden on my sons and their offspring.

I

Setting the Stage

1

Introduction

Few laws have failed so completely as the federal and state statutes designed to create new facilities for the disposal of hazardous and radioactive waste. Despite scores of siting attempts[1] and the expenditure of several billion dollars since the mid-1970s,[2] operating today on new sites in the United States there is only one small radioactive waste disposal facility; only one hazardous waste landfill (in the aptly named Last Chance, Colorado);[3] and a small handful of hazardous waste treatment and incineration units.[4]

In 1981 a leading member of Congress, relying on data from the U.S. Environmental Protection Agency (EPA), predicted that by 1985 the country would need between 50 and 125 new off-site hazardous waste disposal facilities.[5] Numerous commentators said that many more facilities were desperately needed if the nation is to avert an environmental crisis.[6] These facilities have not been built, yet there is no such crisis. The shortage of disposal facilities turns out to be far less severe and more localized than is usually portrayed. Its principal adverse environmental impact is that old, substandard, leaking disposal units stay open because there are no replacements. But there is a genuine political crisis—hundreds of battles around the country, some dethroning elected officials,[7] some verging on violence,[8] over the efforts of the federal and state governments to force hated facilities upon terrified communities.

In this book, I propose an approach to resolve the impasse in siting disposal facilities for hazardous wastes (HWs) and radioactive wastes (RWs). In doing so, I argue that the siting laws are based on a fundamental conceptual error and several factual mistakes and policy blunders.

The conceptual error stems from the way the question is posed. The task is usually framed as finding the best locations for new facilities to dispose of HW/RW.[9] But the fulfillment of this task through government power over land use in effect subsidizes the disposal (and hence encourages the creation) of HW/RW, aggressively ignoring negative externalities and distorting the economics of production. Instead, the task should be posed as how to find the system of HW/RW management that maximizes social welfare, taking full account of social and environmental costs, while achieving fairness. Viewing the problem this way leads to market solutions that reduce the generation of HW/RW, rather than encouraging waste production that results in disposal nightmares.

The most important factual mistakes are the widespread (and erroneous) assumptions that:

• a shortage of disposal facilities increases illegal dumping (when, in fact, illegal dumping has little to do with disposal capacity and can be addressed through targeted enforcement);
• states will cooperate with siting efforts, while willing local communities cannot be found (although the opposite is more often true); and
• monetary compensation can gain acceptance of HW/RW facilities in places that do not want the facilities (although the evidence is that this seldom works).

Three policy blunders have also helped doom siting efforts. The first is separate regulation of all the different kinds of hazardous and radioactive waste streams, thereby dividing the states into victims (those with disposal facilities) and free riders (those without) for each waste stream and dooming regional cooperation. The second is insistence on technically perfect sites; since there aren't any, old facilities remain open in some of the *worst* possible locations. The third is allowing higher levels of government to preempt the authority of lower levels in imposing a site on an unwilling community, a method that not only always fails but is wildly counterproductive.

The siting problem has numerous dimensions—scientific, economic, political, sociological, psychological, legal. In many ways the legal problems are the simplest because if the politics are right, Congress and the state legislatures can change the laws. A political consensus is a prerequisite to a legal solution, and a political consensus cannot be achieved unless the fundamental economic, sociological, and psychological concerns are addressed. Nothing in my proposal requires any new scientific breakthroughs, though advances in waste minimization and other technologies will certainly help enormously.

This book is divided into three parts. Part I sets the stage. Within it, chapter 2 arrays the different kinds of nonradioactive wastes and describes how each is generated and disposed of. Chapter 3 does the same for radioactive wastes. Chapter 4 explains how siting decisions have been made, both historically and today, under federal and state laws, and how old substandard facilities remain open under grandfather clauses. The many permit rules are presented in a typology that reveals the divergence between what underlies most siting disputes and what the environmental laws actually protect.

Chapter 5 considers the effects of the current mechanisms for making siting decisions. It analyzes whether new HW/RW facilities are really "needed"; shows the irrelevance of illegal dumping to the question of need; assesses the fairness of the current system for various regions, economic classes, races, and generations; and erects a framework that strives to explain why certain facilities are so vehemently opposed and why others are accepted. Some of the hidden economics and psychology of the siting process are explored here. Although facility opposition is often trivialized with acronyms like NIMBY ("not in my backyard"),[10] LULU ("locally undesirable land use"),[11] or BANANA ("build absolutely nothing anywhere near anything"),[12] this part will show that even new, "state-of-the-art" facilities pose real environmental hazards.

The prior proposals for addressing the siting dilemma are addressed in Part II. These proposals are divided into those based on efforts to achieve the consent of local communities to new sites (chapter 6); those that rely on governmental coercion (chapter 7); and those that seek to avoid the problem by minimizing waste production, exporting the waste to other jurisdictions, or reducing the problem to legal or linguistic nonexistence (chapter 8). This part also examines the frequent failures and rare successes of experiments with each of these proposals.

Finally, Part III proposes a new alternative, drawing from the many lessons of past siting attempts. It is based upon local control, state responsibility, and national allocation. Under this scheme, all the different kinds of hazardous and radioactive waste streams would be considered together, thereby eliminating much unnecessary regional conflict. The federal government would determine overall national disposal needs and allocate the burdens among the fifty states. The states would then ask communities to volunteer to host facilities—a method that, experience surprisingly shows, can attract numerous offers. The federal government would also make available the multitude of highly (and possibly permanently) contaminated sites that were used for military purposes and nuclear weapons production but have been rendered obsolete by the end of the Cold War. Chapter 9 describes the proposal; chapter 10 measures this proposed alternative against the criteria used in chapter 5 to assess existing siting mechanisms; and chapter 11 discusses certain implementation details. Finally, the Epilogue uses a hypothetical example to show how a siting decision might be made in a particular case.

My proposal is designed to meet the fundamental economic, sociological, and psychological problems of facility siting in a way that leads to a political solution and, ultimately, to a legal framework for implementation. The division of federal, state, and local responsibilities is designed to fit within important values in the American political culture. The reader will have to decide whether the design succeeds.

2

The Origins and Disposal of Nonradioactive Wastes

This chapter and the next describe the different types of waste streams regulated under the HW/RW laws. Chapter 2 focuses on nonradioactive wastes; chapter 3 concerns radioactive wastes. For each kind of waste, origins, quantities, regulation, and disposal are discussed. In an effort to devise a comprehensive approach, I have included here several types of wastes that are not conventionally discussed in this context but that form an important part of the overall disposal picture. As will become clear, domestic and international politics and economics are as important as chemistry and physics in defining which substances are closely regulated.

Approximately 80,000 chemicals are in commercial use today, and about 1,000 more are introduced each year.[1] Only a few hundred chemicals and specified mixtures of chemicals are regulated under the hazardous waste laws. Of the remainder, only a small fraction have been thoroughly tested for toxicity, and there is no doubt that many chemicals outside the regulatory net pose serious hazards.[2] Nonetheless, these chemicals of unknown toxicity are not now part of the HW disposal "problem" because they are not treated as HW and do not consume the same scarce disposal capacity.[3]

RCRA-Regulated Hazardous Wastes

The transportation, storage, and disposal of hazardous waste are regulated primarily by the Resource Conservation and Recovery Act (RCRA),[4] though other federal statutes govern particular disposal methods.[5] RCRA defines "hazardous waste" with general references to threats to health or the environment,[6] but the statute also establishes an intricate system for the EPA to list or characterize the chemicals or types of chemicals that fit within this definition.[7] Industries routinely spend millions of dollars lobbying and litigating to have their particular waste streams excluded from the RCRA definitions because the costs of compliance with RCRA's HW management system are very steep.

The chemical industry generates 88 percent of all the HW in the United States.[8] Primary and fabricated metals and petroleum refining rank next.[9] A small number of sites account for most of the HW. Just 1 percent of all generators create 97 percent of the HW, and three plants—operated by DuPont, Dow Chemical, and Eastman Kodak—generate 57 percent of all HW nationwide.[10]

Despite considerable uncertainty,[11] most estimates of the amount of RCRA hazardous waste generated by the civilian sector are around 250 million tons per year.[12] (This comes to just under one ton for every man, woman, and child in the country, four times the per-capita generation of the country with the second-highest rate, Germany.)[13] This figure greatly overstates the burden on disposal facilities because about 96 percent is disposed of at the point of generation;[14] approximately 95 percent of this 96 percent is water mixed with wastes.[15] Most of it is treated (often by leaving it, with various treatment chemicals, for long periods in large artificial ponds called surface impoundments) and then released into rivers, lakes, and oceans or is allowed to seep underground.[16] Some of this 96 percent is injected underground, and some is burned in about 150 on-site "captive" hazardous waste incinerators.[17] The remaining approximately 4 percent is mostly solid and is sent to a variety of off-site hazardous waste landfills, incinerators, and treatment facilities, all of them privately owned.[18] Landfills receive 26.4 percent of the HW disposed off-site, and incinerators get 4.5 percent.[19]

The commercial, off-site hazardous waste facilities now operating in the United States consist of 103 chemical treatment plants, 95 solvent recovery plants, 60 physical treatment plants, 30 kilns that burn hazardous waste as fuel, 24 landfills, 20 incinerators, and eight deep-injection wells.[20] (Figure 2.1 shows the locations of these facilities.) The 24 landfills and 20 incinerators, and efforts to increase their number, are at the vortex of the current controversy over hazardous waste facility siting.

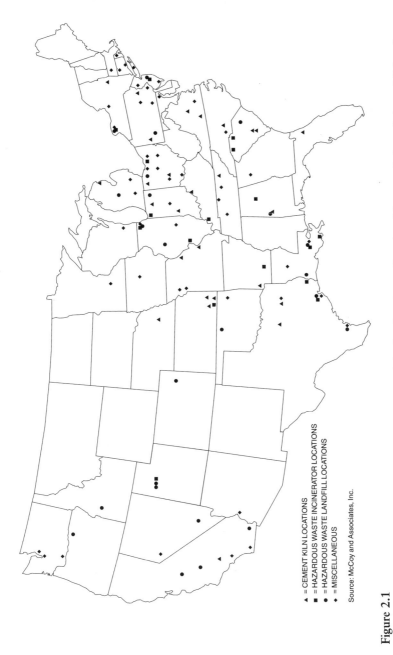

Source: McCoy and Associates, Inc.

Figure 2.1
Commercial hazardous waste treatment and disposal facility locations. (Reprinted by permission of the publisher from *The Hazardous Waste Consultant*, March/April 1993, p. 4.48. Copyright 1993 by Elsevier Science Publishing Co., Inc.)

▲ = CEMENT KILN LOCATIONS
■ = HAZARDOUS WASTE INCINERATOR LOCATIONS
● = HAZARDOUS WASTE LANDFILL LOCATIONS
◆ = MISCELLANEOUS

Despite some regional shortfalls, and deficits in some specialized forms of treatment, there is ample excess capacity at most of the other types of facilities.[21] In fact, during the 1980s at least four hazardous waste facilities received final RCRA permits but failed to open because of insufficient markets.[22] Two states, Kentucky and Georgia, abandoned efforts to site hazardous waste disposal facilities when they concluded there was too little demand.

New laws that discourage the landfilling of untreated hazardous wastes[23] increase the demand for incineration capacity.[24] Other factors (such as waste reduction and on-site treatment) reduce the demand, and the net effect on incineration demand is uncertain.[25] Several new commercial incinerators are now winding their way through the permit process.[26] Incineration capacity has been growing faster than demand,[27] and aggregate projections show no significant capacity shortfalls.[28] In 1993 a trade press report said "[t]oday, there is an overcapacity of offsite hazardous waste treatment facilities and services, especially incineration."[29] In fact, the nation's commercial HW incinerators were running at only about half their capacity.[30] Acknowledging this excess capacity, the EPA in May 1993 announced an eighteen-month "capacity freeze" on HW incinerators while it launches a "national dialogue" on HW management.[31] In the subsequent months several plans for HW incinerators were canceled due to lack of demand.[32]

The existing hazardous waste landfills include several very large ones. The biggest is in Emelle, Alabama.[33] It is estimated by its owner, Chemical Waste Management, Inc. (CWM), to have capacity for another one hundred years of operation.[34] CWM also has a landfill and treatment complex in Model City, north of Niagara Falls, New York, with ample unused land available for further landfill expansion.[35]

In 1990 a leading trade journal concluded that there is no national shortage in hazardous waste landfill capacity;[36] since then, landfill demand has declined sharply.[37] Considering all types of commercial hazardous waste disposal capacity, a vice president of the largest company in the field, Waste Management, Inc. (CWM's parent), wrote in 1991 that "[t]hough some sites will shut down while others add technologies, we are close to a capacity equilibrium unless local political restrictions limit that capacity."[38] In 1993 the Emelle landfill was operating at less than half the rate it did before 1990,[39] and one commentator wrote that "[w]e are currently awash in commercial land disposal capacity."[40]

In 1993 a leading financial analyst, Hugh F. Holman, expressed a gloomy view of the industry's prospects: "The business of commercial hazardous waste management is, in many ways, inherently self-limiting: it is a rare customer indeed that wants to do more business with you."[41] Holman said that trends for more on-site remediation, for treatment rather than disposal, and for waste minimization will continue to erode the market for off-site commercial disposal.[42] A few months later Standard & Poor's Corp. put Chemical Waste Management, Inc. on its "Credit Watch" list because of declining waste volumes, due largely to waste minimization and recycling.[43] Several of the industry's largest companies took major write-offs.[44]

In sum, most legal commentators and some politicians have decried a critical national shortage of hazardous waste disposal capacity and have used this as a basis for wanting to impose facilities on unwilling communities, but the waste management industry, the trade press that covers it, and the financial analysts who study it paint a very different picture. The situation closely resembles that faced by the electric utility industry twenty years earlier; after a long period of calls for government intervention to help it build new power plants, the plant shortage all but evaporated in the 1970s in the wake of surprisingly effective energy conservation measures. Pollution prevention in the 1990s has done much the same thing to the hazardous waste disposal industry.

Civilian Inactive Hazardous Waste Disposal Sites

The explosion of public consciousness about hazardous waste can largely be dated to August 2, 1978, when New York State declared the Love Canal neighborhood of the city of Niagara Falls to be a public health emergency. In the 1940s and 1950s, the Hooker Chemicals & Plastics Corp. had dumped 21,000 tons of liquid hazardous waste into a large ditch. The ditch was covered over, and later a residential neighborhood and a school were built on top. When residents began complaining of illness, officials performed studies that ultimately led to the evacuation of the entire neighborhood.[45]

The EPA, which had for some time been advocating a new statute to control inactive hazardous waste disposal sites, used the massive publicity generated by the Love Canal incident to push through Congress[46] the Comprehensive Environmental Response, Compensation and Liability Act (CERCLA, also known as the Superfund law).[47] CERCLA requires the use of a hazard ranking system to establish the National Priorities List (NPL), a list of the hazardous waste sites[48] posing the greatest dangers.[49] These sites must undergo an elaborate process of investigation and remediation.[50]

The NPL now contains about 1,200 sites nationwide.[51] They are concentrated in the industrial states of the east and west coasts and the Great Lakes region, as shown in figure 2.2. The Congressional Office of Technology Assessment has estimated that the list could readily exceed 10,000.[52] The average cost of investigating and cleaning up these sites is roughly $30 million each.[53] Cleanup involves a broad range of possible actions, ranging from excavating the contaminated material and hauling it to a RCRA hazardous waste landfill or incinerator, to cleaning it up in place using such esoteric techniques as soil gas extraction or bioremediation. Although the practice has now been banned, for several years waste was simply dug up, hauled to another leaky landfill, and dumped without further treatment; several of these receiving landfills later became CERCLA sites themselves.[54] In all, about 26 million cubic yards of soil, sludge, and sediment need to be cleaned up from NPL sites.[55] Many states have their own "mini-Superfund" lists, on which more than 19,000 sites have been placed.[56]

Some of the cleanup actions are extremely controversial. For example, at an abandoned pesticide plant in Jacksonville, Arkansas, tens of thousands of barrels filled with DDT, Agent Orange, 2,4,5-T, and dioxin-contaminated waste had been left to decay since the end of the Vietnam War. In 1992 the EPA, with the support of Gov. Bill Clinton, approved the use of a mobile hazardous waste incinerator to be temporarily erected on the site to destroy the wastes, and the burning began. Litigation and raucous public protests ensued.[57] Though a federal district court enjoined the project on the grounds that the EPA had not demonstrated the incinerator was destroying dioxins as effectively as the EPA's regulations

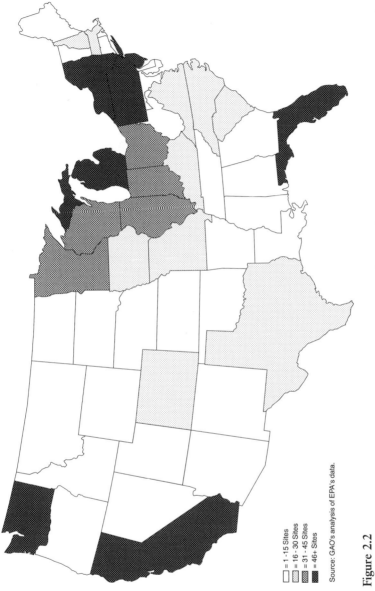

= 1 - 15 Sites

= 16 - 30 Sites

= 31 - 45 Sites

= 46+ Sites

Source: GAO's analysis of EPA's data.

Figure 2.2
Number of sites on the CERCLA National Priorities List, as of September 30, 1992. (Source: U.S. Environmental Protection Agency, Office of Emergency and Remedial Response, 1992.)

required, the U.S. Court of Appeals for the Eighth Circuit reversed and allowed the incineration to proceed.[58]

Local opposition to a local cleanup has become a recurring scenario. In New Bedford, Massachusetts, the EPA decided to put a mobile incinerator on land previously used as a soccer field, to burn contaminated sediments from the Acushnet River. Despite the adamant resistance of the city—and a referendum in which 87 percent of the voters opposed the plan—the EPA insisted on going forward with the incineration, and threatened the city with daily fines of $25,000 if it continued to block the work.[59] In Metamora, Michigan, citizens succeeded in fighting the proposed on-site incineration of thousands of barrels of hazardous waste dug from an industrial landfill.[60] In Niagara Falls, New York, the EPA and the state called for the on-site incineration of wastes dug from Love Canal and other local hazardous waste sites owned by Occidental Chemical Corp.; the mayor opposed the plan, and Occidental decided to seek an off-site commercial incinerator. In all three of these cities, the citizens had called for cleanup of the contamination in their midst but became concerned that digging up and disposing of the material would release more of it into the environment—in other words, that the cure would be worse than the disease.

RCRA Corrective Action Sites

All hazardous waste facilities with permits under RCRA must undertake "corrective action" to clean up any contamination on their sites, including any that has spread beyond the facility boundary.[61] This program, still in its infancy, is very similar to CERCLA's National Priorities List program but addresses contamination at operating facilities rather than at inactive sites. It may ultimately involve somewhere between 1,500 and 3,500 sites.[62] Under some estimates, the cost of the RCRA corrective action program will grow to dwarf that of the NPL program (see table 3.3).

Comprehensive figures are hard to come by, but it appears that, in the early 1990s, on the order of 80 percent of the waste received by off-site commercial HW landfills and incinerators was "livestream"—that is, from ongoing industrial production—with the rest coming from CERCLA, RCRA, and other remedial projects.[63]

Military Facilities

The U.S. military generates approximately 700,000 tons of hazardous wastes annually, including such substances as paint thinner, spilled solvents, hydraulic fuel, aviation fuel, fuel tank and sewage sludges, and herbicides.[64] The Department of Defense traditionally was lax in the use and disposal of such wastes, and defense contractors had little incentive to be careful because they could charge the cost of cleaning up their own messes back to the military.[65] President Carter ordered federal facilities to comply with federal pollution control standards in 1978,[66] but not until 1992 was a statute enacted that allowed the EPA and the states to penalize federal agencies for violating federal or state hazardous waste laws.[67] With the end of the Cold War and the retrenchment of the military, the cleanup of former defense facilities has become a major growth sector.[68]

The Department of Defense has two major cleanup programs for domestic facilities. The Installation Restoration Program covers facilities still in use. It has targeted 1,877 installations for cleanup, with an estimated cost of $24.5 billion.[69] The Formerly Used Defense Sites Program covers 6,786 former facilities such as arsenals, ammunition plants, equipment manufacturing plants, depots, bases, proving grounds, shipyards, forts, and camps; more than half of these will likely require significant remediation, at unknown cost.[70]

Some of this contamination has led to a game of hot potato among federal agencies. The Department of Defense has cleared 20,000 acres on former military sites so they can be returned to the Bureau of Land Management, but the BLM has refused to accept much of this land because unexploded ordnance might remain under the ground. In one large parcel the BLM did accept from the military—15,000 acres at Davis Range in Alaska, a popular hiking area for residents of Anchorage—the BLM must sweep the trails every spring to remove unexploded ordnance that frost action may have brought to the surface during the winter.[71]

Housekeeping was no better at the U.S. military's nearly 400 overseas bases, but there is no integrated program to address their environmental impacts.[72] After the U.S. General Accounting Office surveyed some of these bases in 1986, the Pentagon classified the findings.[73]

Various civilian federal agencies have also identified about 350 sites requiring remediation, but that number seems destined to climb considerably.[74] These include research laboratories, prisons, power plants, properties acquired through foreclosure, and many other facilities.

Chemical Weapons[75]

Though it claims never to have used them in war, the United States has been manufacturing and stockpiling chemical weapons for decades.[76] President Nixon halted U.S. production in 1969 in the wake of a nerve gas leak in Utah that killed 4,300 sheep. The program resumed under President Reagan after then–Vice President Bush cast the tiebreaking vote in November 1985 to allow production of "binary" weapons, which consist of two relatively harmless chemicals that are mixed only upon detonation. Congress insisted, however, that all but 10 percent of the old "unitary" chemicals be destroyed by 1994,[77] a deadline that has been pushed back several times, most recently to 2004.[78] The army decided that shipping all the old weapons—some 27,000 tons in all—to one or two central incinerators was too risky, and several governors opposed transport of the weapons through their states, so the army decided to incinerate the weapons at the places where they are stored. A prototype incinerator at the Tooele Army Depot in Utah, where 42 percent of the stockpile is kept, leaked nerve gas into the atmosphere in 1987; a new incinerator is now under construction.

In 1986 President Reagan promised West German Chancellor Helmut Kohl that the United States would remove some 100,000 chemical-filled artillery shells that had been stored in Germany since the end of World War II. Little happened, but in 1989 Chancellor Kohl secured the promise of President Bush to remove the weapons by 1990. The United States decided to ship them to Johnston Atoll, a U.S. trust territory some 717 nautical miles southwest of Hawaii, to which the United States had removed its chemical weapons from Okinawa in 1971 after complaints from Japan. The shipment from Germany led to protests from the Hawaii state legislature and throughout the South Pacific, but incineration at Johnston Atoll proceeded.[79]

In 1992, in the face of domestic protests, Congress cut off funding for incineration at three U.S. depots.[80] Much of the local opposition was based on the fear that once the chemical weapons had been destroyed, the government would use the incinerators to destroy other defense or commercial waste.[81] Congress also directed the army to explore technological alternatives to incineration.[82] The overall program to destroy chemical weapons is estimated to cost nearly $8 billion.[83]

At an additional eighty-two locations around the country (most but not all of which are on military installations), old chemical warfare materiel was buried or otherwise left behind before 1970. The army is now deciding what to do about these sites, but it estimates the cleanup cost will be roughly $18 billion.[84]

An international agreement, the Chemical Weapons Convention, has been negotiated that would ban the production, use, and stockpiling of chemical weapons worldwide.[85] The former Soviet republics face formidable problems in destroying their own stockpiles.[86]

As a related matter, the U.S. Department of Defense has approximately 345,000 tons of surplus ordnance—bombs, rockets, bullets, shells—that must be destroyed or otherwise disposed.[87] This sort of material was once simply taken into the desert and blown up, or dumped in the ocean, but sounder techniques are now being sought.

Building/Structure Remedial Wastes

Some materials that were used for many years in construction are now recognized as hazardous. The three most prominent examples are asbestos, lead, and polychlorinated biphenyls (PCBs). A federal statute requires the inspection of all school buildings for asbestos, and the performance of abatement where needed.[88] The EPA has estimated that more than 44,000 schools may require asbestos abatement and that 300,000 to 400,000 public and commercial buildings have asbestos that may have to be removed[89] (though no law requires this). An unknown number of residential buildings are also contaminated. National expenditures for asbestos abatement have been running at \$3–\$4 billion per year.[90]

Several federal statutes recognize the dangers of lead[91] (until recently a common component of paint and plumbing fixtures), but there is no comprehensive regulatory program. In several recent incidents, neighborhoods have been contaminated by the lead released by the sandblasting or scraping of large structures, such as bridges and water towers.[92] Many older buildings have been painted with lead-based paint; much of this paint will eventually have to be removed, especially from homes and apartments occupied by young children, who are particularly at risk. PCBs are often present in old transformers and in the ballasts of fluorescent lightbulbs.

All of these substances require special disposal. Lead is a RCRA hazardous waste,[93] and certain debris from its remediation must go to RCRA-licensed facilities.[94] Asbestos, though not a RCRA hazardous waste,[95] is regulated under CERCLA.[96] As a result the operators of many ordinary solid waste landfills, fearful of liability for later cleanup costs, refuse to accept asbestos. PCBs are regulated under the Toxic Substances Control Act (TSCA)[97] and must go to TSCA-licensed landfills or incinerators.[98]

Thus, the demolition or rehabilitation of buildings and structures, when performed properly, adds to the demand for hazardous waste disposal facilities.[99] In all, some 31 million tons of "construction and demolition debris" (both hazardous and not) are generated every year.[100] The waste is typically dumped in unlined landfills constructed for this purpose, even though it often contains hazardous contaminants.[101]

Industrial/Special/Orphan Wastes

Much industrial waste does not fit within the RCRA definition of hazardous waste, or would but for particular exemptions.[102] Under one estimate, 430 million tons of industrial waste—much of it containing high levels of heavy metals and organic compounds—are discharged each year into waste ponds, waste piles, landfills, and other non-RCRA facilities.[103] Most of these facilities lack even the most rudimentary liners or other groundwater protections, and they are barely regulated by either the federal or state governments.[104] A few states—notably California, Illinois, New Jersey, Pennsylvania, and Wisconsin—have required groundwater protection and other controls, but the federal government has no coherent program for these industrial wastes, whose volume overwhelms those of RCRA HW and of municipal solid waste.[105]

Mining and Oil and Gas Wastes

Some 1.3 billion metric tons of mining wastes are produced each year.[106] This material comes from the extraction, beneficiation, and processing of ores and minerals. This waste material is typically placed in a slurry and dumped in tailings impoundments that average 500 acres in size, with the largest exceeding 10,000 acres.[107] More than 24,000 mining waste ponds are in active use,[108] and another 22,300 abandoned mines and processing facilities are expected to require cleanup, at a total cost of $55 billion.[109]

Much mining waste would be hazardous waste, but in 1980—shortly before the RCRA regulations were to take effect—Congress enacted the Bevill Amendment,[110] which temporarily excluded mining wastes from RCRA regulation pending an EPA study and a subsequent rule making. The resulting EPA regulations have engendered a considerable amount of litigation,[111] and much mining waste remains exempt from the RCRA,[112] despite evidence of serious health and environmental risks.[113]

In the oil and gas industry, an estimated 8.6 billion metric tons of brines and drilling muds are discharged each year into some 125,000 oil and gas waste ponds.[114] These wastes have high concentrations of chlorides, barium, and other contaminants but are also exempt from RCRA under the Bevill Amendment.[115] Not exempt is the mercury present in many oil and gas wells and pipelines due to the former use of mercury manometers to measure gas pressure.[116]

Pollution Control Residue

Pollution control devices typically capture the offending material before it can exit the smokestack or drainpipe. Indeed, certain elements of the two main pollution control laws, the Clean Air Act and the Clean Water

Act, have not so much eliminated pollution as consolidated it in residues that require their own disposal. In enacting RCRA, Congress made a formal finding that "as a result of the Clean Air Act, the Water Pollution Control Act, and other Federal and State laws respecting public health and the environment, greater amounts of solid waste (in the form of sludge and other pollution treatment residues) have been created."[117]

One prominent example is sewage sludge—a by-product of sewage treatment plants. In 1990 the nation's approximately 15,000 sewage treatment plants produced some 8.5 million dry tons of sewage sludge.[118] Land application as fertilizer or soil conditioner is the preferred disposal method, but because certain contaminants can make it unsuitable for this use, the EPA regulates sludge disposal.[119]

Sludge disposal is especially difficult for metropolitan areas with little nearby agricultural land. New York City, for example, historically dumped its sewage sludge in the Atlantic Ocean. In 1988, a year when needles and other medical waste washed onto beaches in New York and New Jersey, Congress banned ocean dumping of sewage sludge after December 31, 1991.[120] This increased New York City's annual sludge disposal costs from $20 million to $250 million[121] and led to plans to build five sludge-burning incinerators in New York City and six in New Jersey (for that state's sludge).[122] According to one risk assessment, preventing sewage sludge from being disposed in the ocean would save one statistical life every five years, but incinerating the sludge instead would cause two statistical cancer deaths annually.[123] Local groups have opposed these units[124] and pushed, with some success, for the sludge to be exported to other parts of the country instead.[125] Some regions have willingly accepted this material,[126] but others, claiming the sludge is contaminated,[127] have gone to court to try to stop the exports.[128]

Ash is another residue of certain kinds of pollution control. For every 100 tons of municipal solid waste (MSW) that are incinerated, for example, there remain about three tons of fly ash (the material captured by the air pollution control equipment) and 27 tons of bottom ash.[129] Up to 5.5 million tons of ash are created by MSW generation each year in the United States.[130] Much is so contaminated with heavy metals that it fits within the RCRA definition of hazardous waste.[131] However, since the ash is derived from household trash, the courts were divided on whether the ash is exempt under the RCRA's exemption for household hazardous waste,[132] until the U.S. Supreme Court decided the exemption did not apply.[133]

An even larger volume of ash—about 67 million tons a year—is generated by coal-burning electric power plants.[134] This ash generally does not exhibit the characteristics of RCRA hazardous waste,[135] and the EPA has determined that it should not be treated as RCRA hazardous waste.[136]

Medical Waste

Approximately 500,000 tons of regulated medical waste[137] are generated each year by about 380,000 generators, such as hospitals, clinics, and physicians' offices.[138] The quantity of regulated medical waste is expected to increase considerably, both because many states are requiring much hospital waste (whether or not it is infectious) to be handled specially and because of the increase in in-home health care.[139] (However, greater scrutiny of health care costs, and resistance by insurers, are leading to more reuse of items that previously would have been treated as disposable, such as operating room gowns.)[140]

The Medical Waste Tracking Act of 1988[141] was enacted in the wake of the summer of beach washups noted previously. The special legal attention afforded medical waste seems based more on the fear of AIDS than on any clear evidence that medical waste poses greater dangers in the environmental (as opposed to the occupational) setting than does MSW.[142] In any case, the siting of facilities to incinerate or otherwise dispose of medical waste is enormously controversial,[143] in part because of the psychology surrounding medical waste, and in part because medical incinerators and municipal incinerators tend to have much higher dioxin and furan emissions than do hazardous waste incinerators, largely as a result of the far more stringent regulation of HW incinerators.[144] Existing medical waste incinerators have the capacity to burn about ten times the amount of such waste actually generated, but most of the excess capacity is in hospital incinerators that are used only intermittently.[145]

One related item—wastes containing genetically engineered micro-organisms—has so far escaped much regulatory attention and is usually treated as MSW.[146]

Municipal Solid Waste

Although this book focuses on hazardous and radioactive wastes, ordinary MSW is tied so closely in law, politics, and regulation to HW/RW that it must be discussed as well. Approximately 180 million tons of MSW are generated annually, of which 40 percent is paper; 18 percent are yard wastes; 8 percent is metals; 8 percent are plastics; and 7 percent are food wastes.[147] Of all this MSW, 72.7 percent is landfilled, 14.2 percent is incinerated, and 13.1 percent is recycled or otherwise recovered.[148]

The number of MSW landfills declined from about 20,000 in the early 1970s to about 7,000 in 1991[149] and may fall to only about 1,600 in the year 2003.[150] Strict environmental regulations[151] are leading to the closure of many small landfills and open dumps and to the opening of large new landfills,[152] though haphazard enforcement of these regulations allows many of the other facilities to remain open for years beyond their theoretical closure dates.[153] Old landfills present severe environmental hazards; 21 percent of all hazardous waste sites on the National Priorities List are municipal landfills,[154] and municipalities have been held liable under CERCLA for MSW containing hazardous substances that they have sent to their own and to others' landfills.[155] The leachate from even modern MSW landfills is just as toxic as that from HW landfills, although HW landfills are more rigorously regulated.[156]

Even though the number of landfills is sharply falling, the new landfills (364 of which opened between 1986 and 1991)[157] are so large that total capacity has actually increased,[158] and a bidding war has erupted among landfills looking for more garbage.[159] The EPA has concluded that the nation has adequate MSW landfill capacity, despite some regional shortages.[160]

The siting of MSW-burning incinerators (also called resource recovery plants or waste-to-energy plants) has been very controversial,[161] and many proposed plants have been canceled in the face of public opposition.[162] Still, about ten new plants opened each year from the late 1980s through 1991,[163] and by 1992 there were 190 operating MSW incinerators in the United States. The portion of MSW that is incinerated has steadily increased.[164] Nonetheless, there is a good deal of unused incinerator capacity nationwide,[165] and several municipalities are losing so much money on new but underutilized incinerators that they have imposed special taxes on their residents[166] and passed "flow control" laws to prohibit export of their garbage until such laws were declared unconstitutional by the U.S. Supreme Court in 1994.[167] Two states (West Virginia and Rhode Island) have banned the construction of new MSW incinerators, and several others have imposed temporary moratoria.[168] More than 100 planned incinerators have been canceled in recent years.[169]

Dredge Spoil

Many bodies of water used for shipping must be dredged periodically to maintain their required depth. Especially near major urban areas, this dredged material is often tainted with PCBs, heavy metals, and other toxins. It is usually dumped at sea. From the New York harbor alone, 8 to 10 million cubic yards of dredged material are dumped at sea annually, despite concern by some environmentalists that this practice degrades the ocean.[170]

Some sediments at the bottom of rivers and other bodies of water are so contaminated that dredging and upland disposal in special facilities are necessary. For example, General Electric dumped about 1.1 million tons of PCBs from one of its factories into the Hudson River between the 1940s and 1977. New York State decided that much of the nearby river bottom should be dredged out, but so far the state has been unable to find a place to dispose of this material.[171]

Storage Tank Remediation

In 1984 Congress recognized the threat to groundwater posed by leaking underground storage tanks (USTs), especially those containing gasoline and other petroleum products. The resulting statute requires the upgrading or replacement of many USTs.[172] There are approximately 295,000 contaminated UST sites in the United States, containing at least 56 million cubic yards of soil and debris requiring cleanup.[173] Whether this soil must be treated as a hazardous waste depends largely on the unresolved regulatory status of used oil.[174] More than 166,000 underground tanks had been confirmed to be leaking by mid-1992, but the actual number may be triple that, and the cost of removing all leaking or antiquated tanks and cleaning the contaminated soil and groundwater has been estimated at $41 billion.[175]

3

The Origins and Disposal of Radioactive Wastes

As just shown, hazardous wastes—and materials with hazardous characteristics, regardless of their regulatory status—are produced by virtually every sector of the economy and type of human activity. In contrast, almost all man-made radioactive wastes come from just two enterprises: the generation of electricity by nuclear power plants and the manufacture of nuclear weapons. All other sources, such as medical and industrial uses, account for only a tiny fraction of all radioactive wastes.

Both of these activities are on the decline. No new nuclear power plant has been ordered in the United States since 1979, the year of the Three Mile Island accident, and every plant ordered between 1974 and 1979 has been canceled.[1] The general trend in most of the world today (except for parts of Asia) is away from nuclear power.[2] With the end of the Cold War, the production of nuclear weapons has also all but stopped.[3]

Virtually all radioactive wastes start with one substance—uranium ore. To fuel one typical nuclear power plant for one year, about 125,000 tons of uranium ore must be mined, yielding 175 tons of uranium.[4] In a year the plant will fission about one ton of uranium and convert 1.9 pounds of matter to energy, generating about 7 billion kilowatt hours of electricity. It will also leave about 45 tons of spent fuel and 500 tons of low-level radioactive waste.[5] Eventually, the plant will have to be dismantled and will itself become radioactive waste.

Table 3.1 illustrates the origins of the radioactive waste from power production. It demonstrates the amount of waste, both in volume (cubic meters) and radioactivity (undecayed curies) generated during the forty-year life of a pressured water reactor. (The numbers for a boiling water

Table 3.1
Forty-year lifetime waste production of nuclear reactor

Waste type	Volume (m^3)	Radioactivity (curies)
Once-through fuel cycle wastes		
Mill tailings	4,353,000	37,100
LLRW from uranium conversion	341	9,813
LLRW from uranium enrichment	133	9,716
LLRW from fuel fabrication	3,063	7
LLRW from reactor power generation	30,320	28,660
Reactor spent fuel (HLW)	521	3,270,000,000
Decommissioning wastes		
LLRW	15,100	105,700
Greater than Class C	113	4,070,000
TOTALS	4,403,000	3,274,000,000

Source: Adopted from Robert G. Cochran and Nicholas Tsoulfanidis, *The Nuclear Fuel Cycle: Analysis and Management* (LaGrange Park, Illinois: American Nuclear Society, 1990), p. 278.

reactor, the other major type of commercial nuclear power plant, are not much different.)

This table illustrates that the overwhelming portion of the radiation in RW from civilian nuclear power is contained in the spent fuel assemblies. The hottest reactor components left over when the plant is decommissioned are a distant second. With respect to volume, the great bulk of the waste is uranium mill tailings, with LLRW from reactor operations a distant second.

Table 3.2 combines RW from civilian and defense purposes, and shows the total cumulative volume and radioactivity of wastes generated through 1991. The table also shows that, once again, spent fuel rods from commercial nuclear power plants account for the greatest portion of radioactive waste, measured in radioactivity, with HLW from defense purposes (chiefly weapons production) coming in second. Uranium mill tailings account for most of the volume, followed by defense and commercial LLRW.

Each of the major radioactive waste streams will now be discussed.

Table 3.2
Cumulative radioactive waste generated through 1991

Waste type	Volume (m^3)	Radioactivity (curies)
HLW		
Commercial	1,729	26,210,000
Defense	394,900	970,700,000
LLRW		
Commercial	1,423,000	5,651,000
Defense	2,816,000	13,430,000
Transuranic waste	255,400	2,722,000
Spent fuel*	9,546	23,250,000,000
Uranium mill tailings**	118,400,000	N.A.
Mixed waste	101,400	N.A.
TOTALS	123,400,000	24,270,000,000

Source: Adopted from U.S. Department of Energy, Office of Civilian Radioactive Waste Management, Pub. No. DOE/RW-0006, Rev. 8, *Integrated Data Base for 1992: U.S. Spent Fuel and Radioactive Waste Inventories, Projections, and Characteristics* (Oct. 1992), pp. 9, 14.
* Volume includes spacing between fuel assembly rods. Does not include DOE/Defense fuel to be reprocessed.
** Licensed mill sites only.
N.A. = Not available.

High-Level Radioactive Waste (HLW)

There are two main kinds of HLW. The first is the residue, mostly liquid, from the manufacture of plutonium for warheads. In the United States, most of this residue is now stored in 177 underground tanks at the Hanford Reservation of the U.S. Department of Energy (DOE) in southern Washington State, and 51 tanks at the DOE's Savannah River plant in South Carolina. The rest is located at the Idaho National Engineering Laboratories and at the West Valley site in New York State. In 1957 a tank holding similar wastes exploded in Kyshtym, the Soviet equivalent of Hanford, spreading radioactive contamination over hundreds of square miles.[6] In 1993 a similar but much smaller explosion occurred in a radioactive waste tank in the Siberian city of Tomsk.[7] There is considerable concern that the U.S. tanks may also explode from the chemical and radioactive reactions that constantly occur within them. Several of the tanks are leaking, and sixteen times between 1987 and 1991 they released toxic gases, often injuring workers.[8]

The second kind of HLW is spent fuel from nuclear power plants. There are 109 operating commercial nuclear reactors in the United States, most of them east of the Mississippi River (see Figure 3.1). When the uranium fuel is first loaded into a reactor, it is only mildly radioactive, but after one or more years it becomes too radioactive to use, and it also creates plutonium.[9] When commercial nuclear power began in the 1950s, it was assumed that this spent fuel would undergo reprocessing—a series of physical and chemical operations that separate the uranium and plutonium for reuse.[10] However, reprocessing became an economic and environmental disaster. Only three commercial reprocessing plants were ever built in the United States. One, in Morris, Illinois, did not work and therefore never opened. The second, at Barnwell, South Carolina (near the Savannah River plant), experienced such cost overruns that it was never finished. The third, at West Valley, New York, operated for six years but shut down in 1972, leaving behind hundreds of thousands of gallons of highly radioactive liquid waste and a legacy of fires and accidents.[11]

Reprocessing suffered an enormous political setback in 1974 when India exploded an atomic bomb it had made with plutonium reprocessed from the fuel rods of a nuclear power plant. On October 28, 1976, five days before the presidential election, President Ford, trailing Jimmy Carter in the polls and concerned about the dangers of nuclear proliferation, ordered a temporary ban on commercial reprocessing. After the election, President Carter extended the ban. President Reagan tried to revive commercial reprocessing, but its economics were so unfavorable (partly because of its high cost relative to fresh uranium) that it never resumed.[12] (The military still reprocesses nuclear fuel from submarines and other defense uses at three facilities—Hanford, Savannah River, and the Idaho National Engineering Laboratory.)[13]

Most of the spent fuel rods ever generated by commercial reactors in the United States—9,546 cubic meters through 1991—are stored on-site

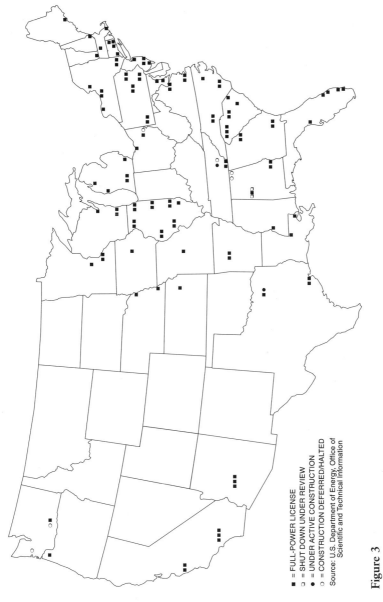

= FULL-POWER LICENSE
= SHUT DOWN UNDER REVIEW
= UNDER ACTIVE CONSTRUCTION
= CONSTRUCTION DEFERRED/HALTED

Source: U.S. Department of Energy, Office of
Scientific and Technical Information

Figure 3
Commercial nuclear power reactors in the United States as of December 31, 1991. (Courtesy of U.S. Department of Energy, Office of Scientific and Technical Information, Oak Ridge, Tennessee.)

at the reactors where they were used. After removal from the reactor, the fuel is initially so hot that it must be kept underwater in spent fuel pools, with the water circulating constantly to cool it.[14] The canceling of reprocessing greatly increased the need for on-site storage capacity, but the utilities have crammed more rods into their existing pools,[15] and it now appears that virtually all the power plants can store their wastes on-site through the remainder of their forty-year operating licenses[16] and perhaps one hundred years or longer with readily achievable facility expansions.[17] Roughly 1,900 tons of spent fuel are generated each year.[18]

With no commercial reprocessing, spent fuel rods are a waste, not a resource. The nuclear utilities, no more eager than anyone else to have nuclear waste stored in their backyard, began pressing for a long-term disposal solution. The Nuclear Waste Policy Act of 1982 (NWPA) required the DOE to establish a system of "long term" or "permanent" deep geologic disposal facilities for both kinds of HLW—waste from bomb production and spent fuel rods.[19] The DOE was told to recommend to the president three sites to be studied in detail.[20] The DOE recommended Yucca Mountain, Nevada; Deaf Smith County, Texas; and Hanford, Washington,[21] and in 1986 President Reagan approved these three sites for study.[22] Just as the studies were about to begin, however, Congress stepped in and ordered the DOE to halt any investigations of the Texas and Washington sites and to put the HLW facility at Yucca Mountain,[23] a place near the Nevada Nuclear Test Site, 110 miles west of Las Vegas, with its nearest neighbor a legal brothel 18 miles away.[24] Nevada then began a long campaign of litigation,[25] raising many serious technical questions about the site[26] and considerably delaying the project,[27] so that opening is not expected until 2010 at the earliest,[28] and may well not occur until after 2020.[29]

Meanwhile, a dissident DOE scientist claimed, based on his geologic studies, that within the next 10,000 years an earthquake could suddenly drive groundwater far up inside the mountain, flooding the nuclear waste and releasing it into the environment. Though much of the scientific establishment has attacked this theory, many Nevadans believe it,[30] and their belief was strengthened by a June 1992 earthquake, measuring 5.6 on the Richter scale, centered 20 miles from the site.[31] The sense of procedural fairness that the DOE sought to cultivate has utterly evaporated, and residents call the statute designating Yucca Mountain the "Screw Nevada Bill."[32] (These difficulties are not unique to the United States; no country has a permanent repository for HLW.)[33] Sensitivities are further heightened by the diseases some residents of this area suffered as a result of open-air testing of nuclear weapons in the 1950s and early 1960s.[34]

The Nuclear Waste Policy Act also calls for establishment of a Monitored Retrievable Storage (MRS) facility,[35] which would prepare spent fuel for emplacement in the geologic repository and act as a central receiving station.[36] The MRS would have capacity for only a fraction of the spent fuel that would exist at the time it opened.[37] The MRS proposal has sparked considerable controversy, especially in Tennessee, which was initially targeted for the facility.[38] One of the concerns was that the MRS facility would become a permanent resting place for HLW because of the difficulties in siting the geologic repository, but Congress in 1987 provided that construction of the MRS may not begin until after a license for the construction of the geologic repository has been issued.[39] (Germany and the United Kingdom already operate centralized, temporary storage sites for HLW.)[40] In 1993 the secretary of energy reversed prior policy and said it was unfair to require the nuclear industry to pay for both the construction of Yucca Mountain and (during the seemingly interminable interim period) also pay for storage; she suggested that the federal government fund a storage facility.[41] Work proceeded on finding a storage site in a volunteer community, and two tribes—the Mescalero Apaches in New Mexico and the Skull Valley Band of Goshute Indians in Utah—stood ready to negotiate a compensation package.[42]

Transuranic Waste (TRU)

Another variety of radioactive waste is TRU—material, such as plutonium, with an atomic number greater than that of uranium.[43] Although most TRU has a relatively low level of radioactivity, it is long lived and highly toxic. Almost all TRU comes from military activities.

To dispose of TRU, the federal government has built, but so far has been unable to open, a facility it calls the Waste Isolation Pilot Plant (WIPP), twenty-six miles east of Carlsbad, New Mexico. (Commenting on the name, one critic has written, "'Pilot,' in truth, is the government's polite way of saying not only 'the first of many' but also 'we hope it works.'")[44] When the project was first proposed in the early 1970s, it was enthusiastically endorsed by local officials, who saw it as a way to replace jobs lost in the declining mining industry.[45] It was also in the state with the first atom bomb test site and with nuclear facilities such as the Sandia Laboratories. Congress authorized WIPP in 1980.[46] Disposal would be in excavated salt formations 2,150 feet underground. The first rooms were mined in the early 1980s, but in 1990 alarmed workers found lumps of rubble lying on the floor and discovered that some of the older rooms were collapsing.[47] With this development, and with changing attitudes toward nuclear power in New Mexico, the state switched to active opposition to the opening of WIPP. A court agreed with New Mexico that explicit congressional authorization was required before the property could be transferred to the DOE and shipments could begin.[48] However, Idaho, where much of the TRU is being stored until WIPP opens,[49] brought counter pressures, and Congress promptly supplied the needed authorization.[50] WIPP remains very controversial and is not expected to start receiving waste before 1998.[51]

For all its size and controversy, WIPP will receive only about 20 percent of the DOE's TRU. Much of the rest lies in shallow burial grounds at nuclear weapons facilities around the country, and there are no clear plans for its final disposal.[52]

Low-Level Radioactive Waste (LLRW)

LLRW is all radioactive waste that is not defined as HLW, TRU, or uranium mill tailings.[53] In the civilian sector, 99 percent of the radioactivity and half the volume in LLRW is produced by nuclear power plants, principally the ion exchange resins that filter radioactivity from reactor cooling water.[54] Other sources include industrial, medical, and research applications.[55] The volume of civilian LLRW generated has been declining rapidly. It dropped by about half between 1980 and 1989, to about 36,000 tons a year, largely due to disposal surcharges that have given generators a strong incentive to produce less LLRW.[56] This is despite an *increase* in the amount of nuclear power generated during the same period.[57]

About 2,816,000 cubic meters of LLRW from nuclear weapons production have been buried around the country, mostly at Hanford, Savannah River, and Oak Ridge.[58] Much of it is simply buried in shallow trenches, usually at the production plants—a practice that continues to this day.[59] Prior to 1970, the U.S. military dumped its LLRW into the sea. Most civilian LLRW has gone to the six commercial LLRW disposal facilities built in the United States after the government banned burial of civilian LLRW at federal facilities in 1962. Of these six, three—those at West Valley, New York; Maxey Flats, Kentucky; and Sheffield, Illinois—have been permanently closed because of water infiltration into the waste trenches and other environmental problems. The three facilities still operating in 1992 were in Barnwell, South Carolina (opened in 1971, near the Savannah River facility); Beatty, Nevada (opened in 1962); and Richland, Washington (opened in 1965, near the Hanford facility).[60]

In 1979 both the Washington and Nevada sites were forced to shut down temporarily.[61] In that year of Three Mile Island, South Carolina was unhappy at being forced to take the entire nation's LLRW, and ordered a 50 percent reduction in the volume of waste accepted by Barnwell, leading Washington and Nevada to announce that they would shut their facilities permanently. This precipitated a national crisis in LLRW disposal. To resolve it, Congress enacted the Low-Level Radioactive Waste Policy Act of 1980 (LLRWPA), which declared that the states, acting alone or in compacts with other states, were responsible for disposing of their own LLRW.[62] The LLRWPA gave South Carolina, Nevada, and Washington the power to exclude other states' waste after 1986. But by 1985, little progress had been made in siting new LLRW facilities, and Congress extended the deadlines, imposed interim milestones, and allowed the three sited states to exclude waste from states that missed the deadlines.[63] Congress also provided that, in 1993, states that had not made provision for disposing of the LLRW generated within their borders would have to "take title" to it and thereby assume liability for damage it causes.[64] In 1990 New York State, acting with (and under pressure by) the two counties tentatively designated as the location for its LLRW facility, challenged the constitutionality of the 1986 amendments.

(I was counsel of record for one of the counties, Cortland County, in this litigation.) In 1992 the U.S. Supreme Court invalidated the "take title" provision as a violation of the states' rights under the Tenth Amendment but upheld the balance of the statute.[65]

The federal requirement that states site LLRW facilities has sparked enormous controversy all over the country.[66] The three states furthest along in developing sites when the Supreme Court ruled were Illinois, California, and Nebraska.[67] However, in 1992 the Illinois siting commission rejected the selected site because of geological problems and other deficiencies.[68] In 1993 a federal court enjoined certain work at the California site due to potential impacts on an endangered species, the desert tortoise,[69] and the Nebraska health and environment departments announced they were denying permits for that state's facility.[70] An elected judge in Texas also rejected the chosen site in Hudspeth County.[71] (Texas subsequently selected an alternative site and agreed to accept LLRW from Maine and Vermont once the site opens, which is not expected before mid-1996 at the earliest.)[72] North Carolina plans to open a new facility in 1996 to replace the facility in Barnwell, South Carolina; North Carolina selected a site in late 1993 and was immediately met with threats of lawsuits.[73]

In January 1993, the Nevada facility shut down completely, and the Washington site closed its doors to all but six western states. The operator of the Nevada site offered to pay the state $20 million for the right to stay open an additional year or two, but that offer was withdrawn after it was disclosed that the senate minority leader, a prime supporter of the proposal, was employed by a law firm that represents the company.[74] This left only Barnwell, which is scheduled to be closed to states outside its compact in July 1994 and to shut down altogether by 1996.

While the states and regional compacts were struggling to build new facilities, a private company managed to take a site that had been used for uranium mill tailings and convert it into a commercial LLRW disposal facility. Envirocare of Utah, Inc. gradually expanded its operations in Tooele County to accept remedial waste from commercial nuclear installations; naturally occurring radioactive material; and other slightly radioactive wastes. This became a competitive threat to the LLRW facility in Richland, Washington, and the Richland operator, US Ecology, Inc., sued to restrict Envirocare's market.[75]

Meanwhile, medical, research, and industrial LLRW generators have been left scrambling for temporary storage,[76] and have petitioned the NRC to allow very low concentrations of short-lived radionuclides to be disposed as nonradioactive.[77] Several states have begun planning to build LLRW storage facilities to handle their own waste, pending the opening of new disposal capacity. All this may reduce generation, because studies show that, if the price of storage is high, the amount of LLRW generated will decline significantly.[78]

Remedial Waste from Nuclear Weapons Production

The sites where the United States manufactured nuclear weapons are, collectively, called the Nuclear Weapons Complex (NWC). The NWC includes fourteen major facilities in thirteen states, on military reservations covering 3,350 square miles and employing more than 100,000 people.[79] The NWC originated with the Manhattan Project in World War II and was greatly expanded in the early 1950s. Among the largest facilities are the Rocky Flats plant in Colorado, which produced plutonium "triggers" for bombs; the Hanford Reservation in Washington, which produced weapons-grade plutonium; the Savannah River site in South Carolina, which made tritium for hydrogen bombs, and plutonium; the Feed Materials Production Center in Fernald, Ohio, which produced uranium metal for weapons; the Oak Ridge Reservation in Tennessee, which fabricated weapons components; and the Idaho National Engineering Laboratory, where fuel from military reactors is reprocessed.[80] The NWC is rapidly phasing down its active operations, many of its units are inactive, and some plans for new units have been canceled.[81] The United States stopped producing highly enriched uranium in 1964 and plutonium in 1988 and has manufactured no nuclear warheads since July 1990.[82]

The NWC has left behind an environmental horror. A former secretary of energy has blamed this contamination on "a 40-year culture cloaked in secrecy and imbued with a dedication to the production of nuclear weapons without a real sensitivity for protecting the environment."[83] At every facility, the groundwater is contaminated with radionuclides, and pollution of surface waters, sediment, and soil is extensive. More than half the sites are on the National Priorities List under CERCLA.[84] They store large volumes of HLW, TRU, and LLRW.[85] At the largest site, the 560-square-mile Hanford Reservation (half the size of Rhode Island), there are an estimated 5 billion cubic yards of solid and dilute radioactive, hazardous and mixed wastes in tanks and other containers, but about 440 billion gallons of liquid wastes have entered the soil, contaminating over 200 square miles of groundwater.[86] At just one DOE facility—in Fernald, Ohio—some 335 million cubic feet of LLRW will be generated in the cleanup process. This is several hundred times the amount of LLRW generated by the commercial sector in a year. The DOE is quietly planning to dispose of the LLRW from the cleanup of the nuclear weapons complex at its own facilities, so as not to overwhelm the few remaining commercial LLRW facilities.[87]

Cleaning up the NWC will be one of the largest public works projects in history. In 1992 the DOE estimated that the work will cost $160 billion[88] and take twenty to thirty years[89]; another recent estimate placed the cost at $240 billion.[90] Included in these costs are some very large waste treatment facilities. For example, the DOE plans to spend $4 billion building a factory to transform the 34 million gallons of HLW in tanks at Savannah River into a glass form for eventual disposal, presumably at Yucca Mountain.[91] Similar plants are planned for Hanford Reservation and the Idaho laboratory.[92] The cleanup at Hanford will not be completed before 2028.[93]

Nuclear Weapons

At the peak of the arms race in the mid-1980s, the Soviets had roughly 33,000 nuclear warheads and the United States about 24,000.[94] The United States has produced nearly 70,000 nuclear warheads since 1945, more than 50,000 of which have now been retired and disassembled.[95] Between them, the United States and the former Soviet republics possess more than 200 tons of weapons-grade plutonium and more than 1,000 tons of highly enriched uranium, either assembled in warheads or held in storage.[96]

The START II agreement and earlier treaties portend a drastic reduction in the number of deployed weapons but do not specify what is to be done with them. There is no coherent national policy on what is to become of the nuclear material contained in these warheads.[97] It can be recycled into new warheads; diluted and used as fuel for nuclear power plants; stored and guarded; or converted into a relatively irretrievable form, such as ceramic or glass blocks, and sent to a repository such as Yucca Mountain or WIPP.[98] America's weapons are being dismantled, and the plutonium cores are stored in a factory in Texas that has experienced many mishaps.[99]

The United States has committed itself to helping the former Soviet republics destroy their nuclear weapons[100] and has arranged to buy some of the material from Russia for dilution into reactor fuel.[101] Meanwhile, there is great concern that terrorists or rogue nations might acquire some of the materials, especially those held by the four republics with Soviet weapons—Russia, Ukraine, Belarus, and Kazakhstan. Already, radioactive materials and bomb parts from these countries have been found trading in the black market.[102]

Another disposal problem—though one that does not (yet) affect U.S. disposal capacity—was the use of artillery shells made of depleted uranium by the United States and its allies in the Persian Gulf War. Roughly 10,000 such shells were used in Iraq and Kuwait in the 1991 war, and there are fears that the shells are causing health problems in the Gulf region.[103]

Decommissioned Nuclear Plants

There are approximately 109 nuclear power plants, 22 nuclear fuel cycle plants (mostly in the NWC), and 54 research and industrial reactors in the United States.[104] All will eventually have to be decommissioned. Nuclear power plants were anticipated to have a forty-year life span, but some are now being shut down much earlier, largely because alternative power sources, especially natural gas, are cheaper.[105] The world's first commercial nuclear power plant, a small unit in Shippingport, Pennsylvania, opened in 1957 and was dismantled in 1989. The reactor vessel was sent whole by barge, via the Panama Canal, to Hanford for disposal.[106] No large commercial plant has yet been dismantled, however. Several have been retired, but they remain intact and their spent fuel pools must be continually monitored, with ongoing security, testing, and training, at an annual cost of several million dollars each.[107] Completely dismantling a large power plant may cost more than one billion dollars[108] and will leave wastes approximately one hundred times more radioactive than the combined total of all the LLRW generated during the reactor's operation.[109] Another option is "entombment"; the plant is encased in a massive structure of concrete and steel, with a structural lifetime of perhaps two hundred years, leaving the task of dismantling a somewhat cooler plant to a distant generation.

Decommissioning can be a risky business. In August 1992, due apparently to sloppy work by Department of Energy contractors, nuclear research equipment exploded during decontamination and decommissioning of a unit at Hanford. Caustic lithium acetate was spread throughout the building.[110]

As with so many other areas of environmental policy, a key question in the decommissioning of nuclear plants is How clean is clean? A coalition of environmental organizations has urged that owners be required to clean up power plant sites to natural background levels of radioactivity; the nuclear industry says that there is no clear definition of "background" and that such a stringent standard would be unreasonably expensive. The NRC expects to promulgate its rule on the required cleanup standards in 1995.[111]

Decommissioning will be complicated by the NRC's practice, which went on for almost twenty-five years,[112] of allowing licensees to bury LLRW on-site without prior NRC approval.[113] In one case, at a Westinghouse plutonium processing plant in Cheswick, Pennsylvania, radioactive waste was found under an employees' softball field. In another, a facility in Pawling, New York, for fabricating and testing uranium oxide, thorium, and plutonium stopped operations in 1972. The land was later sold to the National Park Service for relocation of part of the Appalachian Trail. It was subsequently discovered that parts of the site had soil contamination up to 320 times what NRC guidelines allow.[114]

Decommissioned nuclear submarines pose special problems. Russia has disclosed the locations where four Soviet nuclear submarines sank, laden with nuclear missiles and torpedoes, and where numerous nuclear reactors and other radioactive wastes were intentionally dumped over the last thirty years. The U.S. Navy has lost two nuclear submarines at sea, the *Thresher* and the *Scorpion*, and in 1959 it intentionally dumped one reactor at sea.[115] Approximately three hundred nuclear-powered submarines will become obsolete by the turn of the century, and some nations, including the United Kingdom, are actively considering sea disposal.[116] Any such plan would undoubtedly spark international protests and would arguably violate the London Dumping Convention. It has been estimated that the decommissioning of one large nuclear ship, the *Savannah*, will cost $76 million.[117]

Uranium Mill Tailings

Uranium ore is the basic raw material of both nuclear power plant fuel and nuclear warheads. After the usable uranium is extracted from the ore, finely ground radioactive tailings remain whose radioactivity is reduced to about 6 percent of its initial value after four half-lives, or 300,000 years.[118] More than 230 million tons of uranium mill tailings have accumulated in the United States, representing over 95 percent of the volume of all RW generated in this country.[119] The largest number of adverse health effects from all nuclear-related activities in the United States arise from the mining and milling of uranium.[120]

Historically, these tailings were carelessly managed. Until 1966 they were often used as fill material in building construction, leading to the contamination of thousands of buildings,[121] and, years later, to lawsuits brought by property owners.[122] The great bulk of the tailings, however, was dumped in enormous piles or ponds. In 1979 a tailings dam collapsed at a uranium mill near Churchrock, New Mexico, releasing 100 million gallons of tailings solution that contaminated at least sixty miles of the Rio Puerco along its course through lands in New Mexico and Arizona used by Navajo for watering stock.[123] Contamination from these piles is spread above the surface by wind and below it by groundwater.[124]

In 1978 Congress enacted the Uranium Mill Tailings Radiation Control Act,[125] with a mandate to clean up twenty-four inactive uranium processing sites, containing 24 million tons of tailings.[126] This leaves another twenty-six sites with a total of more than 200 million tons of tailings.[127] Although some of this material will be moved to new disposal sites, the volumes are so great that most will have to be managed in place.[128] Most of these sites are in the Rocky Mountain states.

Naturally Occurring Radioactive Material[129] (NORM)

Numerous activities in such industries as oil and gas extraction, water treatment, mining, and fossil-fired power generation produce NORM (radiation present in nature but brought into contact with humans by these processes). Tens of billions of tons of NORM-containing wastes are generated each year, a volume dwarfing all other hazardous and radioactive wastes combined. Where NORM is produced, workers are often exposed to far higher levels of radiation than would be permitted for workers at a nuclear power plant. However, NORM is virtually unregulated. The EPA has released draft regulations aimed at some NORM wastes,[130] but it has not promulgated them. One state, Louisiana, has regulations on the subject.[131]

There is only one disposal site for NORM in the country, located near Clive, Utah, and established in the mid-1980s to take uranium mill tailings.[132] Recent attempts to site such a facility in Texas were defeated by local opposition.[133] The vast majority of all NORM is simply buried or stored on-site, without any regulatory oversight at all. Its enormous volume, its multiple sources, and confusion over what to do with it have led essentially to regulatory paralysis.

Mixed Waste

Mixed waste is material that is both RCRA hazardous waste and radioactive waste. An example is a solvent-containing rag that was used to clean a radioactively contaminated pump at a nuclear power plant. About 90 percent of all mixed waste produced in the United States is generated by the DOE.[134] Mixed waste is simultaneously regulated as radioactive waste by the NRC and as hazardous waste by the EPA.[135] However, these two schemes of regulation are incompatible. For instance, under RCRA the operator of a disposal facility must verify the contents of a waste package by opening it and taking a representative sample for testing; this procedure, however, could subject facility workers to doses of radiation that violate NRC standards.[136] Partly because of such contradictions, there is not a single disposal facility or off-site storage or treatment facility for most kinds of commercial mixed waste in the United States.[137] The material is simply stored; in 1993 an estimated 589,481 cubic meters of mixed waste were in storage.[138] As a result, the EPA has been forced to grant variances from the RCRA rules for disposal of mixed waste.[139] Even so, the D.C. Circuit Court has ruled that, under RCRA, electric utilities cannot store mixed waste on-site indefinitely while waiting for the EPA to develop disposal guidelines. The court conceded that its decision puts the utilities "in the unenviable position of having no choice but to violate the law."[140] Mixed waste has also become a major headache for states trying to develop their own disposal facilities for low-level radioactive waste.

In 1992 a congressional study revealed that nuclear weapons complex facilities had routinely sent mixed waste to hazardous waste disposal facilities without revealing its radioactive content. The DOE has no firm guidelines on how much radioactive contamination is too much to go to a hazardous waste facility.[141] Much of the DOE mixed waste is now incinerated at Oak Ridge.[142]

Summary

It is a measure of regulatory fragmentation that apparently no one has ever published a cumulative estimate of the costs of cleaning up all the hazardous and radioactive waste sites discussed above. The closest attempt came in a 1991 study from the University of Tennessee. The "best guess" of the costs over the next thirty years of site remediation, assuming a continuation of current policies, was as presented in table 3.3.

Depending on future environmental policies, and on estimation methods, the total cost could be as low as $373 billion or as high as $1,694 billion.[143] These figures do not include such important categories of remedial waste as chemical weapons, asbestos and lead from buildings, non-RCRA industrial wastes, mining wastes, oil and gas wastes, incinerator ash, decommissioned nuclear power plants and weapons, uranium

Table 3.3
Thirty-year site remediation costs

	Billions
National Priority List (Superfund)	$ 151
RCRA corrective action	234
Underground storage tanks	67
Department of Defense sites	30
Department of Energy sites (NWC)	240
State/private cleanup programs	30
TOTAL	$ 752

Source: Milton Russell et al., *Hazardous Waste Remediation: The Task Ahead* 16 (Waste Management Research & Education Institute, University of Tennessee, 1991).
All figures are expressed in billions of 1990 dollars.

mill tailings, and naturally occurring radiation. These categories, together, could well come to a figure of the same order of magnitude—that is, around a trillion dollars over the next thirty years.[144] Whatever the actual figure, it is apparent that society has a very great interest in the sound and economical cleanup of these sites, and in ensuring that even more sites do not become contaminated.

Another attempt to look cumulatively at remedial waste sites was published by the EPA in 1993 and is summarized in table 3.4. It shows that the total volume of disposal capacity that will be required for remedial waste is likely to exceed 100 million cubic yards and may climb even higher.

Table 3.5 arrays the different kinds of wastes from ongoing production processes. It vividly shows that the greatest controversies revolve around some of the waste streams generated in the lowest quantities—such as RCRA hazardous waste disposed of off-site, LLRW, and spent fuel—while much less controversy concerns waste streams generated in quantities several orders of magnitude higher.

It is also instructive to array the different kinds of waste streams according to whether they are "livestream" (that is, the result of ongoing production process) or "remedial" (from the cleanup of previously existing wastes), and by the level of government that is primarily responsible

Table 3.4
Remedial waste: Sites and volumes

Type of site	Number of sites	Material to be disposed (cubic yards)
National Priorities List	1,235	26 million
RCRA corrective action	1,500–3,500	N.A.
Underground storage tanks	295,000	56 million
Department of Defense	7,000	7+ million
Department of Energy	4,000*	200–3.3 million per site
Civilian federal agencies	350	N.A.
State Superfund sites	19,000	N.A.

Source: Adopted from U.S. Environmental Protection Agency, Pub. No. EPA 542-R-92-012, *Cleaning Up the Nation's Waste Sites: Markets and Technology Trends* (Apr. 1993).
N.A. = Not available.
* Discrete sites on 110 major installations.

Table 3.5
Production waste volumes

Waste stream	Annual production (tons)
Naturally occurring radioactive materials	50,000,000,000?
Oil and gas waste	8,600,000,000
Mining waste	1,300,000,000
Industrial/special/orphan	430,000,000
RCRA hazardous waste (on-site disposal)	240,000,000
Municipal solid waste	180,000,000
RCRA hazardous waste (off-site disposal)	10,000,000
Sewage sludge	8,500,000
MSW incineration ash	5,500,000
Medical waste	500,000
Low-level radioactive waste	36,000
Spent nuclear fuel	1,900

Table 3.6
Categories of waste

Livestream wastes

Primary federal responsibility
Nuclear weapons production wastes—HLW
Nuclear weapons production wastes—TRU
Nuclear weapons production wastes—LLRW
Submarine and other military reactor waste
Military RCRA hazardous waste
Commercial nuclear power—HLW
Mixed waste (RW + HW)
Dredge spoil (from Army Corps of Engineers and military)

Primary state responsibility
Commercial nuclear power—LLRW
Commercial RCRA hazardous waste
Mining, oil, and gas wastes
Incinerator ash
Pollution control residue
Industrial/special/orphan wastes
Medical waste
Dredge spoil (often by regional port authorities)

Primary local responsibility
Household hazardous waste
Municipal solid waste
Sewage sludge

Essentially unregulated
Naturally occurring radioactive materials
Wastes containing genetically engineered microorganisms

Remedial wastes

Primary federal responsibility
Nuclear weapons complex waste
Decommissioned nuclear warheads
Decommissioned nuclear submarines
Decommissioned chemical weapons
Decommissioned nuclear power plants
Uranium mill tailings
Abandoned mines on federal lands
Military facilities
National Priorities List (CERCLA) sites

Table 3.6 (Continued)

Primary state responsibility
State Superfund sites
RCRA corrective action sites
Building/structure remedial wastes
Storage tank remediation
Primary local responsibility
Closed municipal landfills
Essentially unregulated
Abandoned mines on private lands

for arranging for or regulating their disposal. This is done in table 3.6. (Because most of the implementation of RCRA has been delegated to the states, I have assigned responsibility over RCRA hazardous wastes to the states.)

Two aspects of table 3.6 are striking. First, federal responsibility for waste disposal is mostly limited to radioactive wastes and to wastes from federal facilities. The states are saddled with most other kinds of wastes, under greater or lesser degrees of federal supervision. Second, 21 livestream and 15 remedial waste streams are listed, for a total of 36. Further subcategorization could have achieved an even longer list. There is very little coordination in finding disposal sites for these 34 different kinds of waste streams. Every one of these different waste streams is embroiled in its own differing and recurring set of siting controversies.

4

How Siting Decisions Are Made

Siting in the Absence of Government Regulation

The private sector in the United States, unlike that in most of Europe and Canada, has been primarily responsible for the disposal of hazardous waste and commercial LLRW. Prior to the onset of modern environmental legislation in the early 1970s, the siting of private disposal facilities was relatively unfettered by environmental constraints. Decisions on where to site disposal facilities were made the same way as decisions on where to put any heavy industry—by companies seeking the best combination of such key factors as proximity to markets and materials; availability of labor; transportation; utilities and infrastructure; and low land and development costs.[1] Even after most environmental regulations took effect, they played only a secondary role in industrial location decisions, with the more traditional factors retaining primacy.[2] Since heavy industry generates most of the HW, disposal facilities tended to be built near industrial concentrations, which in turn tended to be in or near cities.[3] They were also often put on wetlands (then known as swamps) and floodplains, where land was cheap.

In contrast, the facilities in the Nuclear Weapons Complex, where cost was never a controlling factor, tended to be sited more for expediency and sometimes even on caprice. For example, in 1942 J. Robert Oppenheimer personally selected the site of what became the Los Alamos National Laboratory largely on the basis of its grand scenic view and his childhood memories of attending a boys' school there.[4] The same year Gen. Leslie Groves, military head of the Manhattan Project, sent two men scouting for a site for a plutonium production facility. He told them he wanted a place that was remote from population and had ample water and electricity. They found a large spot on the Columbia River in Washington, just one hundred miles from the newly completed Grand Coulee Dam. Groves swiftly bought it, and it became the Hanford Reservation.[5] Almost immediately, concerns were raised about the project's impacts on the Columbia River—concerns that were soon borne out—but wartime exigencies preempted any reconsideration of the site selection.[6] In 1942 as well, General Groves personally selected a site in eastern Tennessee for uranium production, based on its remoteness from population, its cheap land, and its abundant water and power; this site is now known as Oak Ridge.[7] In 1950, with scarcely more study, the bank of the Savannah River in South Carolina was selected for the manufacture of tritium for hydrogen bombs, even though the site sits atop a prolific aquifer,[8] and in 1951, Rocky Flats, Colorado, was chosen for plutonium fabrication with similar inattention to geologic conditions.[9]

Individual entrepreneurs have also played a key role. The most important by far was Frederick P. Beierle, who began his career as a reactor operator at Hanford and who is personally responsible for siting three of the nation's six LLRW facilities and two hazardous waste landfills.[10] In 1963 Beierle and two other men leased a parcel near Hanford for an LLRW landfill; this became the Richland facility, which is still operating. In 1966 Beierle moved to Sheffield, Illinois, and persuaded its citizens to let him build an LLRW facility, which he sold to a company that became US Ecology. (The LLRW facility is now closed, but US Ecology still operates a hazardous waste landfill in Sheffield.) In 1968 he formed a new company, soon renamed Chem-Nuclear Services, and he built what is now the Barnwell LLRW facility near the Savannah River plant, with the active support of the local community and the state of South Carolina.[11] In 1976 Beierle opened a hazardous waste landfill in Livingston Parish, Louisiana; it was later sold to Browning-Ferris Industries. Beierle was also behind an unsuccessful attempt to build an HLW repository in salt caverns in Lyons, Kansas.

Local business leaders have also played major roles in siting facilities. A controversial hazardous waste incinerator was built in East Liverpool, Ohio, at the invitation of businessmen who wanted to attract new enterprise to the economically depressed town.[12] Business leaders in Carlsbad, New Mexico, encouraged the construction of the Waste Isolation Pilot Plant. The nation's largest hazardous waste landfill was sited in Emelle, Alabama, in 1977 by a small group of regional investors who were aided in the regulatory process by the son-in-law of then-Governor George Wallace and by the son of the Speaker of the Alabama House of Representatives. This was before any law required public notice of the proposed facility, and the landfill had been built before most nearby residents realized what it was. (Only after all permits had been obtained did Waste Management, Inc. take over the facility.)[13]

These siting decisions, mostly swift and sure but often environmentally disastrous, stand in marked contrast to what has happened since the federal and state environmental agencies became involved.

Siting Under Government Regulation

Federal Siting Processes

I have already mentioned four federal efforts to site federally owned disposal facilities for HW/RW—Yucca Mountain, WIPP, Lyons, and the incinerators for chemical weapons. Thus far only the last has led to the disposal of a single pound of waste. A fifth effort occurred two decades ago and led nowhere. In 1970 the army began transporting ammunition and nerve gas from depots in the western United States to Florida for disposal at sea. Congress, concerned about where to put such materials,[14] directed the Department of Health, Education and Welfare to study the feasibility of creating a national system of sites for the disposal of hazardous wastes.[15] The resulting report, completed in 1973, recommended a national disposal system with about twenty regional processing facilities. This report was never implemented.[16] (Around the same time, serious efforts were made to establish a national power plant siting authority, but these too failed.)[17]

There have also been two federal attempts to require the states to site facilities. The first was the Low-Level Radioactive Waste Policy Act, which has not yet brought about any new facilities. The second effort has, so far, been similarly unproductive. It arose in the Superfund

Amendments and Reauthorization Act of 1986 (SARA).[18] Congress was concerned that most states were making little progress in siting new HW disposal facilities, and that "Superfund money should not be spent in States that are taking insufficient steps to avoid the creation of future Superfund sites."[19] In an effort "to solve the 'NIMBY'. . . problems that arose because of political pressure and public opposition,"[20] Congress provided that, after October 17, 1989, no state could receive Superfund assistance for remedial actions unless it assured "the availability of hazardous waste treatment or disposal facilities which . . . have adequate capacity for the destruction, treatment, or secure disposition of all hazardous wastes that are reasonably expected to be generated within the State" during the next twenty years.[21] These facilities could be within the state, or outside the state if an interstate agreement for the facility's use was in place.

Acting under this authority, the EPA required every state to submit a "capacity assurance plan" detailing the sources, quantities, and characteristics of the hazardous wastes generated within its borders, and explaining how these wastes would be handled.[22] (The National Governors Association had recommended that, after preparation of the plans, a negotiation process take place among the states to allocate waste disposal responsibilities among themselves, leading to an EPA resolution if the states could not agree, but the EPA did not adopt this suggestion.)[23] Every state submitted a plan, and the EPA approved almost all of them,[24] even when states relied on new facilities that were later rejected or on facilities in other states that opposed importation.[25] There is no evidence that this process has led to the initiation or approval of any new hazardous waste facilities. New York State, which has a large HW landfill on which other states are relying, sued the EPA in 1992 over its refusal to sanction states that are not creating their own facilities.[26]

In 1993 the EPA changed its approach. Each state was directed to submit data on its HW disposal capacity and demand. Based on this, the EPA will determine if there are any *national* shortfalls. If there are, states whose demand exceeds supply in national shortfall categories will be required to submit waste minimization plans and other data. If shortfalls are still projected after that, further measures may be invoked.[27]

State Siting Processes

The federal government's unsuccessful efforts to site federal HW/RW disposal facilities and to require the states to site facilities have been mirrored by the states' efforts. Many states have tried and failed to build state-owned facilities and to encourage private companies to create privately owned facilities.

The two states that came closest to creating their own facilities were Maryland and Arizona. In 1984 the state of Maryland actually built an industrial waste landfill at Hawkins Point in Baltimore Harbor, on the site of an existing chrome ore treatment facility, only to close it after four months of operation because it could not compete financially with commercial facilities in other states that charged lower fees.[28] In 1981 the Arizona Legislature mandated the creation of an integrated hazardous waste treatment facility, including an incinerator and a landfill, in Maricopa County.[29] The state contracted with Ensco, Inc. to build the facility.[30] In 1991, when the plant was nearly complete, the state government, facing rising opposition, canceled the project and paid Ensco $44 million for its trouble, raising the money by selling two state prisons and leasing them back.[31]

Other states have abandoned siting efforts still earlier in the process. In 1975 Minnesota accepted a $3.7 million grant from the EPA to find a site for a hazardous waste landfill. The state identified sixteen sites but met fierce local opposition at each one; in 1978 the state gave up and returned the money.[32] In 1980 an official report to New York State recommended the creation of a privately owned treatment facility on state-owned land.[33] A 2,800 acre site was selected on Lake Ontario, where earlier a nuclear power plant had been proposed. However, there was overwhelming public opposition, and the state canceled the project in 1981.[34] New York State subsequently tried, twice, to construct a secure landfill to receive 250,000 cubic yards of PCB-contaminated sediment from the Hudson River,[35] but both attempts failed, once because of procedural violations[36] and once because of technical inadequacies in the site.[37] Both Georgia[38] and Kentucky[39] took formal siting initiatives but concluded that there is inadequate market for new facilities. Minnesota and North Carolina have begun trying to site their own facilities,[40] and a regional effort is underway in southern California,[41] but the outcomes remain to be seen.[42]

Efforts by states to encourage private firms to locate HW/RW facilities in their borders have fared little better. In the late 1960s and early 1970s, many states adopted legislation that centralized the permitting of power plants and other heavy industrial facilities.[43] With this as precedent, and under the prodding of the EPA,[44] at least thirty-six states enacted hazardous waste facility siting laws, mostly between 1979 and 1984.[45]

No two of these state laws are alike. They have been analyzed and compared many times,[46] and that analysis need not be repeated here. Suffice it to say that most of these laws provide for enhanced public participation and for technical siting criteria. They often create special siting boards to act on facility proposals. The laws vary considerably in the degree of state initiative in the siting process, from aggressively proactive to passively reactive. Some allow the states to preempt local authority, and some preserve local approval power.

Technical siting criteria, especially when combined with siting boards, advance the notion that there is an objectively "best" site, if only people with enough data, expertise, and wisdom can find it. This idea is implicit in many of the state statutes and has been made explicit by some commentators, several of whom see this search as the means to achieving public acceptance of unwanted facilities.[47] The siting criteria most commonly concern depth to groundwater; proximity to wells, surface waters, residences, property lines, and recreational areas; and avoidance of wetlands and endangered species habitat.[48] Elaborate multistage techniques have been devised under which the number of possible sites is progressively reduced by the application of successive "filters" or "constraints."[49]

As shown by the dismal record of siting attempts, these multistage techniques can be counterproductive. Michael O'Hare and colleagues have correctly pointed out that "[t]he general rule seems to be that rationalistic site selection by successive exclusionary judgments serves only to focus political opposition in the relatively small part of the state remaining after the exclusion process, while the broad consensus agreement on the particular criteria being used seems impossible to maintain after its implications become known."[50] Moreover, the criteria themselves, far from being objective, are necessarily laden with value judgments. How close is it acceptable to locate a waste facility to an elementary school—500 feet? 5,000 feet? Can a scenic vista be destroyed, and what precisely is a scenic vista? How much numerical weight should be given to impacts on drinking water versus impacts on endangered species? Experts are of little help in answering these questions.[51]

These screening methods can easily go awry. This occurred, for example, in January 1986 when the Department of Energy announced that one potential site for an HLW repository was the Sebago Lake batholith, a rock formation just six miles north of Portland, Maine, and adjoining Sebago Lake itself, the source of Portland's drinking water. When 3,000 angry people, including the governor, showed up for a public meeting with the DOE that lasted until 3:30 in the morning, the DOE beat a hasty retreat.[52] The DOE's experts had established numerical siting criteria that allowed good geology to overcome proximity to population and drinking water.

Even when siting criteria are agreed upon in advance, their application in particular cases can often be seriously questioned.[53] In fact, in every one of the numerous siting controversies in which I have been involved, the neighbors of the selected site were able to make serious arguments that the selection process was not procedurally or technically sound.

The perfect site is a mirage because the definition of perfection embodies so many contradictions. The perfect site would be far from any population centers, to reduce the risk of health effects, but it would be near a highway and close to where the waste is generated, to reduce transportation accidents; it would be in an area with no development, but it would not be within a wilderness area, a park, an agricultural region, or the habitat of rare species; it would have a high and deep clay layer but no water trapped on the top; and it would be in a region that benefited from the production of the waste to be disposed of, even though the most isolated and dry places in the United States, the western deserts, tend to be in states that generate little nuclear and hazardous waste.

The contradictions in a technical siting approach are highlighted by one peculiar incident that is only slightly beside the central topic of this book. A wealthy businessman, Sterling Clark, accumulated a magnificent art collection, but in the years immediately after World War II he grew concerned that this collection might not survive an atomic war. He resolved to build an art museum in a spot so remote that it could not possibly fall victim to war with the Soviets. He selected the bucolic college town of Williamstown, in the northwest corner of Massachusetts, and built the Sterling and Francine Clark Art Institute there. Ironically, not long after construction began in the early 1950s, the hydrogen bomb was developed, and a strike against the nearby industrial city of Pittsfield might well sweep away Williamstown as well. (Around the same time one of the nation's first nuclear power plants, Yankee Rowe, was also built in the same vicinity.) The institute still stands as a monument to Sterling Clark's artistic taste but not to his technological prescience.

Continuation and Expansion of Existing Facilities

It is a great irony that, at least so far, the principal environmental impact of stringent siting rules, ineffective siting strategies, and the illusory search for the perfect site has been to continue the life of old, substandard, poorly sited HW/RW facilities, so that most of this waste still goes to places picked by Leslie Groves, Frederick A. Beierle, and their counterparts. Of the twenty-one commercial HW landfills operating today, for example, only one is on a site selected since the enactment of RCRA in 1976. The EPA found that about 70 percent of all land-based HW treatment, storage, and disposal facilities would fail the EPA's current siting criteria for protecting groundwater.[54] One sample found potential releases of hazardous wastes from about 90 percent of such facilities.[55] Some old facilities still operate in locations that would be inconceivable under current rules. For example, Radiac Research Corp. operates a commercial hazardous and radioactive waste storage facility in a row building in Brooklyn, New York.[56] One facility in Niagara Falls, New York, was continually used for waste disposal from 1897[57] and received hazardous waste for decades until a state siting board finally ruled in 1990 "this site is not merely marginally unacceptable. . . . [T]he selection of the site is qualitatively so contrary to accepted siting principles that it would necessarily be rejected."[58]

When the waste management industry wants to add new capacity, it is much more likely to seek to expand existing sites than to move to new sites.[59] Both anecdotal experience[60] and formal public opinion research[61] confirm that communities are much more likely to accept expansions of existing HW/RW facilities than the introduction of new ones.

The law makes it immensely easier for companies to continue and expand existing facilities than to create new ones and much harder for opponents to shut down existing facilities than to block new ones. Statistical analysis demonstrates that citizens have seldom succeeded in shutting down existing HW facilities.[62] It is all but universal in regulatory law that products and activities in existence at the time of a rule's enactment receive far more lenient treatment, both substantively and procedurally, than do products and activities that are sought to be introduced later.[63] In an extension of the doctrine in zoning law that "prior nonconforming uses" may continue,[64] hazardous waste facilities have been held to have vested rights to continue their operations,[65] and in some states the "natural expansion doctrine" even requires municipalities to allow landfills and similar facilities to expand.[66]

When Congress enacted RCRA in 1976, it decided not to require existing facilities to meet the new siting and technology standards, for fear that most would have to shut down, leaving hazardous waste with no place to go.[67] Instead, facilities that filed a short form, called a Part A application, and met certain minimal requirements were granted "interim status," which allowed them to continue to operate.[68] A far more elaborate Part B application, typically running many volumes, had to be filed later.[69] By 1984, when Congress reauthorized RCRA, this process was moving so slowly that Congress grew impatient[70] and required landfills to file Part B applications and meet certain groundwater monitoring and financial responsibility requirements by November 8, 1985, or lose their interim status.[71] Only about one-quarter of all the then-existing HW landfills, but fifty of the fifty-nine largest commercial facilities, met this deadline.[72] As a result, most of the smaller landfills shut down, but the larger ones stayed open. Most of the facilities in the Nuclear Weapons Complex are also under interim status.[73] The old landfills still operating under interim status are subject to far laxer rules[74] than are new units.[75] For example, many have no liners to protect the groundwater, since the EPA determined that retrofitting them might do more harm than good.[76] Old hazardous waste incinerators under interim status are also subject to far laxer standards than those with new permits.[77] Certain facilities are also allowed to expand their capacity considerably while still under interim status.[78]

The Overlay of Permits and Other Regulatory Obstacles

The federal and state statutes designed to find sites for HW/RW facilities have received considerable commentary. Much less attention has been devoted to the dense overlay of permits that must be obtained and other legal strictures that must be met by HW/RW facility operators.[79] This focus on the siting laws has somewhat misdirected the academic inquiry, because practitioners know that the highest hurdles faced by project proponents arise in the permit laws, not the siting laws. To build a facility, a developer must obtain each and every required permit; to stop a project, opponents must merely block one. As Benjamin Walter and Malcolm Getz have pointed out, "Dispersing authority among independent veto points strikingly resembles a string of bulbs on a Christmas tree that have been wired in series. When one goes out, so do all the others."[80]

Stringency Spectrum
In an effort to illuminate the nature of these permit and siting requirements, I have arrayed them in what might be called a "stringency spectrum." This list assembles many of the statutes and regulations governing the permitting and siting of HW/RW facilities along a spectrum ranging from the most mandatory to the most prohibitory, starting with those that compel the siting of a particular facility in a particular place (which I call "Must" rules), through those that are essentially neutral, along to those that absolutely rule out a site or type of site ("Must Avoid" rules).[81] After presenting this spectrum, and explaining some ways that the stringency of certain rules is altered, I will discuss the spectrum's implications for the siting dilemma. In an effort to illustrate more fully the range of regulatory devices, I have gone beyond those laws directly applicable to siting hazardous and radioactive waste facilities, and have also discussed certain rules from related areas of environmental and land use law.

Note that not all siting and permitting rules fit neatly within one of these categories and that some rules are hybrids. There is also no indisputable way of ordering the categories; some of the categories toward the middle could arguably be rearranged, though this would not alter the basic point that the rules are quite lopsided in favor of opponents rather than proponents of siting. This list is of primary interest to lawyers; other readers may wish to skip to the discussion of the implications of the regulatory spectrum.

"Must" I am aware of only one "Must" siting law. The U.S. Supreme Court in 1978 ruled that the Tellico Dam in Tennessee could not be built because it might destroy the habitat of a rare fish, the snail darter, in violation of the Endangered Species Act.[82] In response, Congress added a rider to an appropriations bill providing that "[notwithstanding] the provisions of [the Endangered Species Act] or any other law, the [Tennessee Valley Authority] is authorized and directed to complete construction, operate, and maintain the Tellico Dam." The courts ruled that this explicit provision overrode all other federal and state statutes.[83]

There was also an unsuccessful attempt to adopt a "Must" law, though not exactly in the siting context. It was a provision of a 1968 highway law that provided that "[n]otwithstanding any other provision of law or any court's decision or administrative action to the contrary, the Secretary of Transportation . . . shall . . . construct . . . [the] Three Sisters Bridge."[84] It was enacted after a court had halted the bridge (crossing the Potomac River between Virginia and the District of Columbia).[85] Despite the statute's apparently unambiguous language, the Court of Appeals for the District of Columbia, offended that the law was enacted as part of a crude political deal, halted the bridge again.[86] This bridge was never built.[87]

"Must Unless" These laws specifically designate a particular site but do not purport to rule out any other permitting requirements. One such law designated Yucca Mountain for the nation's HLW repository.[88] Another picked a site for Arizona's radioactive waste disposal facility,[89] though it was subsequently repealed. Yet another named a site for an LLRW facility in Texas.[90]

"May Unless"   These rules do not designate a site, but they say that an applicant may take a particular action unless an authority stops it. For example, while discharge of dredged or filled material in navigable waters ordinarily requires a permit from the Army Corps of Engineers,[91] Congress has authorized the Corps to issue "general permits" for categories of activities it determines will have "only minimal adverse environmental effects."[92] Under this authority, the Corps has listed numerous categories and announced it does not even want to be informed when they occur, although the Corps retains the discretion to require individual permits for such activities when it chooses.[93]

A different sort of "May Unless" permit (though one requiring at least a permit application) arises as part of the same program. The Corps' rules for individual dredge and fill permits provide that "[i]n the absence of overriding national factors of the public interest that may be revealed during the evaluation of the permit application, a permit will generally be issued."[94] (This stands in marked contrast to the EPA's guidelines under the same program, which reverse the presumption and state that no discharge shall occur "unless it can be demonstrated that such a discharge will not have an unacceptable adverse impact.")[95]

"May"   Rules in this category create no presumptions either way, but they clear away an otherwise-applicable constraint. The chief example is a zoning ordinance that makes a particular activity an allowed, as-of-right use, not requiring further zoning approvals, but not exempt from nonzoning requirements. Another example is the "declarations of non-navigability" periodically enacted by Congress to exempt the building of structures in particular plots of water from the usual Corps of Engineers permit requirements.[96] Also within this category are statutes exempting named projects from particular permit requirements that have proven troublesome.[97]

Between "May" and "Consider Avoidance" is the point of neutrality. Everything before this point makes it easier to build a facility in a particular spot; everything after it makes it harder to build there.

"Consider Avoidance"   Some laws require decision makers to consider avoiding certain kinds of sites, but do not go so far as to mandate an actual effort to avoid those sites. One example is a provision of the London Dumping Convention requiring contracting parties to consider alternative land-based methods before allowing the incineration of hazardous waste at sea.[98] Another is the New York City Fair Share Rules, which require city agencies to consider whether they are placing undesirable facilities in neighborhoods that are already overburdened by such facilities.[99]

"Try to Avoid"   Many environmental laws contain provisions that are essentially aspirational. They urge the public and private sectors to avoid damaging the environment, but they are neither self-implementing nor enforceable.[100] Professor Harold P. Green has identified fifty-two federal statutes that call for preservation of the environment to protect future generations, references which he says "are more aptly characterized as rhetorical statements of congressional intent rather than provisions of substantive law."[101] Among the laws proclaiming such aspirations are the National Environmental Policy Act,[102] the Clean Water Act (which still declares "the national goal that the discharge of pollutants into the navigable waters be eliminated by 1985"),[103] the Clean Air Act,[104] RCRA,[105] and the Nuclear Waste Policy Act.[106] The contracting parties to the London Dumping Convention also urged each other to prohibit the burial of radioactive waste under the sea floor.[107] Several federal and state statutes express a preference for waste reduction or recycling over incineration or landfilling,[108] but such laws have been held not to inhibit the construction of incinerators.[109]

"May If"   This category covers laws allowing applicants to obtain permits if they meet required conditions. Water pollution permits under the National Pollutant Discharge Elimination System[110] and air pollution source permits under the Clean Air Act[111] fit here, as do requirements under zoning laws for special permits or for variances, although under different enactments the decision makers have varying levels of discretion to deny approvals.

Some legislative bodies have used the subterfuge of "May If" laws to enact what are really intended as "Must Not" laws. For example, the North Carolina legislature, trying to halt a hazardous waste facility, enacted a statute saying that no such facility may discharge effluent without meeting certain conditions that were, in reality, virtually unachievable.[112] As another example, in the statute on siting an HLW repository, Congress required the DOE to phase out research "designed to evaluate the suitability of crystalline rock as a potential repository host medium."[113] The intent of this deceptively bland language was to preclude a repository in any of the eastern states, where the most likely sites have granite, a kind of crystalline rock.[114]

One particular kind of "May If" law requires that certain activities or impacts be disclosed to the public, but does not regulate those activities or impacts, though it often provides ammunition for actions under more prescriptive laws. The principal examples are the environmental impact statement requirement of the National Environmental Policy Act (NEPA)[115] and its state counterparts, and the toxic chemical release forms mandated by the Emergency Planning and Community Right-to-Know Act.[116] The latter has been extremely successful in encouraging waste reduction; the lists of the companies with the highest volumes of pollutants are routinely printed in the newspapers, and, as one commentator said, "[C]ompanies such as du Pont [which is often at the top of the list] are committing to waste reduction programs aimed at getting them off the front pages."[117] (Such rules fit within the long tradition of laws that attempt, through disclosure, to bring legal or moral pressure on those engaging in undesirable activity, such as gambling, large cash transactions, and insurance redlining.)[118]

Another law that fits more or less within this category is the Coastal Zone Management Act, which requires federal actions that affect the coastal zone to be consistent "to the maximum extent practicable" with state-adopted coastal policies.[119]

"May Need to Avoid if Agency Says So"   Several federal statutes say that agencies *may* designate certain areas as unsuitable for development. These laws take on real significance only if the agencies exercise that power. Examples include the Surface Mining Control and Reclamation Act (SMCRA), which allows lands to be declared unsuitable for mining,[120] and the National Historic Preservation Act, which allows the secretary of the interior to designate historic properties.[121] The EPA may declare a wetland to be a critical area, barring the issue of dredge and fill permits under the Clean Water Act.[122] Similarly, the EPA may prohibit dumping in particular sites in the ocean.[123]

"May, at Higher Cost"   Other laws allow an activity to go forward, but extract a financial penalty for doing so (or provide a financial incentive for not doing so). The Flood Disaster Protection Act helps communities obtain flood insurance if, among other things, they avoid siting buildings in flood hazard areas.[124] Impact fees, usually imposed at the municipal level (and allowed in some but not all states), assess developers for the public costs they impose.[125] Requirements that HW/RW facilities compensate neighboring communities would also fit within this category.

"Unless Avoid, Will Take Longer"   Numerous large projects have died, not because they failed to receive all required permits but because obtaining the permits took so long.[126] For this reason, the mere threat of having to get a permit amounts to a strong regulatory disincentive and frequently leads to redesign of projects to avoid regulatory jurisdiction. As one example, the builders of an MSW-burning incinerator in upstate New York changed their air pollution control system from a wet-filter to a dry-filter technology, because the latter would not require a water pollution permit.[127] Numerous developers have avoided intrusions into wetlands and waterways to avoid the notoriously protracted permitting procedures of the Corps. This effect is felt even if the delay involves no required permits, but only a lengthy consultation process, such as that provided under the National Historic Preservation Act.[128] (Where "Unless Avoid, Will Take Longer" laws belong on the spectrum relative to "May, at Higher Cost" laws will largely depend on the length of the likely delays and on the time value of money.)

"Must Avoid Unless"   This large and important category includes numerous requirements with genuinely sharp teeth. The rules here prohibit facility siting unless some (often difficult) condition is satisfied. One such statute is Section 4(f) of the Department of Transportation Act, which prohibits any federal transportation project "which requires the use of any publicly owned land from a public park, recreation area, or wildlife

and waterfowl refuge . . . or any land from an historic site . . . unless
. . . there is no feasible and prudent alternative to the use of such
land."[129] The Clean Air Act has barred significant new emissions into
clean areas unless those emissions are offset, to varying degrees, by
corresponding reductions in emissions from other sources.[130] Other statutes are the SMCRA's ban on surface mining in national forests, unless
certain findings are made concerning the lack of significant recreational,
timber, economic, or other values,[131] and the Clean Air Act's ban, subject
to a state-issued variance, on major new sources that could lower air
quality over certain large national parks or wilderness areas.[132] The Safe
Drinking Water Act's prohibition on federal projects that might contaminate designated sole-source aquifers reads like a "Must Avoid" law, but
the EPA has so much discretion in determining the threat to the aquifer
that it is actually a "Must Avoid Unless" law.[133]

Many "Must Avoid Unless" rules are EPA regulations. For example,
the EPA has promulgated standards under RCRA barring hazardous
waste landfills in floodplains unless certain assurances are provided that
floods will not cause contamination;[134] standards under TSCA preventing PCB landfills from being placed in floodplains, shore lands, groundwater recharge areas, or certain geologic formations, unless the EPA
grants a waiver;[135] and rules inhibiting the siting of MSW landfills in
floodplains, in wetlands, near seismic faults, or in unstable areas, unless
stringent requirements are met.[136] EPA regulations also ban ocean dumping of contaminated material[137] unless there is a strong reason to proceed
with the dumping nonetheless.[138] The NRC bars LLRW disposal facilities
below the water table, unless certain showings are made concerning
radionuclide movement.[139] State laws protecting farmland from destruction are often relevant in HW/RW siting cases.[140]

Finally, a common-law rule, the public trust doctrine, inhibits the sale
or destruction of certain underwater lands.[141]

"Must Avoid"   At the end of the stringency spectrum are those laws
banning, absolutely and without any exceptions, facility siting in particular places or kinds of places. The classic example was the ban in the
Endangered Species Act (ESA) on projects that could drive a species to
extinction by destroying essential habitat. After the Supreme Court ruled

that the construction of the Tellico Dam in Tennessee could not proceed because it might wipe out the snail darter,[142] Congress not only enacted a special "Must" law for this particular dam, but also amended the statute to create a body (dubbed the "God Committee") that can declare exemptions from the ESA in other cases.[143] (This law became an issue again in the 1992 election, when President Bush derided the protections afforded the spotted owl.) Similarly absolute (except for grandfather clauses) are provisions of the Wilderness Act that bar commercial enterprises and permanent roads in wilderness areas,[144] and of the SMCRA that bar surface coal mining in national parks, wildlife refuges, and other protected areas.[145] As noted in chapter 2, in 1988 Congress absolutely barred the ocean dumping of sewage sludge, though the courts found ways to grant limited extensions.

Some "Must Avoid" rules protect particular locations from specific projects. These include the law barring an HLW repository in any location other than Yucca Mountain;[146] a statute prohibiting "dispos[al] of sewage sludge at any landfill located on Staten Island, New York";[147] several Massachusetts statutes barring use of three named locations for HW disposal, after they had been recommended for such use;[148] and a New York State regulation precluding the siting of an LLRW facility at the West Valley site.[149]

The EPA has absolute rules barring hazardous waste facilities[150] and solid waste landfills[151] in areas prone to earthquakes, and prohibiting liquid hazardous waste disposal in salt formations *except* at the WIPP site.[152] The Nuclear Regulatory Commission bars LLRW facilities in one-hundred-year floodplains, coastal high-hazard areas, and wetlands.[153]

Methods of Altering Stringency

Despite the words of an enactment, its effective stringency can be altered in a number of ways. One major device is to restrict judicial review, so that an agency's violation of its mandate cannot be overturned. Though there is a general presumption in favor of reviewability of agency action,[154] several pertinent statutes have partially or totally barred review. These include federal laws on the cleanup of Superfund sites,[155] on preliminary activities in siting HLW facilities,[156] on air pollution over national parks,[157] and on military base closure.[158] Several commentators have advocated such limitations on judicial review as a means of expediting siting decisions.[159] Of similar effect are limitations on the standing of environmental plaintiffs,[160] rules that large bonds be posted to obtain preliminary injunctions,[161] extremely short statutes of limitations,[162] nearly insurmountable standards of judicial review,[163] and other barriers on access to the courts.[164]

Another way to alter the stringency of siting laws is to carve out exemptions. I have already discussed grandfathering. CERCLA exempts from federal, state, and local permitting requirements any remedial action conducted entirely on-site.[165] WIPP is exempt from licensing by the NRC because it will accept only defense-related waste and no spent commercial fuel.[166] Under the "functional equivalence" doctrine, the EPA has been held largely exempt from NEPA,[167] and in some states actions taken pursuant to consent decrees are exempt from environmental review requirements.[168] Certain Clean Air Act requirements are relaxed in designated economic development zones.[169]

Implications

Four things are striking about this stringency spectrum. The first is that it is so lopsided. Far more laws are bunched toward the "Must Avoid" end than the "Must" end. Project opponents have much fuller quivers than do project proponents. This leads to inertia; if new facilities cannot be built, old ones live on.

Second, the extremes are unstable. There was only one successful "Must" law, which arose from a highly unusual set of circumstances, and when push comes to shove—as it did with the Endangered Species Act—the "Must Avoid" laws are vulnerable. This is entirely in keeping with the law's general abhorrence of absolute requirements which admit of no exceptions or flexibility. The failure of absolute laws has often been noted both in the environmental context[170] and in other areas.[171] Thus it is hard to mold the magic bullet—the ironclad guarantee that a facility will or will not be built.

The third striking feature of the stringency spectrum is how often the federal and state legislatures stepped in and made site-specific decisions, often with thinly disguised intent, sometimes based on scientific findings

Table 4.1
Stringency spectrum:
Rules facilitating or hindering facility construction in particular locations

Nature of rule	Examples
Must	Tellico Dam
Must Unless	Yucca Mountain law [but severely delayed]
	Arizona HW site designation [repealed]
	Texas LLRW site designation [pending]
May Unless	Army Corps of Engineers permits
May	Zoning
(Point of neutrality)	
Consider Avoidance	London Dumping Convention
Try to Avoid	Aspirational statutes
May If	Clean Water Act (NPDES) permits
	Clean Air Act permits
	National Environmental Policy Act
	Emergency Planning Community Right-to-Know Act
May Need to Avoid if Agency Says So	National Historic Preservation Act
May, at Higher Cost	Neighbor compensation requirements
Must Avoid Unless	Safe Drinking Water Act—sole-source aquifers
	Section 4(f) (parklands, historic sites)
	EPA location standards for floodplains
	Endangered Species Act
Must Avoid	Wilderness Act
	Ocean Dumping
	NRC location criteria for floodplains, wetlands

and sometimes not. Regardless of the millions spent on studies and experts, in the end it is often the politicians who make the final determinations. Thus the search for scientific rationality in the siting process is illusory as long as important constituencies are unhappy with the scientific results.

Finally, it is remarkable how many of the laws in the strongest negative categories—"Must Avoid" and "Must Avoid Unless"—are aimed at preserving wildernesses, endangered species, wetlands, parks, and historic buildings. These irreplaceable resources are preserved for future generations, in a strong expression of society's moral and aesthetic values,[172] while current community concerns are largely relegated to the political process. Health impacts, the primary basis for most opposition, are implicit in the laws protecting groundwater, air quality, and the like, but they tend not to enjoy the same favored status on the stringency spectrum as does protection of natural areas. Lawyers for project opponents are thus forced to focus on grounds that are often well removed from their original clients' basic concerns—a paradox exemplified by the snail darter that stopped the Tellico Dam and the striped bass that killed the Westway highway.[173] There is only limited overlap between the subjects discussed in the permit hearings and lawsuits about a project, on the one hand, and the community and political meetings about the project, on the other. As Idaho Gov. Cecil Andrus said upon blocking a shipment of HLW: "The legal grounds are not near as important as the moral and political grounds, and I can use the courts till you can step on my beard."[174]

Table 4.1 summarizes the stringency spectrum but limits its listings to only the most important rules directly applicable to HW/RW facility siting. The lopsided nature of the rules becomes immediately apparent: far more laws are on the "Must Avoid" than on the "Must" side of the point of neutrality.

Given the often decisive role of legislatures in selecting sites, the legal processes can be effective at stopping projects by causing one of the Christmas tree lights of the permit process to go out, but they can do little to help build facilities.

5

Evaluating the Current Siting Processes

At the outset of this book, I framed the problem as how to find the system of HW/RW management that maximizes social welfare, taking full account of the social and environmental costs, while achieving fairness. I am using the term "social welfare" as a composite measure of well-being, accounting for the full social benefits and costs accruing to the community.[1] By speaking of both social welfare and fairness, I mean to incorporate both economic and noneconomic factors. The purpose of chapter 5 is to evaluate the current system by these broad criteria. This requires addressing five questions:

1. Does the system allow the sound remediation of waste that has already been created but still lingers, while also providing sufficient disposal capacity for waste that, despite efforts at waste minimization, will be created in the future?
2. Does the system ensure that the full costs of disposal facilities are borne by the users of the facilities?
3. Does the system protect health and the environment?
4. Is the system fair?
5. Is the system politically viable?

In the course of this discussion, I will also explore some of the hidden economic forces and psychological factors at play in the siting process.

Needed Disposal Capacity

Federal and state governments are taking extraordinarily intrusive steps to site HW/RW facilities. The federal government is trying to force facilities on the states, and states are trying to force facilities on municipalities, through incursions on the normal concepts of sovereignty and home rule. There must be some compelling rationale to justify these steps.

The explanations vary between hazardous waste and radioactive waste. For HW, the usual reasons given are that there is a serious shortage of disposal facilities; that without more facilities, illegal dumping will be rampant; and that a shortage of facilities will harm the economy. For RW, the usual reasons offered are that government control is necessary to prevent fissile materials from falling into the wrong hands; that anything less than permanent disposal is unsafe; and that the peaceful uses of nuclear energy require more facilities.

Hazardous Waste

A severe shortage of HW facilities is usually assumed.[2] There is a long history of crying wolf about pressing demands, both for waste disposal facilities[3] and for large unwanted installations in general.[4] It is therefore necessary to look more closely to see if the wolf is really at the door. The question is so difficult that one state siting board conducted a weeklong trial to determine whether a new hazardous waste landfill was needed, and in the end the board threw up its hands and said it could not decide.[5] Many agencies examining the issue take a private company's interest in building a facility as *prima facie* proof of need, reasoning that the company would not otherwise risk its own capital.[6] But this merely proves that the company believes the facility can turn a profit, not that society desperately needs it; that Chrysler still wants to sell cars does not prove there is a shortage of cars. There is an important distinction between market demand (which merely means that someone will buy a product) and societal need (where the absence of something has such acute effects that government should intervene to supply it). The presence of a market demand for Madonna videos and (for different consumers) Ninja Turtle movies does not mean that state intervention to supply either of these products is warranted. The presence of a market demand for HW disposal begs the question of whether government should intervene to increase the supply.

The need for new facilities is also justified by the argument that, without them, illegal dumping will increase. As will be shown, this too has no basis in fact.

A pertinent question is whether people seeking to dispose of hazardous waste can consistently find a lawful place to send it; the answer seems to be yes, as evidenced by the national estimates of an adequate or even excessive supply of disposal capacity. As noted in detail in chapter 2, there are ample hazardous waste landfills and incinerators. By simple reference to the Yellow Pages (typically under the heading "Waste Reduction, Disposal, & Recycling Services"), one can find hazardous waste brokers who will gladly connect waste generators with waste transporters and disposal facilities. Other brokers arrange for the pickup of LLRW for shipment to licensed facilities.

To be sure, prices have soared. Between 1976 and 1991—a period during which producer prices doubled[7]—average waste disposal costs increased from less than $10 to more than $250 per metric ton for landfilling and from about $50 to more than $2,600 per metric ton for sludge incineration.[8] (Since then, prices have dropped somewhat.)[9] It seems likely that the limited number of disposal facilities caused much of this price increase, by enabling the few remaining facilities to charge more for their services. A related major reason for the price increase must be RCRA regulations and CERCLA liability, which have made it much more expensive to provide HW disposal services and at the same time have reduced the supply of such services. Whether these price increases have somehow been a drag on the economy as a whole does not seem to have been studied. But it is clear, as the following discussion will show, that rising prices for waste disposal significantly reduce waste generation. Given the variability of price and the elasticity of demand, there is no "shortage" of HW disposal facilities in strict economic terms, as quantity demanded does not exceed quantity supplied.

Radioactive Waste
The federal government has assumed (though not yet fulfilled) responsibility for off-site disposal of HLW and TRU, for the compelling reason that plutonium must be held securely. Since LLRW is not useful to putative bomb makers, its disposal is a commercial enterprise. There is plainly no surplus of radioactive waste disposal capacity; Yucca Mountain has not been built, WIPP is built but not open, and only three commercial LLRW disposal facilities remain operating. There is,

however, a debate over whether the paucity of RW facilities has major adverse impacts.

The NWPA statutorily determined the need for the Yucca Mountain facility.[10] Critics claim there is no pressing need for a repository because HLW can be safely stored at reactor sites for many decades. They argue, moreover, that the principal impetus behind the Yucca Mountain project was that "[t]he lack of a disposal solution had long been a political albatross around the neck of the nuclear industry."[11] In 1989 the NRC determined that no significant safety or environmental impacts would result from a delay in the availability of Yucca Mountain until 2025,[12] and in 1990 the National Academy of Science concluded that continued at-reactor storage of spent fuel should be safe for at least one hundred years.[13] Both on-site storage, and the development and eventual operation of that portion of the Yucca Mountain facility to be used for commercial HLW, are financed by the nuclear utilities, which pass the cost along to their ratepayers.

The absence of facilities—namely, Yucca Mountain and WIPP—for the disposal of waste from nuclear weapons production poses greater problems. Much of the waste destined for these facilities is kept in crude conditions, posing a significant threat to the environment.

The economics of LLRW resemble those of hazardous waste, except that LLRW appears to be even more price elastic. Nationwide LLRW volumes declined by about half between 1981 and 1989. These quantities are expected to drop still further as the remaining LLRW repositories increase their disposal charges to almost $300 per cubic foot, and waste generators learn that materials substitution and better operational practices can reduce the amount of LLRW created.[14] Temporary shutdowns of disposal facilities have, however, briefly disrupted the operations of some LLRW generators.[15]

Internalization of Costs

Disposal costs significantly affect the demand for facilities for commercial HW/RW disposal. (Waste generation from nuclear weapons production does not seem sensitive to disposal costs.) Under classical economic theory, the market is distorted if the price of a good or service does not fully reflect its social cost.[16] Thus if the price of waste disposal is artificially low, then the amount of waste generated will be inefficiently high.[17]

The current system of HW/RW facility siting, if it succeeded in siting facilities, would keep the price of waste disposal artificially low in two important ways. First, the cost of building facilities (which is ultimately reflected in disposal prices) would be lowered by federal and state override of local zoning controls. Second, under prevailing tort doctrines, facility operators would be (and indeed are) able to escape payment for many of the external costs they impose on their neighbors. I will now discuss these two market distortions.[18]

Zoning Override

One of the favorite legislative techniques in siting HW/RW disposal facilities is to override local zoning and other land use controls. Removing a zoning restriction from a piece of land ordinarily provides a financial benefit to the property owner.[19] Every developer knows that securing the consent of local officials to a project with hostile neighbors is an arduous, expensive process that often requires community compensation, reductions in project size, and changes in design. Eliminating the issue of consent would be a tremendous benefit to the developer. Assuming a competitive market for waste disposal, a zoning override at disposal facilities might well be reflected in lower disposal prices.

While in practice, as shown in chapter 7, zoning overrides have been unsuccessful in HW/RW siting, this analysis suggests that, even if they worked, they would be economically inefficient. In the words of Michael F. Sheehan, "Allowing economic bargaining between the waste facility industry and the communities involved would, within the logic of the economic system, establish a market where the value of a particular site could be weighed relative to other sites and relative to the feelings of the local population. Prices established in this way would internalize a range of local costs otherwise not included in the price of services offered to waste producers or in lower profit rates."[20]

External Costs

If HW/RW facilities are able to inflict costs on their neighbors without compensation—in economists' jargon, if the negative externalities are not internalized—then the neighbors are, in effect, subsidizing the waste generators.

The state of the art in quantifying the externalities from waste disposal facilities is extremely crude.[21] In recent years numerous studies have been performed, however, on the effects of such facilities on one useful measure of externalities—property values.[22] Most of these studies show a strong negative correlation between proximity to an HW/RW disposal site and property values,[23] especially after publicity concerning the site[24] or concerning other contamination incidents.[25] A strong negative effect can result from the mere announcement that a facility will be built.[26] A few studies found no negative impacts,[27] and several studies examined theoretical or practical issues without determining the effects.[28] Overall, evidence exists that in at least some communities HW/RW facilities are lowering property values.[29]

Compensation for these losses is quite limited. The NWPA provides for compensation to state and local governments and Indian tribes for their financial losses in the development of HLW repositories and storage facilities[30] but does not provide for compensation to private parties. The other major federal siting statutes do not provide for compensation at all. CERCLA allows private parties to recover the "response costs" they suffer in investigating and cleaning up hazardous substances,[31] but this does not include personal injury or property damage.[32] A few state statutes provide for damage awards against HW facilities for property damages,[33] while other state statutes arguably preclude such awards.[34]

In general, however, neighbors are left to common-law tort remedies.[35] Although RCRA does not preempt such remedies,[36] both practical and doctrinal problems accompany their use. Among the practical problems are the multiple sources of contamination that are likely to make proof of causation difficult; the long latency periods for most toxic injuries; the resemblance of the illnesses caused by toxic substances to diseases stemming from other causes; the high costs of litigation; and the difficulty in finding a lawyer to take a case on a contingency basis when the prospect of success is so uncertain. The principal doctrinal problem is that most applicable tort remedies, such as nuisance, look not only to the injury suffered by the plaintiff but also to the social utility of the actions of the defendant, thereby denying redress to many people injured by activities deemed by the courts to be socially necessary.[37] On one occasion, plaintiffs proceeding under the nuisance doctrine obtained, after a lengthy trial, an order shutting down a hazardous waste facility,[38] but just as often the neighbors are defeated.[39] The barriers to recovery become even higher when the damage is anticipated but has not yet occurred; the harm must generally be both imminent and highly probable before plaintiffs can prevail.[40] Conversely, if the damage was inflicted long ago but only recently discovered, in some states the applicable statute of limitations will already have expired.

Efforts to stop[41] or to obtain compensation for[42] on-site investigations before final siting decisions are made have been similarly unavailing. (Where efforts in court failed, self-help has been more successful. Citizens in Cortland County, New York, opposing efforts by the administration of Governor Mario Cuomo to test a proposed LLRW site, established a trailer they called "Checkpoint Mario" to sound the alarm if state inspectors were seen; the demonstrations that ensued whenever the inspectors visited were so disruptive that the tests were never completed, and the site was eventually dropped. In Sweden, a nonstop watch has been kept by citizens on the roads leading to Kynnefjall, a proposed radioactive waste repository, since April 21, 1980; as of 1992 it was still in operation.)[43]

As will be discussed in greater detail, among the most common, and severe, impacts of the HW/RW siting process is the emotional anguish suffered by the neighbors of planned facilities. A few courts have awarded damages for such fears,[44] but these cases are very much the exceptions.[45] One state statute calls upon the HW facility licensing agency to consider "community perceptions and other psychic costs,"[46] while the DOE has taken the opposite stance and advised that, in siting an HLW repository, "[p]erceived risk . . . is not an appropriate topic for general repository-siting guidelines; it is a subjective condition that cannot be fairly compared among sites."[47] A West Virginia statute allowed the state to deny a permit for a solid waste facility that was "significantly adverse to the public sentiment." The Fourth Circuit, however, declared the statute unconstitutional because it bore no substantial relationship to the state's legitimate interests.[48]

In a hearing on the expansion of a hazardous waste landfill in Niagara Falls, New York, five days of testimony were taken on the facility's psychological impact on a community that was, the opponents argued, already scarred by Love Canal. The state environmental commissioner ruled that "[a]s a public policy matter, if the Department were to deny an application for a facility after concluding that it met all regulatory criteria and that the risk of its construction and operation was within acceptable limits merely because of fears in the host community, the agency would be abdicating its responsibility. . . . Therefore, I conclude that any psychological impact caused by this facility cannot, standing alone, be grounds for denial of the applications."[49]

After reviewing the law on recovery for psychic distress, Roger A. Bohrer aptly summarized the issue:

At the very least, the issue of emotional distress recovery in the face of techno-logical risk and uncertainty may be seen for what it is—a social choice between subsidy and compensation. A decision to impose liability and to require the internalization of "psychic costs" would *not* stop progress altogether, but it would simply make the products of new technology cost more in the marketplace. By forcing the market to recognize the social costs of technology, a more socially desirable level of consumption of technological products is achieved.[50]

It is notable that psychic costs receive so little recognition in the formal facility siting process, while laws protecting aesthetic and moral values, such as the preservation of wilderness and endangered species, weigh so heavily. In any event, the foregoing shows that the legal system, by denying recovery to many of those injured by HW/RW facilities, subsidizes the generation of hazardous and radioactive waste and would therefore encourage more than the socially optimal number of disposal facilities. That is not to say that economic efficiency is best served by exclusive reliance on a liability system; if the psychic damage of living near a facility were $10,000, but the person could move for $6,000, the socially optimal rule might limit payment to moving costs (plus, perhaps, some increment for the psychic costs of moving). The current siting laws do not provide even that compensation, however. People who are evicted from their property are paid; the neighbors who stay behind are not. The people who own land on the site itself receive fair market value, so, unless they have a strong emotional attachment to their land, they are far more fortunate than those just outside the site boundaries.

In the absence of compensation, the protection for the neighbors lies primarily in the political system, which (unlike the liability system) is exquisitely sensitive to psychic costs, especially when they are played out on the evening news. This protection is often exercised in a blunderbuss fashion. Unable to mount a pinpoint defense, elected officials protect their constituents, the neighbors, best by injecting a paralytic drug into the system. The decision-making process is dragged out for many years; property owners are in long-term limbo because no one will buy or finance homes or land lying under such a large cloud; and the old grandfathered facilities keep chugging along.

Protecting Health and the Environment

Does the current system of HW/RW facility siting protect health and the environment? Much of the pertinent information has already been presented. At this point, I will address two additional issues. First, since the current siting system perpetuates the life of old facilities while sites are sought for modern new ones, I will look at the widespread notion that new, modern facilities can operate with few environmental impacts. Second, I will test one of the key assumptions underlying current siting laws—that a shortage of facilities increases illegal dumping.

Impacts of New Facilities

Much of the siting literature assumes that new HW/RW disposal facilities can be built and operated with a high degree of health and environmental safety.[51] The environmental impact statements and health risk assessments for these facilities (typically prepared by their proponents)[52] usually predict that the risks will be trivial.[53] The actual evidence, however, is far less clear. In part this is due to the enormous uncertainty in the practice of risk assessment, requiring risk assessors to make scores of subjective judgments from inconclusive data.[54] In one not atypical incident, the opponents and proponents of a proposed facility prepared risk assessments that differed by three orders of magnitude.[55] Since there are so few new HW/RW facilities, it is hard to answer confidently whether new facilities will offer much greater protections than old ones.

Epidemiological evidence of actual health effects would certainly be relevant. After an exhaustive review, the National Research Council concluded that there are insufficient data to determine whether or not hazardous waste sites, on a nationwide basis, pose a serious threat to the public health.[56] Another nationwide literature survey found that "residence in counties where hazardous waste sites are located is associated with higher total cancer mortality rates and higher rates of digestive and respiratory cancers," but causation cannot be established based on the present data.[57] Researchers for the California Department of Health Services found that residents near inactive hazardous waste disposal sites complain more often of subjective symptoms (such as headache, nausea, and eye and throat irritation) than do people in control neighborhoods but do not have greater incidences of cancer or birth defects.[58] One survey found that women living within a mile of an inactive hazardous waste site have a 12 percent greater chance of bearing a child with a major birth defect than do women living more than a mile from such a site, but the study acknowledged that cause and effect could not be demonstrated, especially since so many waste sites are in industrial areas.[59] The Congressional Office of Technology Assessment similarly found that the public health impacts from nuclear weapons complex facilities are "plausible, but unproven."[60]

There does seem to be clear evidence that residents near known hazardous waste sites experience a high incidence of headaches, nausea, vomiting, fainting, skin rashes, depression, and anxiety; some or all of these symptoms may be caused by the stress of having such a neighbor rather than by physical pollutants.[61] Illnesses with a psychological origin are no less real, but they cannot readily be addressed through better engineering design of the facilities. One might surmise that a new facility in a volunteer community might cause less stress than an older unit that is hated by its neighbors, but the available data do not allow this theory to be confirmed.

Actual operating experience is also relevant to the impacts of new facilities. The cases reported every week in the *Toxics Law Reporter* attest to the personal injury and property damage caused by prior methods of handling hazardous waste. The hundreds of civil and criminal enforcement cases brought by the EPA under RCRA every year attest to the violations committed by many current hazardous waste operators.[62] About one-third of all hazardous waste is managed in facilities where at least one serious violation was found during the last inspection.[63] It also appears that the enforcement cases actually brought under RCRA are but a small subset of those that could be prosecuted if a more thorough inspection system were in place.[64] The available statistics unfortunately do not differentiate old from new units.

Most indicative of the likely performance of new HW/RW facilities is the experience of the most recently built units—those constructed (usually on existing sites) in the 1970s and 1980s. The largest hazardous waste landfill in the country is the Emelle, Alabama facility of Waste Management, Inc., the largest company in the industry. In a recent lawsuit, it was found to have had releases of hazardous substances and noxious fumes into the environment on several occasions.[65] Waste Management's large HW landfill in Niagara County, New York, has had similar experiences,[66] as have several HW landfills operated by the second-largest company, Browning Ferris Industries.[67]

Events at incinerators have proven even more troubling. In 1992 regulators closed the largest commercial HW incinerator in the country, located in Chicago, after finding improper operating practices such as disconnected pollution monitoring devices, the burning of unpermitted wastes, and false labeling of waste barrels. The experience at many other commercial incinerators is not much better.[68] A court ordered another large incinerator, in North Carolina, to shut down in 1989 following ten years of trouble-filled operations.[69] A brand-new hazardous waste incinerator in New York, built after years of permit proceedings, self-destructed in 1987 during trial burns and never opened.[70] Unannounced inspections by the EPA and OSHA of twenty-nine HW incinerators in 1991 found 395 violations of standards, two-thirds of which the agencies considered "serious".[71] Elevated levels of PCBs have been detected in rodents captured near a new HW incineration/treatment/landfill facility in Swan Hills, Alberta, Canada, and in the blood of about half a dozen waste handlers at the plant.[72] Several epidemiological studies have found elevated levels of respiratory and other disorders near commercial HW incinerators.[73] At a controversial new HW incinerator in Ohio, dioxin emissions during the test burn were up to five times greater than the level used in the risk assessment.[74] Examinations of the results of trial burns of several other hazardous waste incinerators show that emissions of certain heavy metals, particularly cadmium and chromium, may pose significant risk to public health.[75] In 1992 the EPA expressed concern that even well-operated incinerators were having difficulty meeting permit limits for dioxin.[76] In sum, even state-of-the-art disposal facilities can, and do, fail in a multitude of ways.[77]

Occupational, as opposed to public, health risk has been well established. Numerous instances of occupational diseases among workers at nuclear weapons complex facilities[78] and in the hazardous waste[79] and solid waste[80] industries have been documented. Especially perplexing is the widespread practice in the nuclear industry of using large numbers of temporary, often unskilled workers to absorb exceptional doses of radiation. At the West Valley radioactive waste facility, for example, temporary workers were often paid half a day's salary for a few minutes work (such as walking into a highly contaminated room, moving a box or turning a valve, and walking out again), during which they would accumulate a quarter year's allowable exposure.[81] This practice blurs the line between the public and the workforce.

Although these incidents did not necessarily involve injury to public health, they certainly challenge the commentators' assumptions of assured safety. They also suggest that new facilities will be no panacea and that reducing the creation of HW/RW will still deserve high priority. This conclusion is reinforced by the fact that the federal regulations governing the design and operation of HW/RW facilities clearly allow even new units to pose a residual, though slight, health risk.[82] Moreover, environmental risks in the transportation of hazardous wastes are at least as great as those in storage and disposal.[83]

The ambiguity of the evidence about the health effects of existing HW/RW facilities is also an important reason for the opposition to future facilities. If it is difficult to glean any reliable idea of the impacts of an operating facility, where the necessary factual evidence is presumably there for the plucking, it must be virtually impossible to know for sure the effects of a planned but unbuilt unit (especially one using a new technology or an ill-defined waste stream), where a heavy dose of the crystal ball is necessary.

The lessons of history are hardly soothing. Modern landfills and incinerators are built to today's state of the art, but so, in their day, were the Titanic, the Hindenburg, the Three Mile Island nuclear plant, and the Challenger space shuttle. As every Superfund lawyer knows, the recommended waste disposal practices of twenty or thirty years ago have led to today's ruined aquifers. In 1959 the American Society of Civil Engineers declared that unlined landfills would avoid nuisances or health hazards, provided that they were covered daily; by the early 1970s it was clear that liners were necessary, but the minimum specifications for lined landfills in many state regulations until the late 1980s virtually assured groundwater contamination.[84] The regulations of the mid-1990s call for landfills that are far more secure than those built just a few years ago, but since no modern landfill liner is yet more than a few years old, no one can possibly know for years to come whether these structures will function as designed as they age.

As we shall see later in this chapter, this very ambiguity—the unknown effects of an unseen enemy (such as a microscopic quantity of dioxin or a silent dose of radiation)—in itself instills terror, like the ghosts in a horror film.

Illegal Dumping

A major impetus behind facility siting legislation, and the preemption of local authority over siting, has been the fear of illegal dumping of hazardous wastes. Such dumping poses very real health hazards, and is a prime reason it is difficult to be sure that any piece of property, no matter how remote, is uncontaminated. One load of hazardous waste illegally dumped on the side of a road may well release into the environment more waste than would be emitted in a year's proper operations of a licensed disposal facility. The magnitude of the problem, however, has led to a leap of logic as to its causes and cures. Congress,[85] courts,[86] administrative agencies,[87] and many commentators on facility siting[88] have uncritically adopted the idea that a shortage of disposal facilities leads to illegal dumping. In all the siting literature, I have found only one statement questioning this view.[89] Fear of encouraging illegal dumping was also a major reason why Congress has rejected proposals for a tax on the generation of HW[90] and why old grandfathered landfills have been allowed to remain open.[91] However, when the available data are examined, it becomes apparent that illegal dumping has almost no relationship to inadequate disposal capacity and would not be reduced by building more capacity.[92]

As noted earlier, brokers will, for a price, arrange for the shipment of virtually any hazardous waste stream to a licensed disposal facility. The only exception appears to be mixed hazardous/radioactive waste, for which there is virtually no licensed treatment capacity.[93] There are also occasional "stigmatized" loads, which cannot be lawfully disposed because of political problems rather than a lack of disposal facilities. For everything else, there is a legitimate place to go, though the price can be extremely high; the cost of legitimate disposal of hazardous waste is in the hundreds or thousands of dollars per ton, depending on the method used. The price of illegal disposal, on the other hand, is dramatically lower. Information about these prices can be gleaned from past criminal prosecutions. An illegal landfill in Kentucky, known as the Valley of the Drums, accepted up to 100,000 drums between 1976 and 1978 for $0.75 each.[94] An illegal dump in Plainfield, Connecticut, which was closed in 1978, charged $1.50 per drum.[95] An illegal operation near Philadelphia also accepted drums for $1.50 each.[96]

A government investigation in New York provides dramatic evidence of the underground market. In 1992 the district attorney's office in Suffolk County, New York, set up a "sting" operation to catch businesses that were willing to dispose of their HW illegally. Undercover investigators approached businesses and offered plainly unlawful disposal services. They found themselves being forced to reduce their prices to as low as $20 a ton to meet the competition from genuine illegal dumpers.[97] One court has noted the existence of a "vast, unmonitored secondary toxic disposal market—one which . . . weaves across state lines and reaches to every community of this nation."[98]

A survey in the San Francisco area further evidences the extremely low prices in this underground market. The surveyors asked small businesses how much they would be willing to pay for legal HW disposal services; 34 percent said they would pay nothing, and another 18 percent said they would pay no more than $25 a month. The authors concluded that "[i]f the firms are not willing to pay anything, or are unwilling to pay more than $25 per month, their present disposal costs must be very small.

The very small amounts they will pay for waste disposal indicates that they are probably using illegal methods."[99]

Limitations on HW/RW disposal capacity do greatly increase the price of legal disposal, and real capacity shortages could drive the price still higher. It does not follow, however, that this will lead to more illegal dumping. If a four-star restaurant raises its dinner prices from $100 to $120, that will not increase the business at McDonald's; the two establishments serve entirely different markets. The same holds true for HW disposal—there are very distinct legal and illegal markets. If the price of a licensed landfill goes from $250 to $300 per ton, not many of its customers will switch to the $20 method; conversely, a drop from $250 to $200 will not lure the $20 crowd. The restaurant analogy is not perfect because there are many intermediate choices between a four-star restaurant and McDonald's. However, the huge gap between the prices of legal and illegal disposal—$250/ton versus $20/ton—suggests that price shifts of much less than an order of magnitude will not swing many waste generators from the illegal to the legal market or vice versa. Significant changes in the probability and consequences of being caught are likely to have a much more decisive effect on which businesses go to which market.

Several recent studies have revealed the nature of the legal and illegal markets. Of all the HW generated in the United States, 99.6 percent comes from large-quantity generators, and 0.4 percent comes from small-quantity generators[100] (those that generate less than 1,000 kilograms of hazardous waste per month).[101] As seen in chapter 2, the large-quantity generators are overwhelmingly concentrated in a few industries, particularly chemical manufacturing, primary and fabricated metals, and petroleum refining. These companies have much to lose if they are caught in illegal dumping, and they also tend to have sophisticated compliance staffs to advise them on legal requirements. Thus, it is not surprising that several studies have shown that the great bulk of illegal dumping comes from small-quantity generators, and particularly from dry cleaners, auto repair shops, metal cleaners or platers, printers, and pest exterminators.[102] Some have estimated that only about half of all small-quantity generators dispose of their HW properly.[103] Additionally, according to several investigations, organized crime is responsible for much of the illegal HW hauling and disposal.[104]

Thus, the HW most susceptible to illegal dumping is the 0.4 percent from small-quantity generators. It makes no sense, in my view, to distort HW policy and create potentially excess HW disposal capacity for the 99.6 percent of the waste created by large generators in the hopes of luring the small generators. This is especially true when, given the extraordinary discrepancy between legal and illegal prices, this lure is unlikely to be taken. The solution to illegal dumping lies instead in enforcement. On a nationwide basis, very few resources are devoted to inspecting small-quantity generators.[105] More frequent and thorough inspections of small-quantity generators; more cross-checking of toxic release filings; sting operations; and other techniques hold great promise of reducing illegal dumping.[106]

Now to answer the question that began this discussion: The current system of HW/RW facility siting does not harm the environment by creating a shortage of facilities and thereby encouraging illegal dumping. The prevalence of illegal dumping is a failure of the enforcement system, not of the siting system. The current siting system does harm the environment by perpetuating old, substandard facilities, but new facilities offer no panacea.

Affording Fairness

The third issue in evaluating the current system of HW/RW facility siting is fairness. The meanings of the concepts of "fairness", "justice," and "equity" have received considerable attention in the siting literature.[107] Much of the discussion has concerned two ethical systems. The first, the utilitarian system, is historically represented by Jeremy Bentham[108] and John Stuart Mill.[109] Its credo is "the greatest good for the greatest number" and its focus is on the consequences of decisions. In the siting context, it leads to an emphasis on quantitative examination of costs and risks.

The second ethical system is the egalitarian or contracterian. Its greatest historical figure is Immanuel Kant,[110] and its leading contemporary representative is John Rawls.[111] Unlike the utilitarian system, which focuses primarily on the outcomes of decisions and how they affect society at large, the egalitarian system in its modern version focuses more on the procedures by which decisions are reached, with the notion that justice will be achieved by fair procedures and that the effect on individuals, rather than on society at large, is paramount. This tradition above all emphasizes the importance of free and voluntary choice.

The people actually involved in siting facilities have tended toward the utilitarian school, emphasizing the quantitative elements of that tradition. Most of them are engineers and are trained in quantifying observable phenomena. They devise siting mechanisms designed to minimize risk to humans and to the environment with the greatest efficiency. These mechanisms are heavily laden with process—workshops, iterative studies, hearings, reports—but the process is all aimed at achieving the lowest risk at the lowest cost. This result often involves harming individuals (site neighbors), but unless this harm is clear and quantifiable, the standard siting methodologies disregard them as irrational and emotional. The siting professionals would love to find volunteers, but if they cannot, they threaten coercion.

Thus dialogues between siters and neighbors often degenerate into shouting matches. They are speaking different languages. The siters are on a mission for a client (usually a government agency) to find the quantitatively best site; the neighbors want to protect their individual interests, which are not merely economic and medical; and often little common ground is found. The siters argue that everyone should share in the burdens of the technological society from which we all benefit; the neighbors feel that they, as individuals, have had no meaningful say in the technological choices that have led society to be as wasteful as it is. The neighbors feel that the siters are engaging in phantom quantification: the assignment of precise-looking numbers to crude estimates and to the siters' subjective judgments of which decisional factors are more or less important. Even more important, the neighbors feel that what the siters are doing is unfair: unfair to their town, unfair to their children, and, often, unfair to their racial or economic group.

Fairness has elements of both allocation and process. It is unfair to allocate burdens unequally, and it is unfair to reach that allocation through a process that is arbitrary and closed. The discussion that follows focuses on the allocation of the burdens of HW/RW facilities among geographic areas; classes and races; and generations. Unless people feel that these burdens have been fairly allocated, no amount of utilitarian methodology will convince them that they should bear these burdens.

Fairness among Geographical Areas

Fairness among regions is a central theme in siting legislation. The Nuclear Waste Policy Act, written in 1982, contemplated both an eastern and a western HLW repository (although this plan was abandoned in 1987 when the eastern states politically overwhelmed Nevada),[112] and a monitored retrievable storage facility in a third state.[113] The Low-Level Radioactive Waste Policy Act was designed to relieve the burden on the three states with LLRW repositories (South Carolina, Nevada, and Washington).[114] The capacity assurance provisions of CERCLA aimed to assure that every state made provisions to dispose of its own HW. Several states have their own statutes calling for internal geographic equity.[115]

There are three generally accepted principles in achieving regional fairness:

1. The benefits and burdens of waste disposal should be correlated. An area that enjoys the fruits of waste generation should bear the costs of waste disposal.[116]
2. No place should bear a disproportionate share of the region's (or the country's) environmental hazards.[117]
3. Facilities should be placed in the technically best locations in order to minimize adverse health and environmental impacts.

Unfortunately, these three principles are irreconcilable with each other and with other important values. The first and second principles are incompatible because if the first is observed, then disposal facilities will be located near the polluting industries—the chemical waste landfill will be next to the chemical plant, the radioactive waste repository will be next to the nuclear power plant—thereby creating a disproportionate burden on these communities, in violation of the second principle.[118] The first and third principles are incompatible because the technically best locations are usually remote from people and water, and therefore are unlikely to generate much waste (or to enjoy the benefits of its creation).

Observing both the second and third principles would proliferate small waste disposal sites throughout rural America, as each county with suitably dry and remote land would receive a little bit, not much, of the country's waste.

Because of these contradictions, it is perhaps inevitable that all three principles are infringed upon by aspects of the current siting system. The first principle, correlation of burdens and benefits, is badly violated, for example, by the location of the nation's repository for spent nuclear fuel in Nevada, a state with no nuclear power plants,[119] and by the location of the nation's largest HW landfill in rural Alabama, where little HW is generated. The second principle is violated because some communities, such as Niagara Falls, New York,[120] the adjacent Illinois communities of East St. Louis and Sauget,[121] Vernon, California,[122] and Toole County, Utah,[123] voluntarily or not, have major concentrations of disposal facilities, polluting industries, and/or CERCLA sites. The third principle is violated because many of the older waste disposal facilities, which continue in operation while the siting of new units is paralyzed, are in technically inferior locations.

Because it is not apparent which of these three principles should trump the others, these tensions are likely to remain in any siting system. However, they can be significantly reduced in a system that comprehensively addresses all different kinds of HW/RW and requires all states to bear some burdens, as will be discussed in Part III.

The contradictions in various criteria for geographic equity play a key role in the subject that follows—the debate over whether there is a pattern of racial discrimination in the siting of hazardous waste facilities. In a discussion of the nation's largest HW landfill—in Emelle, Alabama—Conner Bailey and Charles E. Faupel write that "a sparse population often is a good indicator of poverty. In the context of Alabama, and the South generally, the population of poor and sparsely populated rural counties is likely to be mostly black. It follows that criteria for siting hazardous waste facilities which include density of population will have the effect of targeting rural black communities that have high rates of poverty," especially since such communities "are likely to be politically marginal at the state and federal levels."[124] Similar complaints have been raised throughout the country as part of a nascent movement against "rural discrimination." In any set of siting criteria based on utilitarian goals, low population density will certainly be considered a positive trait, since it will mean that fewer people will be exposed to risk. But the use of this factor is seen as an admission that the facility is indeed danger-ous—otherwise, why not put it in a city?

Fairness among Classes and Races

The hottest issue in facility siting today is whether HW/RW facilities are intentionally placed in minority communities.[125] Over the years, numer-ous studies demonstrated that poor people are disproportionately ex-posed to pollution,[126] and a 1983 study showed that three of the four commercial HW landfills in the Southeast are in minority communi-ties.[127] The racial issue, however, did not come to the forefront until 1987 with the publication of *Toxic Wastes and Race in the United States* by the Commission for Racial Justice of the United Church of Christ.[128] This study examined the locations of "uncontrolled toxic waste sites" (those on the EPA's "CERCLIS"[129] list of sites with known or suspected contamination) and commercial HW treatment, storage, or disposal fa-cilities. It searched for correlations between the location of commercial facilities and five variables in the community (defined as a five-digit zip-code area)—minority percentage, mean household income, mean home value, number of CERCLIS sites per 1,000 persons, and pounds of hazardous waste generated per person. The study found that "[r]ace proved to be the most significant factor among variables tested in asso-ciation with the location of commercial hazardous waste facilities. This represented a consistent national pattern."[130] In communities with one operating commercial facility, the mean minority percentage in the zip-code area was twice that of areas without such facilities (24 percent versus 12 percent).[131] The mean white percentage in such communities was not revealed. Nor was there any discussion of the racial composition at the time the facility was first built.

The study was less conclusive about the correlation between race and CERCLIS sites, finding that 57.11 percent of the country's black popu-lation, 56.63 percent of its hispanic population, and 53.6 percent of its white population lived in a zip-code area with at least one CERCLIS site.[132] No correlations with income or wealth variables were presented.

Other studies have found a strong racial correlation in the siting of MSW landfills and incinerators in Houston,[133] commercial hazardous waste facilities in Detroit,[134] and hazardous waste incinerators nationwide.[135] A 1984 study found that National Priorities List sites in New Jersey were in communities with high percentages of blacks, low-income people, foreign-born people, and very young and very old people, but it did not compare the strength of the racial and income correlations.[136] A 1992 study found some correlation between community racial composition and the presence of NPL sites. The study found only slight differences in average community racial composition in communities with NPL sites when compared to their geographic regions or to the nation as a whole, on an aggregate basis (meaning adding up the minority populations in the communities with NPL sites and then dividing by the total population of those same communities), but the percentage of blacks and hispanics in communities with NPL sites was found to be greater than the nationwide average. No comparable pattern was found for persons below the poverty line.[137] Likewise, a 1993 study found that counties with higher concentrations of nonwhites have more NPL sites than do counties with fewer nonwhites but also that the more economically advantaged counties are likely to have *more* NPL sites—in other words, race is a much stronger indicator than poverty of the presence of highly contaminated sites.[138]

Additional work has focused on enforcement and cleanup efforts. A 1990 study found that CERCLIS sites in low-income rural communities are being evaluated as quickly as are all sites nationally but that disproportionately few are placed on the NPL, probably because the size of the affected population lowers the hazard ranking score.[139] A 1992 analysis concluded that penalties against polluters are lower when the violation occurs in a minority area, that the EPA takes longer to investigate and clean up NPL sites in minority areas, and that the EPA accepts less stringent remedies.[140]

The impacts of any disproportionate exposure to pollutants would be worsened by the already substandard health status of many minority communities, which stems from such influences as inferior health care, poor eating habits, hazardous occupations, and high consumption of cigarettes, alcohol, and illicit drugs.[141] In one case the Massachusetts health commissioner disapproved a site in Braintree for a new hazardous waste incinerator, largely because in his view the nearest communities have high rates of respiratory disease, due in part to a large elderly population.[142]

Some have concluded that government and corporations make a conscious effort to place HW facilities in minority communities.[143] More prevalent, and more persuasive, explanations for the location of HW sites in minority areas relate to land use patterns and to political power.

According to the land use explanation,[144] factories formerly tended to be sited in center cities, often accompanied by working-class housing. With suburbanization and the decline of central cities after World War II, housing values declined and low-income people, including minorities, were attracted. Most HW is disposed at the factory where it is generated, and when the factory shuts down, a Superfund site is often left behind.[145] Grandfathered HW disposal facilities often persist in these areas, but poor minorities, with limited mobility options, are unable to flee.[146] In this view, these communities are especially attractive to economically rational waste management companies because land and labor are cheap, and residents will accept lower amounts of compensation.[147]

The political explanation (which is not at all inconsistent with the land use explanation) points out that facility siting decisions are often made by the government and that minority communities traditionally are underrepresented in government (which was a major reason, of course, for the Voting Rights Act).[148] Low-income and minority groups have long had low participation rates in political activity in general[149] and in environmental politics in particular.[150] Studies looking for a statistical correlation between political power and siting decisions have been inconclusive,[151] but several particular siting outcomes were blatantly political (although they did not involve minority areas). The bypassing of the normal siting studies and the placement of the nation's HLW repository in a state with little political power, Nevada,[152] is one obvious example. Another came in 1981 when the Arizona legislature designated a spot for the state's HW facility, bypassing the home counties of the state senate's majority and minority leaders.[153] As one proponent of the political explanation has written, "The laws—and the siting of polluting facilities—are products of a political process from which communities of color have been historically excluded and in which today people of color are grossly under-represented. Because the siting decisions are political decisions, the outcome—more facilities in people of color's communities—is neither surprising nor unpredictable."[154]

Several lawsuits have challenged disposal facility siting in minority communities on equal protection grounds. All the suits that have been decided have been dismissed,[155] primarily because the plaintiffs could not meet their burden of proving discriminatory intent or purpose.[156] (One equal protection challenge to a facility siting was successful, but in that case it was because a local elected judge ruled that his county in Texas had so much natural beauty, and that the site was so close to a seismic fault and an aquifer, that an LLRW facility should not be put there.)[157] Several commentators have urged alternative approaches, such as use of Title VI of the Civil Rights Act,[158] use of the equal protection clauses in state constitutions,[159] enactment of a new statute creating a "disparate impact" model for discrimination in HW facility sitings,[160] and judicial adoption of an intermediate level of scrutiny for legislative decisions that have a substantial disparate impact on racial minorities.[161] Other, more cynical commentators have said that "civil rights law has so far miserably failed to combat racism, so why should we think that it will be better able to combat environmental racism?"[162]

Apart from the probably insurmountable obstacle of proving intent to discriminate racially, those challenging proposed sitings on equal protection grounds have another formidable hurdle. A new HW facility will not receive a permit without an administrative finding that it is safe and poses no undue health threat. The plaintiffs will have an uphill battle persuading the court to disregard that finding and to conclude that the proposed facility will endanger the population. For example, in 1993 the Society of Afro-American People in Michigan challenged the issuance of an air pollution permit to a wood-burning power plant. The state of Michigan had allegedly refused to allow a similar facility to be built in a white farming community because of local opposition but permitted this plant in an urban neighborhood with a large black population. The EPA's Environmental Appeals Board found that not only had no racially discriminatory intent been shown, but that the predicted air pollution emissions were far below those that would hurt human health.[163] In general, by the time a facility is built and any dangers become concrete, it will have achieved grandfathered status and will be very difficult to shut down.

Moreover, the theory that race accounts for many siting decisions requires much more factual development.[164] Several groups are now working to replicate the United Church of Christ study to correct some of its acknowledged methodological shortcomings, especially its inattention to timing questions (such as when facilities were sited and when current racial patterns came into being) and to geographic patterns. In 1992 Waste Management, Inc. circulated an analysis showing that 76 percent of its disposal facilities nationwide are located in five-digit zip-code areas with a white population equal to or greater than the host state average; however, this analysis counted white Latinos as white and did not adjust for facility size or type.

Most of the anecdotes and much of the data concerning discriminatory siting come from the southeastern United States. That, however, is a region where, for obvious historic reasons, rural areas have large black populations; in the northeast, where the rural areas are mostly white, most proposed sites have been in white areas. The three sites most often pointed to as examples of environmental racism are HW landfills in Emelle, Alabama; Warren County, North Carolina; and Kettleman Hills, California, but at least superficially plausible explanations have been offered for why all three sites were technically superior.[165] The two counties where new hazardous or radioactive waste landfills were successfully sited in the late 1980s—Adams County, Colorado (including Last Chance) and Tooele County, Utah—were 87 percent and 92 percent white, respectively, at the time of the 1990 Census.

Returning again to our evaluation of the fairness of the current siting system, more work needs to be done to establish whether current efforts to site new HW facilities are racially or economically unfair. There have been few such charges against the proposed new radioactive waste facilities, most of which would be placed in lightly populated areas anyway. One fact does seem clear, however: to the extent that the current siting system perpetuates "grandfathered" facilities and does not allow new sites to open, much HW disposal will continue to take place in poor and minority areas.

Fairness among Generations

A just society will consider the effects of its actions on its descendants.[166] The Constitution states that one of its purposes is to "secure the Blessings of Liberty to ourselves and our Posterity."[167] Few current human activities will have more impact on distant future generations than will the disposal of hazardous and radioactive wastes that may remain dangerous for millenia. Iodine-129, for example, has a half-life of 15.7 million years.[168] A theologian, Ted F. Peters, has stated the question starkly as "how can we morally justify the bequeathal on the part of the present generation of risks and responsibilities that might gravely endanger the health and safety of future generations? How morally appropriate is it for one group to satisfy its own consumptive desires for a few decades and then exact payment from countless as yet to be born civilizations for hundreds of thousands of years?"[169]

As shown by my stringency spectrum, environmental law affords the highest degree of protection to preserving for future generations such items as endangered species, wilderness areas, and historic buildings. Doctrines from completely separate areas of law, such as limitations on the public debt and the rules against perpetuities and against restraints on alienation of property, are also designed to prevent the dead hand of the past from restricting the choices of the present and the future.[170] These principles of protecting the future are constantly violated by today's choices in HW/RW disposal. This is a serious matter because, as Robert Goble has written, "The substances to be disposed of remain toxic for millions of years. Not only is the stuff initially a mixed brew of chemicals, but the mixture changes over time. The wastes have a physical as well as a chemical effect on the material around them, through radiation damage and the generation of heat."[171]

The best way, of course, to protect future generations from HW/RW is to not create it in the first place. If that is not possible, disposal is necessary. The principal disposal options can be arrayed in a spectrum from most to least permanent: destruction; irretrievable disposal; retrievable disposal; long-term containment; and storage. I will now discuss the options in that order.

Destruction   The primary destruction technique is incineration, although bioremediation and other alternative technologies are now being introduced. Incineration works for certain kinds of HW but not for others (such as heavy metals) and not for RW. Unfortunately, incineration of certain wastes creates by-products such as dioxins and furans, which if released into the environment can create their own problems for future generations.[172] Vice President Gore expressed this point of view quite strongly: "The principal consequence of incineration is . . . the transporting of the community's garbage—in gaseous form, through the air—to neighboring communities, across state lines, and, indeed, to the atmosphere of the entire globe, where it will linger for many years to come. In effect, we have discovered yet another group of powerless people upon whom we can dump the consequences of our own waste: those who live in the future and cannot hold us accountable."[173] Thus, although incineration is a permanent remedy in theory, the reality is more complicated.

Irretrievable Disposal   Permanent shielding of exceptional items has been an aspiration of many civilizations, from the builders of the pyramids (who relied on secret passageways and the curses of the gods to keep out trespassers) to the believers in the legend of the Holy Grail (who looked to a race of knights to guard the cup). Geological repositories for radioactive waste, such as Yucca Mountain and WIPP, continue that hope. Recognizing the danger that some future society might forget about these facilities and inadvertently drill for oil or water there, the DOE has spent several million dollars designing a "keep out" sign for WIPP that would be effective for 10,000 years and recognizable by any future earthling.[174] A perpetual care fund is also being established for WIPP, with the theory that the income from a permanent endowment will allow the DOE to pay for the monitoring and security for, say, the life of the sign.[175] Perpetual care funds are a well-established feature of cemetery finance[176] and have been set up for some HW landfills as well.[177] The Uranium Mill Tailings Radiation Control Act requires perpetual surveillance of tailings disposal facilities,[178] which the EPA expects will be designed to last at least 1,000 years.[179]

No great cynicism is required to scoff at the notion of maintaining a sign, or a bank account, or a federal department for 10,000 years. (After all, the oldest continually operating organization in the Western world, the Catholic Church, is less than 2,000 years old.)

The term *repository* suggests passive repose, but the insides of some of these facilities might be anything but dormant. Scientists are now studying the issue of "thermal loading" at Yucca Mountain, for example. Under one scheme, HLW would be emplaced in such a way that the temperature of the rock around the waste packages would remain above the boiling point of water for 300 to 1,000 years; the heat would presumably drive away any moisture that might otherwise reach the waste packages and prevent or greatly retard their corrosion for several centuries. Under another scheme, the repository would be loaded to keep the temperature of the waste and the rock near it above the boiling point of water for 10,000 years. This would allow much more waste to go into Yucca Mountain, but it would also require spent fuel to be "aged" (like cheese or wine) for about 60 years to reduce its high initial temperatures and then to be repacked more densely in the repository to maintain conditions above the boiling point.[180] If, as seems likely, Yucca Mountain is initially loaded with waste around the year 2025, then this repacking would be performed around 2085 by the generation of the great-grand-children of those making the decision today. This multigenerational task seemed to delight one Los Alamos chemist—otherwise depressed by how the demise of the Cold War had dimmed his career prospects—who was quoted in *Science* magazine as saying, "There's no better job security than in the storage of plutonium. Its half-life is 25,000 years."[181]

Retrievable Disposal   The idea behind retrievable disposal is that waste is kept just as environmentally secure as in irretrievable disposal, but if a future generation decides to do something different with the material, it can obtain access. For example, before the signing of the Antarctic Treaty, a proposal was made to emplace a radioactive waste canister in a shallow hole in the Antarctic ice sheet, and allow the canister to melt its own way to the bottom; under one variation, a cable would be attached that would allow retrieval.[182] Earth-mounded bunkers, which are increasingly used for LLRW worldwide, are a less exotic example. The principal negative aspect of retrievable disposal for RW is that the wrong people might do the retrieving. Fear of this escalated after workers at the West Valley, New York, facility stole radioactive tools and sold them at a public auction a few miles away. A similar incident occurred at the Beatty, Nevada LLRW facility.[183] Retrievable disposal is generally disfavored for HLW and TRU for fear of theft of plutonium. However, some have favored it because it would allow future generations to utilize the energy embodied in these materials.

Long-Term Containment   With geologic disposal, it is assumed that the waste will remain isolated from the environment, so that if future generations forget about the facility, the odds are that nothing bad will happen. Conversely, with long-term containment, it is assumed that, some day, the waste *will* reach the environment. The principal example is the most common waste disposal method of all—the landfill. The EPA has repeatedly stated that, regardless of sound construction and operation, all landfill liner systems will eventually fail, and the hazardous constituents of the waste will migrate into the broader environment.[184] The Department of Energy and the Nuclear Regulatory Commission have said the same thing.[185] This is the principal reason that the geology of sites selected for landfills is so important, and why Congress moved in 1984 to ban the land disposal of untreated waste. Residuals repositories—landfills that will accept only the residuals of HW treatment and that are equipped with special protections against water infiltration[186]—promise a significantly longer intact life, but just how much longer is unknown.

The remedy selected in many CERCLA cleanups offers another important example of long-term containment.[187] At many sites, especially those being cleaned up by their owners rather than by the EPA, the contamination is left in the ground but is enclosed in a liner and a cap, much like a landfill. Such "containment" remedies have been used in about half of the NPL sites where the cleanup work has been completed. Eventually, possibly decades or centuries later, the site may have to be cleaned up all over again, after the liner leaks.[188] The methods used to address leaking underground storage tanks are also often temporary, requiring later cleanup.[189] Most of the cleanups now underway at the nuclear weapons complex involve either containment of the contamination in place or excavation and storage of the waste in containers, often under "marginal conditions."[190] Disputes over containment versus off-site disposal have involved high stakes; at one chemical plant in Newburgh Heights, Ohio, which left behind radioactive waste when it closed in 1972, on-site burial would cost $7 million and off-site disposal would cost $50 million, and the dispute (as well as other factors) has left the site unremediated for more than twenty years.[191]

Storage   The least permanent method of disposal, and the one which most explicitly shifts the burden to future generations, is storage. There is every sign that storage is becoming the de facto method of dealing with radioactive waste and some hazardous waste. The TRU destined for WIPP has long been sitting in storage awaiting shipment, much of it since the 1950s; at the Idaho National Engineering Laboratory there are 129,000 drums and 11,000 boxes of sludge, clothing, glass vials, and other debris contaminated with plutonium.[192] The civilian HLW waiting for Yucca Mountain is stored at the nuclear power plants;[193] spent fuel rods from federal government reactors have been sitting for years in storage pools intended to hold the rods for only eighteen months;[194] as seen in chapter 3, the HLW from nuclear weapons manufacture is stored in large tanks at Hanford and Savannah River, some of which are leaking. The plutonium from dismantled nuclear warheads seems destined for storage for years to come.[195] The proposed monitored retrievable storage facility may store the waste for a very long time. Some leading commentators have advocated leaving much of this material in storage for another one hundred years or so, awaiting improvements in technology and changes in the public attitude toward permanent repositories such as Yucca Mountain.[196] Other nations have also, more or less explicitly, adopted a policy of long-term storage of HLW.[197] The Minnesota Public Utilities Commission was so concerned that a proposed storage facility for spent fuel from a nuclear power plant would become a de facto permanent repository that it allowed the plant to build only a small building with a few years' storage capacity.[198]

The siting of LLRW facilities is going no more smoothly. Many nuclear power plants are building storage facilities for LLRW until repositories open. These facilities are often little more than steel-frame, metal-siding buildings on a concrete slab, sometimes with poured or precast concrete walls.[199] Many environmental groups advocate storage of LLRW (medical and industrial as well as utility) at the nuclear power plants.[200]

Congress has prohibited the extended storage of HW banned from land disposal.[201] Such storage poses real hazards beyond the ever-present (and often-realized) danger of leakage. In 1984 a fire at an HW storage facility in Jacksonville, Florida, destroyed several tanks containing PCBs and other organic chemical wastes, spreading oily droplets of HW onto nearby homes, cars, and vegetation; a similar event also occurred in Quebec.[202] The U.S. General Accounting Office has stated that, despite the expenditure of $187 million on emergency preparedness, the eight communities where old chemical weapons are stored pending destruction could not effectively respond to a leak, and the residents could be in severe danger.[203] Nonetheless, several environmental groups have called for aboveground storage of HW until better disposal or destruction technologies are developed.[204] The Dutch, who cannot build landfills because of their high water table, have already largely adopted this approach for HW treatment residues.[205]

Some communities fear that what starts as storage will gradually transmute into permanent disposal. Exactly this is happening in the town of Lewiston in Niagara County, New York. During World War II the government shipped some of the radioactive waste from the Manhattan Project to an ordnance works there for "storage," keeping some of it in a 165-foot concrete silo. In the mid-1980s the silo was demolished, and the waste was buried in a shallow engineered bunker. In 1993 the Department of Energy decided it was safest to leave the waste on the site and build a permanent cap over it. Many in the community are advocating, instead, that the material be excavated and hauled to a permanent repository once one is built.

The discussion so far should make it clear that, in answer to our basic question of fairness, the current system of RW/HW disposal is pushing many costs onto future generations.[206] It is also further depressing the cost of HW/RW disposal today (and thus subsidizing the creation of waste) by imposing many of the costs of final disposal on our distant descendants.

There might be justification for imposing these costs on future generations if, in creating the wastes, we also created things that will benefit future generations enough to cancel out the costs. If we limit the inquiry to tangible goods that will survive and be part of a future society's capital supply, there will certainly be some dams, roads (or at least road cuts), and maybe a few buildings constructed today that will still be useful in a few hundred years, but the waste created in the course of their manufacture will be but the tiniest fraction of all the waste we leave behind. There will also be enduring intellectual creations; twenty-fifth-century citizens will presumably read books written in the late twentieth century on personal computers made with plastic components and powered by nuclear reactors. Here, too, only the tiniest fraction of today's lasting waste will have gone to the service of today's few lasting works of the mind. And what benefit will the twenty-fifth century realize from a nuclear warhead, on standby in a silo for a few years in the 1980s, to be disassembled and then entombed in Yucca Mountain for millenia?

The existence of all these waste-generating items today will, of course, shape the nature of all future societies and determine the identities of all future people, just as were it not for the coal-guzzling steamship that transported my grandparents across the Atlantic many years ago, my parents would not have met and I would not exist. Am I an ingrate if I do not look with gratitude at an ash heap that might have come from that ship?

This attempt to balance intergenerational accounts leads nowhere. Each generation creates goods and services overwhelmingly for its own purposes, and each generation should bear responsibility for the by-products of that production. Our generation, more than any that came before, fails to take responsibility for its own wastes.

Political Viability

The final question in evaluating the current HW/RW facility siting system is whether it is politically viable—that is, whether it can achieve enough political and public consensus for facilities actually to be built. One clear lesson of the past two decades is that adamant, sustained citizen opposition, when backed by local government, almost always wins.[207] Since the passage of RCRA in 1976, not a single hazardous or radioactive waste disposal facility has opened, and stayed open, on a new site in the United States in violation of this principle.

One new HW facility did open on a new site in 1993, but whether it will remain open is still in doubt. Its travails illustrate how difficult it is for waste management companies to operate in the current climate. The story began in 1980, when a group of business leaders in the economically depressed city of East Liverpool, Ohio, invited Waste Technologies Industries (WTI) to build a commercial hazardous waste incinerator on a twenty-two-acre site along the Ohio River, across from West Virginia. WTI agreed, and it received permits from the EPA and the State of Ohio in 1983. The incinerator was built in the late 1980s at a cost of $160 million, but before it could begin operations WTI needed several permit modifications, for which it applied in 1990. By then the incinerator had become highly controversial, largely because of its proximity to an elementary school and to a residential neighborhood. WTI had been sold to Van Roll America, Inc., a U.S. subsidiary of a Swiss company, and questions arose concerning the circumstances of the sale and whether it had been fully disclosed.

The controversy achieved such prominence that during the 1992 presidential campaign, the Democratic vice presidential candidate, Sen. Al Gore, traveled to East Liverpool several times and vowed that the incinerator would never open. The state of West Virginia sued, unsuccessfully, to block the plant's opening; the U.S. Court of Appeals for the Fourth Circuit ruled that West Virginia should have raised its technical objections during the permit process years earlier.[208] The new EPA administrator, Carol Browner, recused herself from the matter because her husband worked for an environmental group involved in the case. In March 1993, over the objections of the new vice president, the EPA regional administrator bowed to the federal courts and allowed the plant to open. Though protesters chained themselves to the White House fence, limited commercial operations began on April 12. In July 1993, however, the state of Ohio said it was reopening its investigation of the circumstances of the sale of WTI. Ohio also announced that preliminary tests revealed higher-than-expected mercury emissions, and tightened the emissions standards applicable to the plant. The EPA then announced that dioxin emissions were also higher than anticipated, leading to further disputes. Though the plant did begin commercial operations, during the controversy it lost several of its major customers, and as of early 1994 it remained in both legal and economic limbo and experienced temporary shutdowns due to equipment malfunctions.[209]

In East Liverpool and everywhere else, public sentiment shapes siting decisions. Thus the factors that create or diminish public opposition must be explored.

What Creates Public Opposition

Several public opinion polls have shown that nuclear power plants, RW facilities, and HW facilities are all lumped together as the most feared land uses, far more feared than, for example, chemical plants, oil refineries, or coal-fired power plants.[210] In some of the surveys, people said they would not want a nuclear power plant or an HW/RW facility closer than one hundred miles from their homes;[211] in others, ten miles was an acceptable distance.[212]

This opposition has not always been so. In the first third of this century, radium-laced patent medicines were hawked as cures for rheumatism, diabetes, and lagging sexual powers; a toothpaste containing radium was sold to "brighten the teeth," and tap water was bubbled through radium.[213] In 1962 the Consolidated Edison Co. seriously (though unsuccessfully) proposed siting a nuclear power plant in the middle of New York City.[214] In 1975, 60 percent of Americans told pollsters they favored the construction of new nuclear power plants; by 1983, 60 percent were opposed.[215] Years of polling showed that concern over RW was imperceptible until 1973, when a leak of liquid HLW at Hanford received wide publicity.[216] HW was also a matter of little public concern; a 1973 survey for the EPA found that most people had positive attitudes about HW facilities and would accept one in their county.[217] Public opinion took a dramatic swing in the late 1970s, however. Love Canal came to light in 1978. In March 1979, the movie *The China Syndrome* appeared, and two weeks later its warnings were eerily confirmed at Three Mile Island. A member of the Nuclear Regulatory Commission wrote that Three Mile Island shifted the burden of proof in the public debate from the nuclear opposition to the industry and its regulators, by demonstrating that the catastrophic potential of nuclear power was more than the obsession of a few fanatics.[218] Since then, public opposition to nuclear power and to HW/RW facilities has solidified.[219]

Many theories have been offered for the widespread public opposition to HW/RW facilities, but I believe that the reasons can be summed up with two words: *dread* and *intrusion*. These have important implications for the siting dilemma.

Dread Polls confirm that by far the most important reason behind opposition to HW/RW facilities is concern over the impact on health,[220] particularly the health of one's children.[221] This is so not only in the United States but in other societies as well.[222] Many educated people (though few if any of the experts in the field) attribute a large percentage of adult and childhood cancers to hazardous waste sites.[223] Worry about health effects is also at the root of the emergence since about 1980 of thousands of grassroots groups of facility opponents and "toxic victims," organized into two nationwide coalitions, the Citizens Clearinghouse for Hazardous Wastes (CCHW) in Arlington, Virginia, and the National Toxics Campaign in Boston.[224] Each issue of CCHW's newsletter, *Everyone's Backyard,* contains reports from regional correspondents about successful efforts to stop HW/RW facilities. These groups differ markedly in style, agenda, and constituencies from the mainstream national environmental organizations.[225]

Lois Gibbs, who founded CCHW after being evicted from her home near Love Canal, tells a story (possibly apocryphal) that dramatically illustrates these differences. At a hearing in Louisiana concerning a hazardous waste site, she recounts, citizens set up an aquarium filled with contaminated drinking water from their wells. They loudly announced that the fish they were about to place in the tank would be dead by the end of the hearing. When the government officials and traditional environmentalists in the room protested, the crowd began to chant "kill the fish." Gibbs said, "If we have to kill the fish to make the point, we'll do it. We're sacrificing our children."[226]

Every era has had its own dreads. The Israelites cast out Moses' sister, Miriam, when she contracted leprosy.[227] Successive civilizations have had ample reason to be terrified of plague, smallpox, and polio, until each was eradicated or controlled. These dreads have manifested themselves in disputes over siting; Louis Pasteur's effort to find a laboratory to develop a rabies vaccine was hampered by residents of Paris who feared they would contract the disease.[228] We now have cancer and AIDS.[229] Often people who have been exposed to agents that might cause these diseases, or who simply fear such exposure, become preoccupied with health problems.[230] When wastes are the source of these agents, a further layer of revulsion occurs, at least in the view of Freudians.[231] When the waste is burned, as in incinerators, the complex psychological reaction to fire comes into play.[232]

Yet another layer of horror accrues when the waste is radioactive. Since Hiroshima and Nagasaki, fear of nuclear war has become embedded in the culture. People over forty well remember the fallout shelter craze of the 1950s, the Cuban Missile Crisis of 1962, films like *On the Beach* (1959), *Fail Safe* (1964), and *Dr. Strangelove* (1964), and any number of grade-B movies about post-nuclear-war mutants.[233] These horrible images of nuclear war have become melded with the image of nuclear waste in the public mind. Polls conducted around the world during the 1980s showed that nuclear power and nuclear waste were regarded at the extreme negative end of almost every attribute of risk perception, such as dread, lethality (likelihood that a mishap would prove fatal), potential for catastrophe (multiple fatalities), involuntariness, and uncontrollability.[234] The mental linkage of nuclear weapons and radioactive waste greatly impedes the siting of RW facilities; although most RW is not prone to atomic explosion,[235] many members of the public may not draw that distinction.[236]

The distinctions drawn by most members of the public tend to be rather sharp. This phenomenon was well described by a physician writing of a different context (the process by which patients decide whether to undergo risky medical procedures):

One feature of human judgment is the tendency to categorize an entity as either "dangerous" or "safe" without recognizing that low and high levels of exposure can have different or even opposite effects. For example, many respondents in a recent survey believed that a teaspoon of ice cream has more calories than a pint of cottage cheese. . . . Presumably, ice cream is considered as inherently high in calories and cottage cheese low, regardless of the amount consumed. The same type of thinking that brands an activity as healthy or dangerous independent of dose can lead individuals to believe that a person cannot ingest too many vitamins, that infinitesimal amounts of salt or sugar are harmful, or that even the most trivial contact with anyone having [AIDS] is dangerous.[237]

The same might be said about public reaction to HW/RW facilities. Many people tend to regard them as horribly dangerous, with no meaningful shades of gray between danger and safety; thus most proposed mitigation measures to reduce dangers will sway few minds.

Protests against hazardous and radioactive waste facilities are able to draw upon powerful symbols. In bumper stickers from siting controversies, the skull and crossbones appear repeatedly. Michael R. Reich made a similar observation in his study of protests by the victims of chemical contamination around the world:

Victims chose symbols of life and death not simply because the symbols expressed the pain of their suffering but because they represented ultimate forms of power: to restore the dead to life and to threaten the living with death. In choosing symbols that represented unjust death, the victims challenged a basic legitimacy of the social order: the legitimation of death and its integration into social existence. The symbols around death also served to unite and mobilize the victims, to help them overcome feelings of despair and hopelessness. The choice of these images reflected a broader pattern of using death symbolism in movements for social and political change.[238]

The sociologist Kai Erikson has written that "[m]aybe we should understand radioactive and other toxic substances as naturally loathsome, inherently insidious—horrors, like poison gas, that draw on something in the human mind."[239] He says that toxic emergencies possess two distinguishing characteristics that add to the dread they induce. First, they are unbounded and have no frame or end; the "all clear" is never sounded. Second, "they are without form. You cannot apprehend them through the unaided senses; you cannot taste, touch, smell, or see them. That makes them especially ghostlike and terrifying. Moreover, they invert the process by which disasters normally inflict harm. They do not charge in from outside and batter like a gust of wind or a wall of water. They slink in without warning, do no immediate damage so far as one can tell, and begin their deadly work from within—the very embodiment, it would seem, of stealth and treachery."[240] An invisible, ambiguous threat tends to induce what Irving Janis calls hypervigilance—a complete, sometimes even obsessive attention to possible risks and ways to avoid them.[241] There can be no less fertile soil for a proposal to site a waste disposal facility.

Intrusion   The insidiousness of their threat is closely related to the second major reason HW/RW facilities are so hated. The facilities themselves are seen as imposed on communities without consent, and once they arrive they do their damage silently. When forced on unwilling localities, the facilities are seen as colossal intrusions, almost as foreign invasions. Attempts to override local siting authority almost invariably backfire and *increase* local opposition, partly by intensifying the community's perception of risk.[242] Some studies have shown that people will accept voluntary risks approximately 1,000 times more hazardous than risks they perceive as involuntarily imposed[243]—they will parachute out of an airplane or smoke a cigarette, but they don't want anyone to build a waste plant miles from their house.[244]

The sanctity and security of the home have a central place in American law and culture.[245] The Constitution places a high value on guarding against intrusions in people's homes; the Third Amendment states that "[n]o soldier shall . . . be quartered in any house without the consent of the owner," and the Fourth Amendment guarantees "[t]he right of the people to be secure in their . . . houses." An unwanted facility is another kind of intrusion. As Michael R. Edelstein has written in his study of communities that had experienced toxic accidents, "[r]ather than a place to escape to, with the contamination home had become a place that residents could not escape from. Parents particularly feared the consequences of continued residence for themselves and their children. Thus, home was inverted in the sense that it now was accompanied by a strong sense of fear and insecurity. Rather than buffering the family from the dangers of the outside world, home embodies these dangers."[246]

This sense of intrusion is magnified when the waste is imported from other areas. The EPA has acknowledged that public opposition is greater when the facilities would accept out-of-state waste.[247] (This is an especially difficult problem, as all fifty states export some HW to out-of-state treatment facilities, and forty-eight—all but Alaska and Montana—import such waste.)[248] Approximately 8 million tons of HW are shipped off-site every year, and about half of that crosses state lines.[249] The Supreme Court has made clear that the Commerce Clause precludes states and localities from excluding or discriminating against out-of-state waste.[250] Yet, the fear or the reality of importing waste from other areas has been a major impetus behind public opposition to waste disposal facilities of all sorts—HW landfills,[251] HW incinerators,[252] sewage sludge landspreading,[253] and municipal solid waste landfills.[254] This is a major reason why several states refused to enter into compacts with others for regional disposal facilities for LLRW[255] and HW.[256] Opposition has also been raised to acceptance of waste from elsewhere in the same state,[257] county,[258] and city.[259] In several celebrated incidents, barges[260] or train-loads[261] of waste (one of which, bearing Baltimore sewage sludge, came to be known as the Poo-poo Choo-choo)[262] were forced to wander aimlessly after their intended destinations refused to take them, and no one else would accept them. A centerpiece of Daniel R. Coats's successful run for the Senate seat vacated by Dan Quayle was opposition to MSW imports, especially from New Jersey. One of his television ads featured a fat, cigar-chomping man wearing a Yankees cap and a Cape May T-shirt, littering the steps of the Indiana State Capitol.[263] After his election, Senator Coats sponsored a bill, which passed the Senate (but not the House), allowing states to ban or tax out-of-state shipments of MSW.[264]

The EPA has promulgated a regulation under RCRA targeted against any state action "which unreasonably restricts, impedes or operates as a ban on the free movement across the State border of hazardous wastes."[265] Nonetheless, many states have attempted to restrict imports.[266] There has been extensive litigation in both the federal[267] and state[268] courts challenging, and usually striking down, these attempts.

The desire for local autonomy in the use of land, and freedom from outside interference, is an element of human nature not to be underestimated. Both history and the present day are full of examples of bloody conflicts—whether colonial revolts, civil wars, or guerilla resistance to imperial adventures—over whether a given patch of ground would be controlled by the natives or by outsiders. Lately many of these fights have involved environmental issues. In the words of a leading scholar of the former Soviet Union, Marshall I. Goldman, "[A]lmost all the nationalist and ethnic stirrings that have occurred since Mikhail Gorbachev came to power have originated within the environmental movement. . . . [E]nvironmentalists increasingly have begun to attribute the pollution affecting their regions to decisions that Moscow had made. . . . In almost all cases, the orders appeared to stem from Moscow, and their implementers appeared to be Russians. This made it all the easier for environmentalists to link up with local nationalists and echo their calls for more local autonomy and decisionmaking."[269]

Similar fights also occur in the United States. Legislation that would expand federal control over local land use consistently meets a chilly reception in Congress. Movements have arisen around the country to prevent Washington or the state capitals from telling local governments how to control the use of land.[270] These movements have some of the same powerful psychological wellsprings as the resistance to the forced importation of outsiders' waste.

The operation of this resistance is seen in Illinois, the home of the nation's largest commercial nuclear utility, Commonwealth Edison. The company has twelve nuclear power plants at twelve different sites, all in the northern third of the state. Each plant has its own stockpile of spent fuel rods. This seems to cause relatively little concern to the neighbors. However, if Commonwealth Edison attempts to transfer spent fuel rods from one plant to another, there is a loud public reaction against the importation of "outsiders' waste." Thus every plant must provide its own spent fuel storage pool, and inter-plant transfers are not allowed (though shipments of spent fuel rods to and from nuclear facilities in other states routinely pass through Illinois without notice).[271]

Another incident in Illinois illustrates the same point even more starkly. A hazardous waste landfill in Wilsonville opened in 1976 and operated quietly for about four months. However, the local newspaper then reported that the landfill would receive PCB-contaminated soils collected from a spill in Missouri. Protests immediately mounted, and demonstrators blocked the facility entrance to stop the trucks coming in from Missouri. The village government dug a culvert across the access road, effectively closing it. Soon thereafter the facility was shut down altogether.[272]

A concept closely related to intrusion is trust. Opposition is magnified when the community does not trust the people or institutions seeking to place HW/RW in their midst. The political prospects for several proposed HW facilities were severely damaged when it became known that their

proposed developers had histories of environmental or other violations.[273] Other proposals were defeated after the public came to believe
that the government was negotiating them in secret with corporate spon-
sors.[274] Overall trust by Americans in their institutions has been declining
since the Vietnam War and Watergate, and some have said this is a major
reason for the great loss of public support for nuclear energy—a technology that relies on large organizations.[275] Two major motion pictures,
The China Syndrome (1979) and *Silkwood* (1983), portray the nuclear
industry as venal and deceptive, and idolize its whistle-blowers. As Ivan
Selin, the chairman of the Nuclear Regulatory Commission, has said
(quoted in an article by Allan Pulsipher):

I would be delighted if the public would look at the facts on nuclear safety in an
objective manner. But, to be fair about it, our government . . . has lied to our
country quite often in the area of nuclear work. . . . I'm not really sanguine that
public information, public frankness, public openness in and of itself is enough
to overcome a very understandable cynicism of "Why should I believe you
now?"[276]

Especially since Three Mile Island, large segments of the public mistrust the entire nuclear industry, adding seriously to the difficulty in siting
RW facilities.[277] The U.S. Department of Energy faces widespread distrust in its cleanup of the Nuclear Weapons Complex.[278] This distrust
can expand from individual companies or agencies to entire industries
and even to the technological society. In the words of Professor Erikson,
"If science and technology have become the source of risk . . . it is
because toxic peril has moved people so far up the scale of suspicion that
they come to distrust not only public officials and experts, not only the
social order and the natural world, but also the very ethos of science and
technology."[279] Indeed, the antinuclear movement (much like the anti–
Vietnam War movement before it) is in no small measure a protest
against what is seen as lethal corporate and governmental gigantism.

Mistrust is often accompanied by despair. The forces behind hazardous
facilities are seen not only as evil, but as invincible. As Peter Sandman
has said, "Ironically, nearly everyone is impressed by the community's
power of opposition—except the community, which sees itself as fighting
a difficult, even desperate uphill battle to stop the siting juggernaut."[280]
During this battle, residents feel themselves trapped because the threat

of the facility often makes it impossible for them to sell their homes at full price—an injury for which there is no legal remedy.

The sense of intrusion is further magnified if the community feels it is being treated unfairly—if, for example, there is no obvious reason why it is being singled out to be exposed to a waste facility. A sense of random victimization increases the outrage felt by the neighbors.[281]

To be sure, dread and intrusion are not the only factors at play in public opposition to HW/RW facilities. The tremendous public outcry against ocean incineration of hazardous waste seems to stem more from the sense of the oceans as a precious resource to be preserved for all of humanity; but even here, it does not stretch my categories too far to say that all of us regard ourselves as having a stake in the oceans, so that any insult to them is an intrusion on our own property.

What Reduces Public Opposition?
Although a complex of reasons, which can be grouped together under the headings "dread" and "intrusion," intensify public opposition to HW/RW facilities, several other factors reduce opposition.

Local Waste and Local Jobs   While importation of waste is mightily resisted, storage or disposal of waste at the point of generation often proceeds smoothly. Nuclear power plants around the country have expanded their capacity for storing spent fuel, for example, with very little public opposition.[282] Many companies have been able to build on-site storage or disposal capacity with little or no controversy.[283] This lack of opposition is attributable to several factors: no importation of waste (and thus no intrusion into the community) is involved; the waste generators are often industries that create many local jobs;[284] inertia and familiarity blunt opposition to what could be seen as just more of the same; and transportation risk is eliminated. Some of the largest chemical companies have adopted on-site disposal as a matter of policy.[285] Even Lois Gibbs's Citizens Clearinghouse for Hazardous Waste has advocated on-site disposal.[286] When faced with the continuation of a polluting industry or the loss of many local jobs, communities have often agreed to keep the jobs and sacrifice the environment, even at discernible risk to public health.[287]

Another reason for the favorable reception of on-site disposal is its strong appeal to fairness. There is an obvious equity in having each location take care of its own waste—an equity that has a major impact on public reaction to siting proposals.[288]

Unfortunately for the waste disposal industry, off-site facilities tend not to create many jobs.[289] Only rarely will a hazardous waste facility create so many jobs that it has a major impact on the local economy. One such unusual example is the nation's largest hazardous waste landfill, the Emelle facility in Alabama; with a payroll of more than 400, it is Sumter County's largest employer, and as such it enjoys widespread support in the local business community.[290] More typically, the nation's largest LLRW facility, at Barnwell, has fewer than 200 employees.[291] The West Valley nuclear facility at its peak had an operating staff of 170.[292] A new HW incinerator in Ohio would employ 104.[293] An integrated waste management facility in Alberta, Canada, employs 94.[294] The Envirocare radioactive waste landfill in Utah employs 60.[295] These numbers are not trivial, but they are not so overwhelming—as those in an automobile assembly plant, for example—that they can overcome intense local opposition.

In several instances the prospect of large numbers of local jobs did succeed in mobilizing local support for HW/RW facilities, although not always for long. At one time the Yucca Mountain facility in Nevada enjoyed local support, in part because the community believed that many of the 6,800 jobs at the Nevada Test Site (one of the state's largest employers) would be jeopardized by a nuclear test ban treaty.[296] When WIPP was first conceived in the early 1970s, the nearby city of Carlsbad, New Mexico, which was facing major layoffs in the local potash industry, strongly supported the facility.[297] The West Valley nuclear project, which began in the early 1960s, received strong support from local residents, who were promised that it would lead to strong growth in the local economy.[298] It is important to note, however, that both WIPP and West Valley were initiated before the onset of major opposition to nuclear power and RW.

Local Control   The opposite of intrusion is invitation. When a community invites a facility into its midst, the risk is voluntary, not involuntary.

Voluntary siting elicits a far different psychological response, as the discussion in chapter 6 concerning the Canadian experience in facility siting will evidence.

Regardless of whether the facility comes to town by invitation, local control over the facility's operations has a powerful impact on local reaction. Numerous public opinion studies have shown that the ability of a municipality to monitor a facility, to participate in its management, and to shut it down if necessary, has a far greater impact on local acceptance than any other measure—considerably greater than financial compensation, for example.[299] Some grassroots activists share this view.[300]

This finding was empirically confirmed by the success of Browning-Ferris Industries (BFI) in siting a HW landfill and treatment facility in Last Chance, Colorado—the only HW landfill siting success since the passage of RCRA in 1976. BFI initially applied for county approvals in November 1980. The county denied the permits in June 1982. In early 1983 the state amended the siting law to confirm that local governments could veto facilities but also to provide that, if no locality approved a site within two years, the state could step in and site one. The county then extracted several concessions from BFI, including a reduction in volume; prohibitions on certain types of wastes; specified engineering techniques; and county specification of haul routes. The county also secured the right to inspect the facility; $100,000 per year from BFI to pay for inspectors plus 2 percent of gross revenues; the right to review and approve detailed engineering and construction drawings; and the requirement of county approval of all fee schedules and hours of operation. Having obtained this degree of local control and compensation, the county approved the project in August 1983. Opposition from some environmental groups and neighboring counties continued, but the state granted the necessary permits in 1987, and the facility finally opened in 1991 (although by then BFI had sold it to another company).[301]

Local Culture   As Mary Douglas and Aaron Wildavsky have written, "[E]ach culture, each set of shared values and supporting social institutions, is biased toward highlighting certain risks and downplaying others."[302] They go on:

The sudden appearance of intense public concern about the environment can never be explained by evidence of harm from technology. Weighing the balance between harm and help is not a purely technical question. Technology is a source of help as well as harm. Objective evidence about technology in general is not going to take us very far. Acceptance of risk is a matter of judgment and nowadays judgments differ. Between private, subjective perception and public, physical science there lies culture, a middle area of shared beliefs and values.[303]

After studying chemical contamination incidents in Michigan, Italy, and Japan, Michael R. Reich came to a similar formulation. "Culture provides a context within which victims approach the task of organization and from which they select appropriate symbols," he wrote. "As a system of interacting symbols that give shared meanings to a social group, culture plays an essential role in our cases of toxic contamination, providing people with a universe of meaning, within which they define, implement, and legitimate reality."[304]

Different communities obviously have different histories, educational and income levels, racial and ethnic mixes, physical settings, political systems, leadership styles, predominant religions, and dozens of other characteristics that combine to produce each community's own unique local culture of "shared beliefs and values." Some communities have cultures that are not averse to HW/RW facilities. Their view of the world does not see these structures as demons. Survey research has demonstrated that public trust of power generation and waste disposal facilities increases with greater proximity to and familiarity with such facilities.[305] Experience confirms this.

A prime example is Richland, Washington, a city near the Hanford Reservation, which dominates the local economy. The local high school basketball squad, "the Bombers," plays on a gym floor inscribed with a mushroom cloud; local residents can bowl at the Atomic Lanes or have a massage at the Atomic Health Center. There is strong local support for RW management, and the community is now turning into an environmental boom town (no pun intended) with thousands of new jobs devoted to cleaning up the mess from weapons production.[306] Similar local cultures strongly supportive of RW or HW facilities have been described in Barnwell, South Carolina, home of the Barnwell LLRW facility and near the Savannah River Reservation;[307] in Oak Ridge, Tennessee, whose economy is based on nuclear technology and whose local government volunteered for the monitored retrievable storage facility for HLW;[308] in Los Alamos, New Mexico, a community dominated by a nuclear weapons laboratory;[309] in Nye County, Nevada, the home of Yucca Mountain, the Nuclear Test Site, and the Beatty LLRW facility;[310] in Fall River County, South Dakota, an area with extensive uranium mining, which indicated it would welcome an LLRW facility;[311] in Hermiston, Oregon, site of an army depot where munitions and toxic chemicals had long been stored, which was amenable to receiving nerve gas from Okinawa;[312] in Pinawa, Manitoba, the site of Canada's Whiteshell Nuclear Research Establishment, where the local government volunteered for Manitoba's HW facility;[313] in the Bruce and Chalk River areas of Canada, which have nuclear facilities;[314] and in Swan Hills, Alberta, a community with a major oil and gas industry and the volunteer for Alberta's HW facility.[315]

If any generalization can be drawn from these experiences, it is that communities that already have risky facilities tend to have local cultures that have accepted such risks and often will accept still more risks. Those who participate in the political and economic life of the community perceive the benefits as exceeding the costs, and do not share the view, prevalent in so many other places, that wastes are unacceptably repugnant. It appears that this is mostly because we are less likely to dread what we know. People who work in an oil refinery are less likely to fear chemicals than are people who work in an office. Similarly, as we have just seen, one of the major factors in public reaction to hazardous and radioactive waste is the degree of trust in the institutions handling the waste. People who are familiar with and trusting of such institutions as the military or the chemical industry will more readily accept these organizations' waste management efforts; people who see these as foreign hostile forces will rebel. Based on survey research, Aaron Wildavsky and Karl Dake concluded that people with hierarchical personalities (who tend to have high degrees of trust in large institutions) and people with individualist personalities (who believe in market principles and individual decisions) are far less likely to fear technological innovation than are egalitarians (who do not trust large institutions and who believe markets are unfair).[316] Thus the reaction of a community to a proposed facility may also be related to which personality types predominate there. A

community's economic base, its cultural milieu, and its dominant personality types are certainly related—a college town and a steel mill town will
have different characters in many ways—and for these purposes we need
not seek cause and effect.

In gauging the origins and nature of local consent, it must also be
borne in mind that many communities are not homogenous. Different
individuals have different perceptions, different economic situations, and
different degrees of proximity to the proposed site. Robert D. Benford
and colleagues, in their study of the attempted siting of an LLRW facility
in Boyd County, Nebraska, made this trenchant observation:

> To the extent that the NIMBY label is used pejoratively, the implication is that
> those who oppose the location of hazardous waste sites in their backyards are
> selfish people who are unwilling to accept some risk for the common good. This
> fails to examine the structural situations of NIMBYs. Some respondents are more
> able than others to appear unselfish. They are the well-educated, economically
> secure, upwardly and hence geographically mobile who, due to their advantaged
> position, have a number of options other than being saddled with the label
> NIMBY or having to live near a noxious facility. First, they may welcome a
> hazardous waste facility to their community (to their poorer neighbor's back-
> yard), cloaked in the mantle of good citizenship, while using their status to ensure
> that they accrue a disproportionate share of the facility's benefits. Second they
> may avoid having the hazardous wastes in their backyard by using their resources
> or power to affect decisions. If those strategies fail, they can move elsewhere. For
> the poor, elderly, uneducated, and relatively powerless, such options are less
> available. They are, in short, more vulnerable.[317]

The dichotomy in risk perception within a community was also noted
by Francoise Zonabend in her sociological study of the community
around a nuclear fuel reprocessing plant at la Hague in Normandy. The
workers in the plant, she reports, divide themselves into two categories:
the *rentiers* (cautious people) and the kamikazes. The *rentiers* take great
precautions about safety, rigorously follow procedures, and pay close
attention not to exceed the permissible doses of radiation. The kamikazes
are fearless in exposing themselves to radiation. Plant management takes
candidates' attitudes into account when assigning jobs.[318]

These local divisions in public opinion do not mean that genuine local
consent cannot be achieved. They do suggest, however, the importance
of (1) a referendum of the entire electorate, rather than reliance on the
approval of the community's governing body, (2) an explicit and well-

thought-out plan for use of compensation funds, and (3) a compensation package that allows the closest neighbors of the site to move out if they wish.

Achieving Public Acceptance

The Case of Niagara County   Several of the above factors—especially dread, intrusion, local control, and local culture—came together in one siting controversy in Niagara County, New York, in the northwestern corner of New York State, across the Niagara River from Ontario.

The county has long played a central role in the politics of hazardous waste. The city of Niagara Falls, at the southern end of the county, is the site of Love Canal, the impetus for the Superfund law. The natural beauty of the Falls made the city a honeymoon capital, but the same natural forces attracted heavy industry that required abundant water and electricity. Major chemical companies built factories along the Niagara River's banks above the Falls. One of them, Hooker Chemical (now known as Occidental Chemical), dumped tens of thousands of gallons of liquid hazardous waste in Love Canal, and similar quantities in another spot, known as the S Area, adjacent to the city's drinking water plant, eventually requiring the plant's relocation. A company called CECOS International, which became a subsidiary of Browning Ferris Industries, built a succession of commercial hazardous waste landfills on a site in the city, and over the years, as portions of the site filled up, sought and obtained approvals to build more landfills on other portions. In 1987, I was retained by the city and the county to oppose the latest expansion of CECOS. In 1990—after a four-month trial before a state administrative law judge and a state siting board, and after several rounds of appeals on numerous issues—the state denied the expansion application, largely because of the poor geology of the site (it sat atop permeable rock, and thus contamination could readily reach the river, and it was very close to a trailer park and other residential communities). With that, and with other defeats in Louisiana and Ohio, CECOS went out of the hazardous waste disposal business.

Twelve miles north of Niagara Falls are the towns of Porter and Lewiston. They are predominantly rural communities, and their only population center, the village of Lewiston, was 98.9 percent white at the 1990 Census. They have a long history of waste disposal. Within their borders are a shallow bunker containing low-level radioactive waste left over from the Manhattan Project; a large commercial landfill for municipal solid waste, called the Modern Landfill by its owners; and a 710 acre site, known as Model City, that was once occupied by the U.S. Army for the Lake Ontario Ordinance Works; was later the scene of extensive illegal dumping; and was subsequently bought by a commercial hazardous waste company, SCA Services, which built a series of hazardous waste landfills there. In 1984 SCA was acquired by Waste Management, Inc., the largest company in the business,[319] and the site came under the control of the buyer's subsidiary, Chemical Waste Management, Inc. (CWM).

In 1988 CWM received approval to build a large new HW landfill, called SLF-12 (for Secure Landfill No. 12), on a twenty-two-acre portion of the site, with a capacity of one million cubic yards. The permit had been only feebly opposed by the community, and only on traffic grounds, and the state siting board found that this was an excellent location for a landfill, given its natural geology. SLF-12 quickly became the second largest commercial HW landfill in the country (measured in terms of annual tons of waste received), behind only CWM's landfill in Emelle, Alabama. As a condition of building SLF-12, however, CWM was required by the EPA, under RCRA's corrective action provisions, to clean up the prior contamination on other portions of the site. The EPA told CWM to consider construction of an incinerator at the site to destroy some of this remedial waste.

In 1990 CWM announced its desire to build two hazardous waste incinerators on the site, with a total capacity of 100,000 tons a year. These would not be for the on-site remedial waste, but rather would be commercial incinerators that would accept waste from all over the country and from Canada. This time, the community was outraged. I was retained by the towns and the county to oppose the incinerators.

The incinerators presented a classic combination of dread and intrusion. The community had long lived with the landfills in their midst, and though the landfills were not popular, they were grudgingly accepted. The landfills were well screened so they were not visible from the road, and since the community's water came from the Niagara River rather than from wells, there was little concern about danger to drinking water. The principal concern about the landfills was that the trucks carrying waste there passed in front of a school complex where all the community's children spent their lives from kindergarten through high school.

The incinerators, however, were a different story. This was a new technology with which the community had not become comfortable. There were serious concerns about dioxins and other contaminants escaping from the smokestacks and wafting over the school, less than two miles away, and over the homes. These concerns were heightened by the operating problems at CWM's incinerators in Chicago and Sauget, Illinois. Moreover, the incinerators would accept liquid as well as solid wastes, posing a greater danger in case of a truck accident near the school. At community meetings there was palpable fear about the danger the incinerators would cause to the community and, especially, to its children. Mothers demanded, but could not be given, absolute guarantees that their sons and daughters would not get cancer from the emissions.

There was special resentment that the wastes for the incinerators would come not just from western New York but from the whole country and Canada. SLF-12 had a similarly large market area, but the community had come to feel itself the nation's "dumping ground"—a complaint heard in siting controversies everywhere—and that enough was enough. Thus the smoke from and the trucks to the incinerators were dreaded, and the international wastes from a multinational company were intrusive.

Our legal strategy on behalf of the community reflected these concerns. We retained technical experts who could advise about the physical dangers of the emissions. (Finding qualified experts in the United States was difficult because most of the specialists in HW incineration had, or aspired to, consulting contracts with Waste Management, Inc., so we had to go to Canada.) We also pressed the case that western New York should not bear such a large load of the continent's waste disposal needs. The state had attempted to impose a permit condition on SLF-12, limiting the importation of wastes from other states, and we were active in defending this condition against the inevitable challenge under the Commerce Clause of the Constitution. New York State also sued the EPA for not penalizing states that had failed to develop their own HW disposal capacity, despite the capacity assurance provision of SARA. We also pressed the EPA to adopt a regional approach in the state HW plan that had been mandated by the state legislature.[320] The strategy was to fight CWM at every turn, in every possible forum, whether or not directly related to the incinerators. Ironically, much of the legal and technical opposition work was paid for by funds received by the towns from CWM as part of the gross receipts tax on CWM's landfill operations.

Meanwhile, SLF-12 was filling up at a remarkably fast rate, and before it was full CWM wanted to build yet another landfill unit, which it called RMU-1 (for "residuals management unit," emphasizing that the waste there would be the residue of treatment processes). It would cover forty-seven acres and have capacity for 2.8 million cubic yards, making it one of the largest HW landfills ever built. The communities were unhappy about this proposal, but in view of the high siting score that had been received by SLF-12, they were pessimistic about stopping it.

At that point the germ of a settlement idea was born: let CWM build the new landfill in exchange for dropping the incinerators. The twelve-member community advisory committee had several meetings with CWM representatives. The chief negotiator for the citizens was a local resident, R. Nils Olsen, Jr., a law professor at the University of Buffalo. Olsen, with my participation and that of the advisory committee, negotiated with CWM's lawyers starting in 1992, and in June 1993 achieved a settlement. The towns, the county, and the citizen advisory committee would not oppose the permit for RMU-1, and in exchange CWM would not apply within the next ten years to build a hazardous waste incinerator (except possibly one that would burn only on-site remedial waste) anywhere in Niagara County. Since the application and permitting process is so lengthy, this would ensure no commercial incineration for at least fifteen years, by which time incineration technology may have improved significantly.

Importantly, the agreement also provided for a large decree of local monitoring and even some local control over RMU-1. CWM agreed to a detailed "site operations plan" that strictly limited the hours when trucks could travel to or from the site, banning trucks carrying HW during hours when school buses would be transporting children and also banning most trucks during sleeping hours. CWM agreed to give the county an equipped emergency response truck with a computer link to the facility; to train emergency response personnel; to work on developing alternative haul routes; to accept a perpetual care condition for the landfill; to provide advance notice to the community of expected unusual traffic activities; to participate in a study of the air intake system at the school, to guard against catastrophic accidents; to report all violations of transportation regulations to the District Attorney and the local police; to contribute $16,000 a year to the County Health Department to support county oversight of the facility; and to other similar mitigation measures.

The agreement was widely, but not universally, praised by the community. Lee Simonson, chairman of the county legislature, hailed it as a David and Goliath victory.[321] Timothy P. Henderson, president of the leading local environmental group, Residents Organized for Lewiston-Porter's Environment, Inc. (ROLE), which was a signatory to the agreement, wrote, "We went eyeball-to-eyeball with a billion-dollar industry and . . . they blinked. . . . This agreement is about a community taking responsibility for the operations of a hazardous waste facility in its own backyard. I for one am not comfortable leaving those decisions to a group of nameless and faceless technocrats."[322]

On the other hand, a local resident, Lisa Aug, accused ROLE and the towns of appeasing CWM in the same way that Neville Chamberlain appeased Hitler. She wrote on the op-ed page of the local newspaper:

ROLE is blinded by its tunnel-vision opposition to incinerators. No, I don't think they're safe, and even if they were I wouldn't trust CWM to operate one properly. But an incinerator does destroy toxic waste. Toxic chemicals dumped in a landfill remain deadly forever, eventually leaking out to poison everything around them. Not just for 10 years, not just for the seven generations for which we must be responsible, but for all time. . . . [The money that CWM brings to Niagara County] is blood money, dripping with the gore of children yet unborn who will have to cope with the legacy of this "great deal."[323]

A few days after the agreement was reached, the state had its hearing on RMU-1. Only nine people spoke, and most of them supported the application.[324] Soon thereafter CWM received its permits for RMU-1.

The Case of Tooele County   Other forces have come together to achieve successful siting in Tooele County, Utah, with very little controversy.[325] Tooele (pronounced "too-willa") County is emerging as the nation's hazardous and radioactive waste capital.

The county covers 7,000 square miles, roughly the size of Connecticut, but has only 28,000 residents. Its population was 91.5 percent white in 1990. In 1941 the War Department acquired 2,100 square miles for what became the Tooele Army Depot, for the manufacture and storage of conventional and chemical weapons, and the Dugway Proving Ground, for the open-air testing of nerve gas. These remain the County's main employers. Within its borders are also the Salt Lake Bombing and Gunnery Range, used for training F-16 pilots from Hill Air Force Base; the Bonneville Raceway, used for automobile speed trials; and one of the nation's four or five largest emitters of toxic pollutants, the AMAX Magnesium Corp., which extracts metal from evaporation ponds on the western shore of the Great Salt Lake. The Tooele Army Depot is the only place (other than an atoll in the Pacific) where the Army has managed to build an incinerator for chemical weapons, and the county's leaders have offered to accept even more chemical weapons for disposal in exchange for $20 million to build a new county hospital. Within the county, the Skull Valley Band of the Goshute Indians (with fewer than 200 members) has also considered hosting the monitored retrievable storage facility for high-level radioactive waste.

The siting episode of greatest interest here began with a 4.3-million-ton pile of uranium mill tailings that had been sitting for years in what is now downtown Salt Lake City, the residue of a uranium processing mill operated for the old Atomic Energy Commission from 1951 to 1964. The U.S. Department of Energy, AEC's successor, began looking in 1975 for a place to dispose of these tailings. A total of thirty-seven potential sites in Utah were identified. The state picked three of these sites as its top choices, and then the DOE, after extensive analysis, chose one of the three, on state-owned land.

The formal address of the site is Clive, Utah, but Clive is actually a Union Pacific railroad siding, not a town. The nearest neighbors (eight of them) are fifteen to twenty miles away. The site is on the eastern edge of the Great Salt Lake Desert, and the underlying shallow aquifer is so saline that its water is undrinkable. The area has very little species diversity and is inhabited primarily by jackrabbits, deer mice, and desert horned lizards. There are no known cultural resources in the area, though the cannibalistic Donner Party may have passed nearby in 1846. The only activities on or near the site were grazing, hunting, and occasional recreational vehicle use.

Most of Utah is divided into 640-acre "sections." The DOE acquired 100 acres of one section for the mill tailings. Then a private company, Envirocare of Utah, Inc., acquired the remaining 540 acres and obtained licenses for the disposal of low-level radioactive waste, naturally occurring radioactive waste, and similar materials. The DOE agreed to take title to the site upon the termination of Envirocare's license. Tooele County, which has chronically high unemployment, was happy to have these facilities. After a series of public meetings, it established a 100-square-mile West Desert Hazardous Industry Area for such activities, and barred residential uses in this area. Later United States Pollution Control Inc. (USPCI), a subsidiary of Union Pacific, built a hazardous waste incinerator one mile to the west of the Envirocare site, and Aptus, a Westinghouse subsidiary, built a hazardous waste incinerator seven miles to the northeast. USPCI also built a hazardous waste landfill nearby, called the Grassy Mountain Facility, though there is not a single blade of grass for many miles, and an adjacent PCB landfill called Grayback Mountain. The district has created more than nine hundred new jobs and brings in $2 million in annual "mitigation fees," which have allowed the county to freeze its property taxes. Speculators hoping to build still more facilities have increased the price of a 640-acre section from $320 to $750,000.

The only apparent controversy concerning Envirocare has come from US Ecology, which has argued that it would take away business from its LLRW landfill in Hanford, Washington. There are no neighbors to object. Though the site began with just Utah waste, it has become nationally important and is receiving slightly radioactive waste from civilian sites around the country.

Implications

As the preceding discussion has shown, much opposition to HW/RW disposal facilities is caused by some combination of dread (an emotion largely derived from the nature and physical origin of the waste) and intrusion (a concept that considers whether the facility is imposed or invited, whether the waste is local or imported, and whether the facility's builders are trusted). A number of factors increase public acceptance of such facilities: the treatment of local waste; the provision or protection of local jobs; the degree of local control; and the nature of the local culture.

This framework helps explains a number of apparent anomalies:

• There have been few siting successes with new commercial HW incinerators because they are both dreaded (the by-products of incineration, such as dioxin, are much feared) and intrusive (they import waste from a wide area). MSW incinerators attract much public opposition (because their emissions are still dreaded), but, because they are not so intrusive (they are usually built by municipalities to burn local garbage), there have been many siting successes.

• Although experts proclaim their safety, medical waste incinerators have met with ferocious local opposition largely because of their psychological association with today's most dreaded disease, AIDS.

• Radon, a naturally occurring radioactive gas emitted by many rock formations, has stirred relatively little public concern, largely because it is not at all intrusive—that is, it comes from nature, rather than being imposed by some outside, mistrusted human agency—and, if it gets into a house, it is eminently controllable.

• Some activities that statistically pose a far greater risk of injury than does HW/RW disposal, such as the trucking of gasoline around the country,[326] stir little concern because they are neither dreaded nor intrusive (and also because their risks are spread over a larger population).

• To obtain gratification or other benefits, individuals engage in personal activities (such as smoking and car racing) that subject them to considerably greater risks than do HW/RW disposal facilities, but those activities are voluntary and therefore not intrusive.[327]

• On-site disposal of HW/RW is generally accepted, because it is not intrusive.

This discussion of public opinion also has important implications for the siting process. Most important, it shows that the widespread practice of trying to preempt local control and force disposal facilities on unwilling communities is much like the medieval practice of bleeding the sick: it is exquisitely counterproductive. Not only does it never work, it actually increases opposition exponentially by turning what might be a voluntary risk into an involuntary, highly intrusive risk. As I showed earlier, preemption (if it worked) would also amount to a hidden subsidy for the creation of hazardous and radioactive waste. Unfortunately, as I demonstrate in chapter 7, preemption is deeply embedded in national siting policy and law.

A second implication is the flip side of the first: the search for volunteer sites should be central to a sound siting policy. It should be possible to find communities whose cultures will induce them to volunteer. I will discuss prior successful attempts to find volunteer communities in chapter 6.

A third implication is that the opposition to importation, and the relative ease of on-site disposal, are likely to lead to a proliferation of a very large number of very small disposal facilities. This situation presents its own environmental hazards, but my proposed solution to that dilemma must await chapter 9.

Finally, to address the fifth question at this chapter's opening: The current system of HW/RW facility siting is not politically viable. It is based on a lack of understanding of the psychological and sociological dynamics of siting, and it fosters public opposition that is almost always fatal to siting attempts.

II

Prior Proposals for Reform

6

Consensual Proposals

Part II describes and assesses prior proposals to reform the siting process. I have divided them into three categories—those that seek to achieve local consent to build a facility (consensual) (chapter 6); those that seek to impose a facility (coercive) (chapter 7); and those that seek to avoid local siting altogether by eliminating, redefining, or exporting the waste (avoidance) (chapter 8).

Compensation

Strictly speaking, three separate concepts are involved here. "Compensation" aims to recreate the status quo and to make communities whole by paying for actual damages. "Mitigation" prevents, reduces, or eliminates adverse impacts before they occur. "Incentives" or "rewards" are positive inducements, beyond any actual or predicted damages, to reward communities for accepting risk.[1] In the discussion that follows, I use the word *compensation* generically to cover all three concepts.

Drawing on the work of Frank Michelman,[2] the seminal paper on compensation in HW/RW facility siting was published by Michael O'Hare in 1977,[3] and many writers have since embraced and embellished the idea.[4] The basic notion is that "[s]iting of noxious facilities tends to concentrate costs within an area proximate to the site while providing diffuse benefits over a wide area. Compensation measures can rearrange this distribution of costs and benefits."[5] Under this theory, compensation serves three purposes: it induces localities to accept facilities; it makes the victims whole; and, by internalizing the external costs of these facilities, it increases economic efficiency.

The idea of compensation has been widely embraced in HW/RW siting legislation. The HW siting statutes of at least thirteen states mandate some kind of compensation or economic incentives.[6] Twelve states offer compensation packages for communities getting LLRW facilities.[7] The NWPA provides for payments of $20 million per year for the host state (or tribe) for the HLW repository, and $10 million per year for the host of the monitored retrievable storage facility,[8] with additional "impact aid" to the state and affected local governments.[9] The state or tribe can also receive several million dollars a year prior to the opening of the facilities, but only if it waives its rights to object to the siting.[10]

Despite all this support, compensation has never been used successfully in siting an HW/RW disposal facility in a reluctant community in the United States.[11] The reason is clear: the opposition to these facilities stems mainly from concern over their impact on health, particularly children's health, and there is no amount of money that people will accept to allow others to endanger their children. That is not to say that people will not endanger their *own* children, through everything from passive smoking to disuse of car seats to overt abuse. But these dangers are imposed by parents, not by an intrusive outside force. I have seen clients of mine chain smoke while bemoaning the possible effects of a nearby toxic dump on their children, though their own smoking poses a far greater danger.

Individuals who perceive these facilities as dangerous will not change these perceptions when offered money,[12] and they view the offer itself as immoral,[13] "bribery,"[14] or "blood money."[15] Health, like love or salvation, does not belong in the marketplace; to buy or sell any of these is a travesty.

The moral problems are amplified when money is offered to low-income communities.[16] In the words of one writer, "Garbage tends to concentrate in depressions; it rolls downhill until it hits those places most desperate to deal."[17] Poverty makes choice less meaningful. It can drive people into lives of prostitution, mugging, and drug dealing as superior alternatives to starvation. If a community needs the money from a landfill to pay for a clinic, a fire station, and a school, is that community able to make an enlightened choice? On the other hand, does a landfill opponent have a right to say the community must go *without* those things it desperately needs, especially if the clinic and the fire station will likely save far more lives than the landfill will endanger? For well-off environmentalists fighting projects in impoverished areas, the line between education and paternalism can be hard to draw.

Except for the unusual case of Tooele, Utah (discussed above), the closest that compensation has come to achieving success in siting an HW/RW facility in the U.S. was in 1981, when the City of Baltimore agreed to locate an HW landfill in an industrial neighborhood surrounded by other landfills. Under the proposal, the city would receive a redeveloped park and $5 per ton of waste, and the twenty-two nearby families would be relocated. The city, however, also insisted on a ban on out-of-state wastes—a condition that helped doom the project economically.[18] (As noted above, compensation also contributed to the successful siting of the HW landfill in Last Chance, Colorado. Local control, however, seems to have played a much more decisive role there.)

Thus compensation works when, and only when, the community does not believe the proposed facility poses an undue hazard. In the voluntary siting processes in Canada, discussed below, communities that were already amenable to hazardous waste facilities were made even more so by the offer of compensation.[19] Compensation has been quite successful in siting MSW landfills and incinerators, which have much lower perceived risks than do HW/RW facilities.[20]

In those situations where compensation is effective, a number of practical issues arise:

What Kind of Compensation?
Cash payments to the municipality are only one form of compensation. Long menus of possible measures have been prepared, such as the purchase of fire trucks and ambulances; promises of jobs to local residents; purchase of buffer zones around facilities; funds for street improvements; and free use of the facility by local businesses.[21] Where the neighbors fear that a facility will contaminate their drinking water, a guarantee of replacement water is often well received.[22] One popular measure is "value protection," a form of insurance against declines in property values.[23]

How Much?

Since most of the anticipated effects (such as odors, noise, and lost aesthetic value) are hard to quantify,[24] several schemes have been proposed for auctions and other mechanisms to derive an appropriate and acceptable compensation amount.[25]

Who Benefits?

It is easy to make payments to the host municipality but much harder to devise a formula for allocating the payments to those who are adversely affected—landowners, tenants, businesses, and so on. Moreover, facilities often have spillover effects on neighboring areas.[26] Decisions on how the money flows are often made politically. One HW landfill in California pays $7 million a year to the county, but none of this money makes it back to the remote (and mostly Latino) town where the landfill is located. In contrast, at a proposed LLRW facility in Illinois, the town would have gotten all the compensation and the county none, sparking great resentment by the county.[27]

When?

Most compensation schemes (except for that under the NWPA)[28] pay nothing until the facility is operating. However, neighbors can begin experiencing losses (especially to property values) as soon as the site is proposed.

Ongoing compensation—such as per-ton payments—can become addictive. If a community grows accustomed to having no property tax because of facility fees, then the closure of the facility becomes a feared event, and permit renewals garner active support. Several communities with landfills and nuclear power plants have experienced this phenomenon.[29]

Some communities have tried clever methods—too clever by half—to increase their compensation. Since the NWPA provides for payments to local governments in lieu of the taxes they would receive from comparable private activities on the land,[30] the Nevada Legislature created a new county around Yucca Mountain, called Bullfrog County, with 144 square miles but not a single resident. The state then set an extremely high local tax rate for the county, and provided for the reversion of the in-lieu-of-tax payments to the state if the local government (which did not exist) did not use them. The Nevada state courts struck down this scheme as a violation of the state constitution.[31]

Negotiation and Mediation

Many commentators have urged that siting disputes be resolved through negotiation, usually with the assistance of a mediator.[32] Several handbooks have been written to guide the process.[33] Nineteen states have procedures for negotiation or mediation between facility developers and proposed host communities.[34] The NWPA established a presidentially appointed position, the nuclear waste negotiator, whose job is to "attempt to find a State or Indian tribe willing to host a repository or monitored retrievable storage facility at a technically qualified site on reasonable terms."[35] He has one of the most thankless jobs ever conceived; local officials cringe at the thought of being seen with him, for it could destroy their careers.[36]

Massachusetts has pioneered mediation in HW siting disputes. In 1980 that state enacted a siting law conceived by three scholars from MIT and Harvard—Michael O'Hare, Lawrence Bacow, and Debra Sanderson—who had written widely on compensation mechanisms. The statute involved a formal process of mediator-aided negotiations among the state, the locality, and the facility developer, leading to a compensation agreement; if no agreement could be achieved, one would be decreed by an arbitrator. This statute served as the model for laws soon enacted in Rhode Island, Connecticut, Wisconsin, and Virginia, and it received extensive scholarly commentary.[37] By now it is clear, however, that the law has not succeeded. There have been six serious siting attempts under the Massachusetts law since 1980, the most recent one ending in 1992; every one of them failed.[38] Nor have any new HW disposal sites been created in the other states that adopted the Massachusetts model (although an on-site wastewater treatment facility was permitted in Providence, Rhode Island).[39]

In 1993 Michael O'Hare and Debra Sanderson wrote a retrospective on their experiment. They criticized state and local officials for not pushing through facility proposals in the face of "irresponsible behavior" by opponents and for allowing the process to "degenerate . . . into a hostile and occasionally ludicrous circus." They concluded:

The Massachusetts negotiated compensation model of facility siting is so complicated in practice, and so contingent on local factors, that no one can judge confidently whether it is on the whole hopeless, flawed by correctable, or merely unlucky. We believe its principal liability is that it offers two fatal temptations: to public officials, it appears to offer an alternative to taking leadership risks; and to frightened citizens, it appears to offer a way to avoid, rather than confront and control, physical risks and anxiety. Specific features of the process . . . might be corrected with good effect. But we see larger and more pervasive forces as the real obstacle. The NIMBY problem is, at heart, symptomatic of the pessimistic expectations that citizens, industry, and government all hold for each other and themselves; raising those expectations is not a task that can be accomplished by any legislated decision process.[40]

Negotiation suffers from precisely the same shortcoming as compensation: outside threats to the health of one's children are nonnegotiable. Many Massachusetts residents feared the proposed facilities were such threats, though O'Hare and Sanderson believe the threats are tiny and much overshadowed by the benefits. Negotiation and compensation work in the same situations—where the community does not fear for its children's health. Indeed, the principal purpose of the mediation process is to negotiate a compensation package.[41] In Wisconsin, thirty-four solid waste facilities, but not a single new hazardous waste disposal facility, have been sited using negotiation.[42] Negotiation does little to address the usual underlying causes of public opposition, and citizen activists warn against even entering into the process—"industries already have the battle more than half won when they can get their citizen opponents to sit down with them and speak their language."[43] Low-income communities are also concerned that they do not have sufficient bargaining power to extract necessary concessions.[44]

Public Participation and Public Information

Public participation is an important element of most environmental permit processes.[45] Sherry Arnstein has formulated a "ladder of citizen participation," ranging (top to bottom) from citizen control to delegated power, partnership, placation, consultation, informing, therapy, and manipulation.[46] Participation methods currently used in HW/RW siting run the shortened gamut from placation (negotiated concessions) to manipulation (public relations campaigns). The most common forms are public hearings, citizen advisory committees, and membership on state siting boards. Eleven states also give technical assistance grants.[47]

A strong undercurrent running through much of the facility-siting literature hints that public ignorance is at the root of opposition and that the mission of citizen participation and public information is to correct or neutralize this ignorance.[48] There is not much evidence, however, that public participation serves this purpose, and it appears that, especially when poorly handled, it can actually increase public opposition.[49] Moreover, the undercurrent seems based on a false assumption; facility opponents have been found to be just as knowledgeable, on average, as proponents.[50]

Numerous methods have been used to try to increase public acceptance. In many siting programs (especially for radioactive waste facilities) the federal or state government provides funding for local "citizen participation" efforts, but in localities that start out being opposed, this money is often used to hire lawyers and experts to defeat the proposal. Euphemisms are rampant; just as trash collectors became "sanitation engineers," now an incinerator is a "thermal treatment unit" and a landfill is a "residuals management unit." One common technique is to stage household hazardous waste cleanup days, in which citizens can take old paint remover, pesticides, and other waste for proper disposal, and in the process learn that they, too, contribute to the HW problem.[51] Another is advertising campaigns,[52] though sometimes these can backfire. (The American Nuclear Energy Council ran television ads in Nevada promoting the HLW repository, but counter-ads soon appeared in which two Las Vegas disk jockeys climbed into a huge pair of overalls to create a two-headed mutant, "Yucca Mountain Man." After the campaign, 75 percent of the public opposed the facility—the same as before.)[53] There is evidence that more information or educational efforts about nuclear power, for example, only mobilize latent fears about the atom.[54]

All in all, public information campaigns have failed in their purpose. Although such campaigns can increase perception of risk,[55] there is little or no evidence that they can reduce this perception.[56] Media blitzes have increased public concern over AIDS, radon gas, drunken driving, and other threats, but there are few examples of successful campaigns to allay preexisting fears. (An exception is the swift debunking of a clear hoax, such as the Tylenol tampering scare, but even this can be a steep uphill climb.) An alarm bell is far harder to ignore than a lullaby.

One reason for the failure of public information efforts is the filtering effect of the news media. Except in very small communities, facility proponents cannot hope to speak directly to everyone; they must convey their message through the media. The media, in turn, are more apt to report negative than positive news: "X is dangerous" is a better story than "X is safe," just as there are few stories about the thousands of airplanes that make it safely to their destinations every day (and any such stories would find few readers). Thus reports about the danger of a proposed incinerator or landfill will tend to receive more prominent play in the newspaper or the evening news than will assurances of safety.[57]

Volunteers

The usual practice in siting has been called "decide, announce, defend" (often followed by "surrender"). The opposite approach is to ask for communities to volunteer to host facilities.[58] This method has three principal advantages and four principal disadvantages. The advantages are: (1) it decreases intrusion—by making the risk voluntary it reduces the perception of risk; (2) it draws out those communities with cultures that will accept these facilities; and (3) it tends to lead to payment of the full social costs of a facility, since the hidden subsidies of preemption are eliminated. The disadvantages are: (1) it would be coincidental if volunteering communities also happened to have favorable geology and other physical conditions; (2) equity problems can arise because low-income communities may be more prone to volunteering; (3) it is difficult to define the borders of the relevant community, and people just beyond may object; and (4) it is generally assumed that few communities will volunteer.

Several Canadian provinces have managed to avoid all of these disadvantages and successfully site HW facilities in volunteer communities. The methods they used are instructive. In the late 1970s, two attempts by private corporations to site hazardous waste facilities in Alberta failed in the face of strong public resistance. The provincial government declared a moratorium on private siting efforts and decided to take over the process itself. In 1982 Alberta established a Crown (government-owned) corporation, the Alberta Special Waste Management Corp. (ASWMC), to build a centralized facility with HW incineration, treatment, landfilling, and deep-well injection. In 1984 the ASWMC asked all the municipalities in the province if any would like to be considered for the facility; 70 percent responded positively. After a series of information presentations and community meetings, several towns dropped out (as they were free to do). The ASWMC selected five finalists based on technical criteria. Referenda were then held in the five towns to ensure that local support existed. The ASWMC chose the town of Swan Hills, a community founded in the 1950s as a camp for rig hands for the oil drilling companies that still dominate its economy, and where the referendum had passed by a 79 percent margin. A champagne celebration was held in Swan Hills after the selection, and one of the losing communities protested with newspaper ads. The facility (which is owned 60 percent by a private company and 40 percent by ASWMC) opened in 1987 and became the most comprehensive waste disposal facility in North America. In 1990 it received local support for quadrupling its capacity.[59]

The Manitoba Hazardous Waste Management Corporation (MHWMC) was established in 1986. It went through a very similar process. About fifty municipalities expressed interest in an HW facility, and thirty-five open houses were held to determine initial community support. Five communities became actively involved in the siting process—four rural towns plus the city of Winnipeg. Two of the towns dropped out after referenda, and one was disqualified on technical and economic grounds. That left one rural town, Montcalm, and Winnipeg. Montcalm is an agricultural area with a population of about 1,700. A major north-south highway to the United States runs through it, accustoming the community to truck traffic. Several neighboring communities have supported Montcalm's efforts. Winnipeg, which has 625,000 of the province's 1.1 million people, would put the facility adjacent to its MSW landfill, about three kilometers from the nearest residential area.[60] In 1992 the MHWMC chose Montcalm and issued a construction permit.[61]

A new hazardous waste facility has also been sited in a willing community in Quebec,[62] and Ontario is now attempting to site an LLRW facility using the same process as Alberta and Manitoba, with similarly positive responses. Some twenty-six communities expressed initial interest; twelve withdrew shortly before or after community meetings, but fourteen stayed in the process. As of 1993, the list had been narrowed to two communities, both of whose municipal councils have voted in favor of the LLRW facility.[63]

A voluntary process failed in British Columbia, and the reasons for its failure are revealing. Ten communities advanced themselves as possible sites for a hazardous waste management facility, but the facility proponent—a consortium of private firms—kept public involvement to a minimum, selecting both the technology and the actual sites almost unilaterally. The communities were not given the technical or financial resources they needed to participate meaningfully in the process. The provincial government was also eager to have the facility in operation before the next election three years hence, leaving too little time for the necessary public involvement. The necessary public support did not materialize, and the process ended in failure.[64]

Other than Tooele, Utah; Last Chance, Colorado; and Baltimore, Maryland, there have been no locally desired HW/RW sitings in the United States since 1976. This is, however, largely the result of state rather than local opposition. The voters of Fall River County, South Dakota, in a uranium mining region, voted in favor of an LLRW facility in 1985, but the state rejected the proposal.[65] Fremont County, Wyoming, expressed interest in an HLW storage facility in 1992, but the governor vetoed the idea;[66] the same thing happened in 1992 in Apache County, Arizona,[67] and in 1987 in Oak Ridge, Tennessee.[68] In 1969, 95 percent of the residents of Hermiston, Oregon, favored an army nerve gas storage facility, while 90 percent of the state's residents were opposed.[69] Many residents of Nye County, Nevada, the home of Yucca Mountain, support the construction of an HLW repository there, but the state has vigorously fought the proposal.[70] Martinsville, Illinois, a rural town of about 1,200, volunteered to host the LLRW facility for the Midwestern states, and 68 percent of its residents voted their support in a 1988 referendum. However, a state board refused to issue a permit because the site was geologically unsuitable.[71] (One town in Michigan and two in Wisconsin had also volunteered for this facility.)[72] Other communities—Naturita, Colorado; San Juan County, Utah; Edgemont, South Dakota; Barnwell, South Carolina; and Richland, Washington[73]—have supported HLW or LLRW facilities, only to be thwarted by state opposition.

Some other communities initially volunteered but later changed their minds. This occurred in Boyd County, Nebraska, which first supported and then opposed an LLRW facility;[74] Woodland, North Carolina, which changed its mind about a hazardous waste incinerator;[75] and Carlsbad, New Mexico, which initially invited the WIPP project.[76] The town of Ashford, New York, the location of the West Valley nuclear facility, had a referendum in 1991 on hosting New York's LLRW repository. The vote was 702 opposed to 533 in favor, but the town board soon thereafter voted to approve the facility anyway. The following November the town board was reelected. The ultimate outcome in Ashford remains to be seen.[77]

Several MSW facilities have been sited after calls for volunteers,[78] sometimes following favorable votes in local referenda.[79] Fights have arisen between volunteer communities and their neighbors,[80] but this is an unusual result.

Volunteer communities typically see a waste facility as a vehicle for economic development.[81] (Given the strategic importance of this issue, it is surprising that so little research has been performed to test this belief.) For example, the Industrial Development Corporation of Lea County, New Mexico, has placed advertisements in the hazardous waste trade press inviting new facilities. The ads show a dirt road through a desert, with the headline, "This isn't the photo we'd use to attract tourists, but hey, you guys are in waste management!"[82] People responding to the ad receive a packet of information about the county's geology and its proximity to Los Alamos, WIPP, and "the fastest growing market for waste technology in North America: the Republic of Mexico." The material adds that "Lea County residents have lived with oil and gas and potash mining for years, and maintain a practical view of economic opportunities presented by toxic substances."

This hope for economic development seems to have borne fruit in Blainville, the volunteer community that received Quebec's facility. In conjunction with the new facility, a cloverleaf exit was built linking the town to the main road to Montreal and eliminating a bypass that had been seen as a deterrent to economic growth. The Canadian military also transferred an unused 2,000-acre camp to the town for use as an industrial park. The community's economy has indeed flourished with the highway bypass and the industrial park.[83]

The quest for volunteers is less likely to succeed if it is in the wake of a failed attempt to impose a site. Connecticut's LLRW program is one example of an effort thwarted by local opposition and is now attempting to induce communities to come forward on their own. However, the mistrust, antagonism, and fear of radioactive waste that resulted from the prior attempt have shaped public opinion, and finding volunteers is proving very difficult. Once an area has become fearful of a site, regaining public confidence is extraordinarily difficult. This is illustrated by the case of the Nyanza Superfund site near Boston. The site, a dye manufacturing plant, was added to the Superfund list in 1981. In 1982 a local citizens group conducted a door-to-door survey of the nearest neighbors and found eight cases of cancer and other illnesses among the forty-seven households it covered. This survey was widely publicized in the news media. Health officials said the survey was not scientifically valid. Many residents believed the survey over the officials. This incident became a source of lasting distrust among those who believed that the government was indifferent to, or was even covering up, the health threats. The testy relationship that ensued between the citizens and the officials strained further cooperative efforts.[84] This community would hardly have been receptive to a call for facility volunteers. Thus the search for volunteers may succeed in fresh territory, but it is not likely to work in places where coercive siting has already been tried and failed.

Special questions arise with attempts to site HW/RW facilities on Native American reservations.[85] In recent years tribal governments have attained increasing autonomy in environmental regulation over their reservations, and under several federal environmental statutes (though not RCRA) they are treated as states.[86] Different tribes have very different attitudes toward HW/RW. In 1991, for example, tribal police of the Shoshone-Bannock Tribes in Idaho intercepted on a highway in their reservation a truck carrying spent nuclear fuel that was on the way to the Idaho National Engineering Laboratory.[87] The Sioux challenged an HLW facility in Minnesota and tried to prevent trucks with nuclear materials from crossing reservation land.[88] The Mississippi Band of Choctaw Indians in 1991 rejected, by a 786 to 525 vote, a proposal for an HW landfill near their reservation.[89] On the other hand, five tribes have expressed interest in hosting the monitored retrievable storage facility for HLW.[90] The governors of Wyoming and New Mexico are trying to block these attempts, but it is not clear whether they have the authority.[91]

These events have aroused heated emotions on several fronts. Some environmental advocates have charged that the federal government wants to take undue advantage of Native Americans.[92] Others say the environmentalists are being paternalistic. Two Albuquerque lawyers—one a member of the Pawnee Tribe, the other a member of the Cherokee Nation—have written:

Too often, the environmental community appoints itself the officious protector of the Indians. . . . To people like ourselves, Indians who have devoted our careers to the defense of Indian rights, this is unspeakably arrogant. . . . Much of the environmental community seems to assume that, if an Indian community decides to accept such a project, it either does not understand the potential consequences or has been bamboozled by an unprincipled waste company. In either case, the clear implication is that Indians lack the intelligence to balance and protect adequately their own economic and environmental interests. This is clearly a racist assumption; the same assumption that guided the federal policies that very nearly eradicated Indian people in the late nineteenth and early twentieth centuries.[93]

Traditional Western notions of democracy are confounded when assessing the nature of tribal consent to a hazardous facility. In some tribes, the governing body is elected by the members of the tribe, but other tribes are governed at least in part by hereditary leaders or by theocracies (with officers selected by religious leaders).[94] Outsiders who try to challenge the legitimacy of a decision made by a tribal council find themselves in the awkward, if not untenable, position of attacking the way a different culture has come to govern itself.

The tribe that has been the most aggressive in seeking facilities is the Mescalero Apaches, the descendants of the legendary warriors Geronimo and Cochise. Their 460,000-acre reservation is in New Mexico, near the White Sands Missile Range and the Alamogordo test site. Their long-time president, Wendell Chino, is popularly elected by the 3,000 members of the tribe. Under his leadership, the reservation has attracted a $30 million ski resort (Inn of the Mountain Gods), a sawmill, and a bingo parlor. He is supporting the siting of the monitored retrievable storage facility on the reservation as a way of remedying the tribe's still-high unemployment rate. In the most recent election for tribal president, conducted during the midst of the MRS controversy, Chino won against a facility opponent by a vote of 391 to 176. The MRS issue remains very controversial within the tribe, and New Mexico's governor, its legislature, and all its representatives in the U.S. Congress oppose the plan.[95]

Risk Substitution

In 1991 two commentators independently suggested the risk substitution approach—that is, allowing a new HW facility to be built if there is a corresponding reduction in the risk from some other nearby facility.

Kent E. Portney, after reviewing the public opinion evidence, posited that most siting schemes had failed because they attempted to reduce people's risk perceptions, an approach, he soundly concluded, that is all but futile. To address this problem, he suggested searching for existing facilities, such as chemical plants, ammunition factories, and nuclear power plants, that neighbors regard as dangerous; buying the old factory and shutting it down; and building the new HW facility on the same site or nearby. This, Portney reasoned, would be acceptable to the neighbors because there would be no net change in the risks to which they were exposed.[96]

Bradford C. Mank had a similar notion but focused on just one type of old facility—contaminated sites such as orphaned Superfund sites or MSW landfills. He suggested that waste management companies be allowed to build new disposal facilities in exchange for remediating an existing contaminated site in the community.[97]

I believe that these proposals, although innovative, miss two important psychological dynamics. First, as discussed in chapter 4,[98] people react differently to old risks than to new ones. An existing facility next to where a person has been living for years may appear less threatening than a new one, even if objectively the latter poses far less risk. Second, if people are really concerned about an existing risk, they may demand that it be abated regardless of any plans for new facilities. People may consider it unjust to be forced to accept a new risk in order to eliminate an old one, especially since, once they have the new facility, they will never be rid of it. Additionally, Portney's proposal will foster considerable opposition if it costs jobs in the community, particularly since HW disposal facilities create relatively few jobs. There are also economic problems with both proposals. The cost of buying out an operating factory—even assuming that existing contracts with suppliers, labor, and customers allowed it—would likely be in the tens or hundreds of millions of dollars. The average cost of cleaning up a National Priorities List site is $30 million.[99] These are very large costs to bear in acquiring a site to conduct a financially risky business; regulatory or political vagaries may shut down the waste disposal operation prematurely, preventing full amortization of the site acquisition costs.[100]

Nonetheless, several recent events show that risk substitution does have some application in the real world. In 1992 the New York City Council approved a comprehensive solid waste plan that included a major new incinerator, but only after the mayor had agreed to shut down two old incinerators; that allowed the city council speaker to say the plan would, on a net basis, improve air quality in the city.[101] The same year in New Mexico, the Mescalero Apache tribe announced it would be willing to spend as much as $150 million to clean up contaminated uranium mining sites in the state in exchange for hosting the monitored retrievable storage facility for HLW.[102] And in 1993, as shown earlier, communities in Niagara County, New York, agreed to the expansion of an HW landfill in exchange for a moratorium on the construction of HW incinerators.

7

Coercive Proposals

This chapter describes past proposals for reform of the siting process that would attempt to use coercive means.

Preemption

In chapter 5, I demonstrated that preemption of local authority is counterproductive as an HW/RW siting strategy because it greatly increases perception of risk by making siting involuntary (intrusive), and it in effect subsidizes the creation of HW/RW. Moreover, it has never succeeded in actually siting a new facility.[1] Nevertheless, preemption remains a common feature in federal and state siting law and policy.[2]

RCRA provides that nothing in it "shall be construed to prohibit any State or political subdivision thereof from imposing any requirements, including those for site selection, which are more stringent than those imposed" under the EPA's regulations.[3] Nonetheless, the EPA announced in 1980 that "the process of site selection [should] not be hampered by blanket vetoes."[4] In 1983 the EPA promulgated a regulation disapproving "[a]ny aspect of State law or of the State program [under RCRA] which has no basis in human health or environmental protection and which acts as a prohibition on the treatment, storage or disposal of hazardous waste in the State."[5] In 1985, during consideration of what became the capacity assurance provisions of CERCLA, a Senate committee declared that "the process of site selection should find a way to transcend blanket political vetoes. No community should be able to remove itself from consideration on political grounds alone. Everyone must take responsibility for assuring that adequate sites are available."[6]

In 1986 the Eighth Circuit held that a county ordinance that prohibited the storage, treatment, or disposal of HW conflicted with RCRA because the ordinance did not allow the wastes to be handled in the manner deemed safest by Congress and the EPA.[7] In 1988, after North Carolina enacted a law inhibiting the siting of a particular HW facility, the EPA began proceedings to revoke the state's authority to implement RCRA.[8] Although some environmentalists claimed at the time that the EPA's proceedings were motivated by improper influences from the waste management industry,[9] the EPA ultimately found that North Carolina's actions were legal because they did not amount to a statewide prohibition on HW facilities.[10]

For obvious reasons of national security, federal primacy in the disposal of radioactive waste is well established,[11] although the states have been given the responsibility to find sites for LLRW.

Meanwhile, at least twenty-four states have provided specifically for state override authority of local zoning laws in the siting of HW facilities,[12] and very few states leave facility siting decisions entirely in the hands of localities.[13] The judicial decisions span a broad range of views on the degree of authority remaining to local government.[14]

Several commentators have argued that local authorities should be overridden when they unduly restrict HW facility siting,[15] but others recognize that this is futile.[16] As Gail Bingham and Daniel S. Miller have written, "Simply preempting local controls . . . is unlikely to resolve the siting dilemma because it does not address the causes of opposition. Rather than just disappearing, the opposition just surfaces somewhere else—in administrative challenges that complicate permit proceedings or in lawsuits that tie permits up in court."[17]

Moreover, the list of required permits is so long and the process is so protracted that election cycles are likely to come and go before final resolution, allowing a dedicated community to pressure candidates for office to support their side, or at least to call for "reevaluations" of prior decisions. Few candidates can stand in front of a room full of voters who are angry and fearful for the safety of their children, and refuse at least to take a second look at a decision to put a hated facility in their midst. That would take a degree of political courage (or arrogance, depending on one's point of view) seldom seen these days, as can be attested by anyone who has recently observed Congress trying to deal with the federal deficit. Each reevaluation demanded by the politicians further delays the process and gives opponents an opportunity to bring forth new facts and new laws. Thus, as long as the community remains cohesive and adamant, its resolve will likely neutralize a nominally preemptive siting law.

Penalties

Orlando E. Delogu has proposed federal legislation requiring states to site HW/RW facilities, with a provision that "any state that will not fashion an effective siting mechanism will lose all forms of direct and indirect federal financial support . . . in those program areas in which NIMBY-type activities [preclude] needed sites."[18] The Low-Level Radioactive Waste Policy Amendments Act of 1985 (LLRWPAA) imposes financial penalties on states that do not make provisions for disposal of LLRW, and originally required such states to take title to the LLRW.[19] However, the latter provision was struck down by the Supreme Court as a violation of the Tenth Amendment.[20]

While mechanisms of this sort require states to site facilities, they do not dictate any particular sites.

Governmental Facilities

The federal government, with requisite congressional authorization, theoretically could pick a site, take it by eminent domain if necessary, and build a facility on it. As we have seen, a comparable approach was taken at Yucca Mountain and has been stymied at every turn. States could do the same, but all such attempts to date have failed.

In addition to the obvious political problems, states may be reluctant to build their own HW/RW facilities because doing so would make them liable under CERCLA for contamination caused by the facilities.[21] The states would, in effect, take title to the waste—the same consequence that the Supreme Court held was unduly intrusive when imposed by Congress under the LLRWPAA.

8

Avoidance Proposals

This chapter describes various proposals for reform of the siting process that attempt to avoid siting facilities altogether, at least for particular kinds of wastes or within the geographical area of concern to the person making the proposal.

Waste Reduction

Few would disagree that the ideal way to solve the HW/RW siting dilemma would be to avoid generating the waste in the first place, if that could be done with acceptable economic consequences. Reducing generation of HW/RW has numerous benefits in addition to reducing the need for, and the environmental impact of, disposal sites: it increases the efficiency of raw materials utilization; it reduces the potential liability of generators; it reduces accidents in the transportation of the waste; it reduces leakage of waste at the locations of production, storage, transportation, and disposal; and depending on the technology used, it can reduce worker exposure to hazardous materials and reduce the presence of such materials in consumer products. Moreover, facility opponents consistently demand greater waste reduction.[1]

Despite these undisputed advantages, the law is strikingly weak on waste reduction. The Clean Air Act and the Clean Water Act are full of command and control mechanisms to reduce the production of air and water pollutants, and the Clean Air Act contains explicit marketplace incentives to reduce pollution. But the laws governing hazardous waste and radioactive waste have no comparable provisions.[2] Some HW statutes urge waste reduction, but there is nothing with any teeth. The RW statutes have nothing at all that requires or even encourages waste reduction.

RCRA declared the objective of "minimizing the generation of hazardous waste and the land disposal of hazardous waste by encouraging process substitution, materials recovery, properly conducted recycling and reuse, and treatment"[3] and announced a national policy "that, wherever feasible, the generation of hazardous waste is to be reduced or eliminated as expeditiously as possible."[4] The only enforcement mechanism (added by the Hazardous and Solid Waste Amendments in 1984) is that each generator must certify, in the manifests that accompany all shipments of HW, that it "has a program in place to reduce the volume or quantity and toxicity of such waste to the degree determined by the generator to be economically practicable."[5] This certification is contained in fine print on the standard manifest form.[6] A small quantity generator must merely certify that "I have made a good faith effort to minimize my waste generation and select the best waste management method that is available to me and that I can afford."[7] A lawyer would be hard pressed to find anywhere else such a short sentence with so many loopholes. The biennial reports each generator must file with the EPA must also state "the efforts undertaken during the year to reduce the volume and toxicity of waste generated."[8] It does not appear that any generator has ever been sanctioned for inadequate or false certification of waste minimization efforts. In mid-1993 the EPA published guidelines on the elements of a waste minimization plan that would allow a generator to issue a proper certification;[9] the results remain to be seen. President Clinton also directed all federal agencies to develop and implement pollution prevention plans.[10]

Despite the name of the statute, the Resource *Conservation* and *Recovery* Act, the EPA devoted such little funding and so low priority to conservation and recovery that one disgruntled official said RCRA should more aptly be called DRIP, the Dump Regulatory and Investigatory Planning Act.[11] One study of the EPA's implementation of RCRA in the late 1970s remarked that "EPA could have looked for ways to encourage product and process redesign and recycling in order to lower the volume of waste requiring disposal. As public opposition to all disposal options has built, one is struck by how little attention this approach received from EPA during the Carter years."[12] CERCLA contains no provisions concerning waste minimization at all. The EPA's directives to the states concerning capacity assurance reports stated that waste reduction was the preferred method of addressing hazardous waste problems,[13] but again no enforcement mechanism exists.

In 1990 Congress enacted the Pollution Prevention Act, declaring in even stronger rhetoric the national policy of reducing pollution at the source.[14] This statute calls on the EPA to develop and implement a strategy to promote source reduction, to give grants to the states, and to collect data and establish an information clearinghouse, but it has no regulatory punch whatsoever.[15] Congress authorized $16 million per year to implement the act.[16] This amounts to half the cost of cleaning up an average National Priorities List site. Several states enacted their own pollution prevention statutes, but few of these laws are any stronger than the federal laws.[17]

The environmental impact statement (EIS) process is ordinarily an important mechanism for examining alternatives, but there is no requirement that the builders of waste-generating facilities (such as chemical plants) discuss waste minimization in their EISs.[18] EISs for waste disposal facilities, such as incinerators, typically devote a page or two to waste minimization, compared to dozens of pages on recycling and hundreds on technological alternatives.

Many industries are working to reduce the hazardous waste they generate, not because of the toothless laws just mentioned but because of the high price of waste disposal[19] and because of fears of liability at disposal sites.[20] Since waste generators have been held jointly and severally liable under CERCLA for cleanup of landfills to which they sent waste,[21] large generators have become very skittish about using landfills. This has led to an increased preference for HW incineration[22] and to the formation of consortia to inspect disposal facilities. In the words of one facility operator, explaining why waste generators increasingly choose incineration over landfills: "[T]he liability goes up the smokestack."[23] Fear of liability has also led to a degree of overcompliance; many risk-averse companies send to HW landfills, rather than to ordinary solid waste landfills, material that technically may not be RCRA hazardous waste, but that (it is feared) may one day be listed as such, or might be contaminated with some HW, or material whose regulatory status is uncertain.[24] The price of guessing wrong is very high; a company or manager can be convicted of a felony for treating as ordinary solid waste a load that is really RCRA hazardous waste, and the associated bad publicity and loss of customers can drive a company out of business. If it is clear that a load is technically non-HW, but the chemical looks like it may soon be listed as an HW, a generator may wish to play it safe and send it to an HW landfill, because if it goes to an ordinary MSW landfill, and the landfill later becomes a Superfund site (which many MSW landfills do), then under CERCLA the generator will have to share in the costs of cleaning up that landfill. New York State has estimated that 40 percent to 60 percent of the waste disposed at its hazardous waste landfills is not legally hazardous.[25]

Whatever the reason, the amount of HW reduction and recycling is reported to be steadily rising.[26] Many estimates of the potential for waste reduction exist. Because the various studies that offered these estimates employed different methods and terms, their results are not comparable. Several of them, however, calculated that more than half of all the hazardous waste generated in the country could be eliminated through technological measures.[27] The situation is similar to that faced in 1973 when oil prices soared in the wake of the first OPEC oil embargo, forcing many industries to rethink their production methods. In just over a decade—between 1973 and 1985—American industry cut energy requirements per unit of output by one-third.[28] As Ronald T. McHugh wrote, "the economics of waste minimization has frequently been the driving force in process engineering evolving from the early time-and-motion studies, through material shortages in World War II, to the energy crisis of the 1970s, and finally to today's emphasis on compliance and materials cost savings tied to the direct and indirect costs of environmental requirements."[29] Among the methods available for reducing generation of hazardous waste are changes to process inputs; improved plant management or housekeeping; changes in process equipment or technology; recycling and reuse of materials within a process; and changes in the design of end products.[30]

The amount of hazardous waste generated depends critically on the price of waste disposal. This has been demonstrated both theoretically[31] and through extensive interviews with plant operators.[32] The experience at the Borden Chemical Plant in Richmond, California, illustrates this dynamic. The plant formerly generated approximately 350 cubic yards of phenolic resin sludge per year. When the cost of landfilling the sludge increased from $50 to $150 per yard, the plant changed its procedures for rinsing filters, rinsing reactor vessels, and increasing employee awareness of how to prevent small but significant losses of materials. The amount of sludge produced fell 93 percent.[33] Conversely, low waste disposal costs can discourage waste minimization. USS Chemicals disposes of its liquid wastes through deep well injection; although it could recover phenol from the wastewater prior to injection, or reduce phenol wastes at the source, the company continues to use injection because it is cheaper.[34]

Waste minimization efforts can greatly affect the need for disposal facilities. For example, plans to build a new HW incinerator in Washington State have been called into question as a result of the vigorous waste minimization program of Boeing, without whose waste the incinerator would not be economically viable.[35] DuPont's waste reduction efforts allowed it to cancel an agreement to supply one-third of the waste that was to feed a new HW incinerator in Ohio.[36] Large chemical and petrochemical plants are rebuilt on a ten-to-fifteen-year cycle. As new plants, designed since waste minimization became a major issue, are developed, waste generation at these facilities—which are by far the largest generators of HW—can be expected to decline significantly.[37]

The United States lags far behind Europe in the use of waste minimization technologies. Bruce Piasecki and Gary Davis, who have compared waste management on the two continents, attribute this not only to the relative scarcity of land and materials in Europe but also to the contrast between dominance of chemical engineers in the European waste management industry and dominance of former landfill operators in the U.S. waste management industry. They write that "[t]he evolution of hazardous waste management from garbage collection also helps explain America's peculiarly long reliance on landfill disposal. Both the practitioners and regulators had long experience with landfill disposal; thus both parties were slow to address hazardous waste management as a chemical engineering problem instead of a dirt-moving one."[38]

An important opportunity is available to reduce the amount of high-level radioactive waste generated, but the pricing mechanism for disposal of such waste provides no incentive to utilize it. If the fuel rods in nuclear power plants are used longer than originally contemplated—say, eighteen months instead of twelve months—less waste is created. This practice, known as "extended burnup," would be especially attractive to utilities if the price of uranium increased (which has not been the trend in recent years). Some utilities practice extended burnup anyway because it translates into less down-time during refueling, but the reduced amount of waste generated does not reduce disposal costs. That is because, under the Nuclear Waste Policy Act, utilities are charged for spent fuel disposal based on the amount of electricity generated by their nuclear plants, not based on the amount of waste created.[39]

Health and the environment ordinarily benefit from waste minimization, but not always. For example, there are two primary kinds of smoke detectors. One uses a photoelectric cell; the other uses a tiny amount of a radioactive element and is called an "ionizing unit." Because of the radiation, some have advocated banning the ionizing units in favor of photoelectric cells. However, detectors with photoelectric cells cost about four times as much as those with ionizing units. Quadrupling the cost of smoke detectors will undoubtedly reduce the number of people who buy and install them, especially in low-income neighborhoods. Every year fires kill several thousand people who might have been saved by smoke detectors—largely low-income families and the elderly. The number of fire-related deaths might well increase if fewer people buy smoke detectors because of the higher price. Thus the health effects of this particular kind of waste minimization may well be quite negative.[40] This sort of trade-off is rare, but when it occurs it should be explicitly considered.

Taxes and Charges

Much current discussion about reforming environmental law centers on economic incentives that would use such mechanisms as marketable permits and effluent charges. Reducing the generation of MSW by use of disposal charges has often been proposed,[41] and numerous municipalities have actually adopted similar approaches,[42] often with dramatic results in reducing the amount of waste generated.[43]

The notion of financing the Superfund through a tax on waste generation or disposal of hazardous waste (called a "waste end tax") was discussed when CERCLA was first enacted in 1980 and also during the reauthorizations of the RCRA in 1984 and CERCLA in 1986. In each case the idea was rejected, largely because of concern that such a tax would increase illegal dumping.[44] (As shown in chapter 5, this concern was misplaced.) Instead, the Superfund is financed largely through a fee on chemical feedstocks,[45] plus a surcharge on corporate income taxes.[46] This method creates little or no incentive to reduce waste generation.[47] Several states have adopted waste-end taxes, however.[48]

Some commentators have described an intricate system in which waste taxes or charges are fine-tuned to the external costs of the disposal methods used.[49] However, these systems require quantification of external costs in a way that is simply beyond the current state of the art.[50] Any lawyer who has handled a toxic tort trial knows that it can take weeks of testimony to prove that a particular incident of waste disposal led to a particular illness. Quantifying the future adverse health and environmental effects of numerous methods of disposal of thousands of different kinds of wastes, and then defending that quantification in front of some tribunal (for the taxed industries will surely demand a due process right to challenge their assessments), would create a litigation industry rivaling that which has arisen around CERCLA.

There is, however, another way to use economic incentives to reduce hazardous waste generation: rather than creating artificial market mechanisms, eliminate the hidden subsidies and allow the market itself to create those incentives. I will return to this in chapter 9.

Recycling[51]

In the public image, recycling involves such environmentally benign activities as bundling newspapers and separating cans and bottles. At the plants where those newspapers, cans, and bottles are actually processed into new products, however, much pollution is created. Proposed facilities to recycle and compost ordinary MSW have attracted considerable community opposition.[52] On New York's Long Island, a hazardous waste treatment facility and a plastics recycling factory are directly across the street from each other; the residential neighbors complain bitterly about the HW facility, but odor studies have shown that the plastics recycler is responsible for most of the bad smells in the area. There is a debate whether, at least for certain materials, secondary manufacturing (making products from recycled material) produces less pollution per ton of material processed than does primary manufacturing (though when the entire system is considered—from resource extraction through disposal—recycling is more likely to have the edge). Solid waste recycling indisputably, however, preserves virgin materials, conserves landfill space, saves energy, and reduces the need for incineration.[53]

When the material to be recycled is hazardous waste, the problems are considerably more severe. Some HW/RW recycling operations, such as smelting of lead from automobile batteries,[54] reprocessing waste oil,[55] and reprocessing spent nuclear fuel,[56] although often worthwhile, are highly polluting. A hazardous waste recycling facility is not much more desirable as a neighbor than any other kind of hazardous waste treatment plant. Indeed, the line between recycling and treatment is often blurry. Extracting silver from a load of used photographic chemicals, for example, would be considered recycling under some circumstances and treatment under others. Which side of the line a company falls on makes enormous legal difference, because a full set of RCRA permits is needed on one side, while little or no governmental permission is needed on the other side; the determination of where the line is located involves frequent battles between companies and regulators over the fine points of incomprehensible regulations. The actual environmental impacts of different options are often lost in the game of regulatory chess.

On-Site Treatment and Disposal

As discussed in chapter 2, 96 percent of all hazardous waste is disposed at the plant where it is generated. This is mostly wastewater, but on-site disposal of solids is also in growing favor, in view of its legal and psychological advantages. One prime example is the use of mobile incinerators. These units travel in a convoy of several large trailers and are set up at Superfund sites for a few months at a time to destroy on-site wastes. They are then disassembled and moved to another location.[57] These units attract less opposition than do commercial incinerators because they do not involve intrusion of waste from other regions, and they are seen as temporary. They are far from universally accepted, however. On-site incineration at NPL sites peaked in 1988 and has been declining ever since.[58] Some areas have expressed concern that, after installation, the mobile unit will be made permanent and will accept off-site wastes.[59] This concern is not entirely without basis. Congress initially directed that the on-site incinerators for destroying old chemical weapons be dismantled after all the material had been burned. Later, however, Congress rescinded this requirement, and the military is reportedly studying using these incinerators to destroy remedial waste from other sites.[60] This plight resembles that faced by the states of South Carolina, Nevada, and Washington, which thought that the Low-Level Radioactive Waste Policy Act of 1980 would quickly relieve them of the burden of the entire nation's LLRW. In fact, a former governor of South Carolina has coined what has been called "Riley's Law"—"Nuclear waste tends to remain where you first put it."[61]

For remedial waste, on-site disposal (as opposed to treatment) has the disadvantage of preventing the property from obtaining a "clean bill of health," thereby scaring away potential purchasers who are concerned about future liability.

On-site disposal is still in growing favor for waste from ongoing production processes.[62] One reason is its public acceptance. For example, to avoid public opposition to large-scale medical waste incineration, twenty-two hospitals on Long Island, New York, have worked out a regional system under which regulated medical waste is treated in on-site autoclaves at each hospital, and the residue is then burned at an existing resource recovery incinerator.[63]

There are other reasons as well for the increase in on-site disposal:[64]

• The EPA prefers permanent remedies in the cleanup of contaminated sites, and many new treatment methods are best performed on site.

• Some cleanups, especially those at the NWC and military facilities, are so large that it would be uneconomical to transport the waste off-site; instead, treatment facilities are built on-site.

• Off-site land disposal leaves the generator vulnerable to later CERCLA liability if the landfill (or the transporter that takes the waste to the landfill) mishandles the material.

• On-site treatment and disposal units require fewer permits than do off-site facilities.

Another on-site alternative is long-term storage, which was discussed in chapter 5.[65]

Alternative Treatment Technologies

Hazardous waste landfills and incinerators have a very bad reputation in many communities. New treatment technologies are being developed that have not, at least so far, acquired the same negative image and thus do not attract the same opposition. One engineer cataloged sixty different kinds of treatment technologies.[66]

These alternative technologies are typically used on-site to treat remedial wastes. One prime example is bioremediation—the use of bacteria or fungi to destroy waste. (A similar approach is used in most sewage treatment plants.) Bioremediation comes in many varieties.[67] One can be described in the manner of a recipe: Dig up the petroleum-contaminated soil from around a leaky tank. Spread it on a flat paved surface, like a parking lot. For every eight parts soil, add one part sheep manure (to supply the microorganisms to eat the oil) and one part wood chips. Mix with bulldozer. Let stand for two or three months, mixing frequently. Yield: usable top soil or mulch.[68] Bioremediation is a young technology, and its effectiveness and environmental impacts are still unclear. Genetically engineered organisms are being created to attack particular kinds of wastes, and some have expressed concern over the release of these organisms into the environment.[69] Others have suggested that, if the wrong methods are used, certain carcinogens could be created.[70]

Another newly fashionable technology is thermal desorption, in which organically contaminated soils, sludges, and sediments are heated (but not burned) in a chamber that volatilizes the organics and collects them for further treatment or recycling. Between 1990 and 1992, the use of thermal desorption in hazardous waste cleanups doubled, while the use of mobile incinerators at HW sites lost popularity. This reversal of fortune has been attributed to the stigma attached to incineration. It is not clear, however, that thermal desorption involves lower risks to the community.[71]

These alternative technologies are being advanced by two technology-forcing provisions of the hazardous waste laws. One of them, adopted in the 1986 amendments to CERCLA, requires the EPA every year to test and demonstrate a certain number of new technologies for use at Superfund sites.[72] The resulting Superfund Innovative Technology Evaluation (SITE) program had, through the end of fiscal year 1991, looked at a total of 76 different technologies, and innovative technologies had been selected for 228 different cleanups, although significant economic and institutional barriers remain.[73]

The other technology-forcing law had its origins in 1981 in California Gov. Jerry Brown's Office of Appropriate Technology. That state began a program to phase out the land disposal of hazardous wastes by requiring their treatment instead.[74] This approach was adopted by Congress in HSWA, the 1984 amendments to RCRA.[75] The EPA was required to promulgate technological treatment standards for all the different kinds of HW under RCRA. In the laborious process of doing so, the EPA has required the use of incineration and a broad range of other technologies.[76] This has greatly increased the use of treatment technologies but not necessarily decreased the demand for hazardous waste landfills, since the treatment residuals are still landfilled, and they are often bulkier than the original waste. Another impact, ironically, was to impede the construction of new HW treatment facilities in the late 1980s because HW companies had to wait for the EPA to promulgate its new standards before knowing what the demand for their equipment would be.[77]

Further technological advances in the coming years may decrease the need for off-site facilities for the disposal of remedial wastes, but it is impossible to predict the impact of these future developments.[78] One difficulty is that, unlike landfilling and incineration, which are suitable for a broad range of wastes, many of the new technologies are effective with only a few specific kinds of waste streams. Methods that work wonderfully for certain chemicals are useless, or even harmful, with others. Waste streams from ongoing industrial processes tend to be fairly uniform, but many contaminated sites have an undefined, heterogenous chemical soup, making it hard to know with confidence that a selected technology will be effective.[79]

Several economic forces have inhibited the adoption of innovative HW treatment technologies. As just noted, the EPA's long delays in promulgating technology standards under HSWA's land disposal ban left the industry with great uncertainty over what technologies would be approved. The continued operation of old grandfathered landfills has depressed the price of HW disposal, making new technologies less attractive; ironically, one of the reasons these landfills have been allowed to stay open is because of the unavailability of alternative treatment capacity, creating a vicious circle.[80] As an added irony, successful waste minimization efforts also made treatment technologies less attractive by reducing the market.[81] The major waste management firms were reluctant to push for new technologies that would allow their profitable landfills to be shut down, and the major industrial generators, despite having the requisite technical expertise, were reluctant to go into the waste management business for fear of CERCLA liability.[82] Partly as a result of these forces, the United States is far behind many other advanced nations in developing HW treatment technology.[83]

Deregulation

One common but seldom-discussed way to avoid the HW/RW siting problem is to define the material as nonhazardous. This form of deregulation has allowed many kinds of wastes to avoid the rigors of the RCRA program and to be disposed of as ordinary garbage, burned as fuel, flushed down the sewer, or kept on-site indefinitely. The EPA has a formal procedure for generators to petition to have a particular waste stream "delisted" under RCRA,[84] but the principal impacts of deregulation have been outside that relatively narrow and rigorous provision.

Government exemptions have placed large waste streams beyond the pale of HW regulation. For example, the 1980 Bevill Amendment exempted huge volumes of mining waste and other materials from RCRA. RCRA's exemption for hazardous waste used as fuel presents another significant example.[85] Such exemptions invite manipulation of waste categorization. RCRA's fuel exemption, for instance, led to a practice known as "sham recycling" in which cement kilns, lime kilns, blast furnaces, and the like burned huge quantities of liquid hazardous waste at prices far below those charged by the better-regulated HW incinerators. Until the EPA finally closed the loophole in 1991 by imposing stringent air pollution regulations, these units were burning nearly twice as much HW as were HW incinerators. Even now these units—often sited many years ago in what are now residential areas—account for a significant portion of the nation's HW incineration capacity.[86]

Another example of attempted deregulation occurred in the wake of the D.C. Circuit's decision in late 1991 invalidating, on procedural grounds, aspects of the EPA's definition of HW.[87] The EPA, under the prodding of Vice President Dan Quayle and his Council on Competitiveness, took that opportunity to propose an entirely new set of definitions that would have had the effect of exempting broad groups of materials from RCRA.[88] By September 1992 this proposal had become an issue in the presidential campaign, and the White House chief of staff, James A. Baker III, ordered its withdrawal.[89] Similarly controversial was a 1990 proposal by the NRC to declare certain radioactive materials as "below regulatory concern," so that they could go to ordinary MSW landfills rather than to LLRW facilities. This proposal drew storms of protest, and NRC withdrew it in 1991.[90] Even so, low concentrations of uranium and other source and by-product nuclear materials can be incinerated without a license.[91]

Often these exemptions arise from economic imperatives; if it is frightfully expensive to dispose of a common waste as hazardous, the path of least resistance is often to declare it nonhazardous. One instance concerns "shredder fluff"—the material left over from the shredding of discarded automobiles and appliances after the ferrous metal has been removed. Unfortunately, older cars and appliances—still being discarded and shredded today—also have electrical or electronic devices that contain PCBs. These PCBs contaminate a whole load of shredder fluff. Under normal rules, the entire load would have to be sent to a hazardous waste landfill. The cumulative amount of shredder fluff produced every year nationwide is enormous. Instead of requiring special disposal of all this material, several states have found an exemption from the rules and declared shredder fluff nonhazardous, even if it does contain small amounts of PCBs.

Another important example is the "domestic sewage exclusion," which exempts from RCRA regulation most wastes that become mixed with sewage; this allows huge quantities of liquid industrial waste to be poured into the sewers and be treated by sewage plants that are not required to meet RCRA's rigorous standards.[92]

These examples involved attempts to reclassify something as ordinary solid waste rather than as hazardous or radioactive waste. There are also instances where material is moved even further down the regulatory ladder and deemed not to be waste at all. Some states allow what are called "beneficial use determinations"—rulings that a particular use of what would ordinarily be considered a solid waste is so beneficial that it does not really amount to disposal and therefore requires no disposal permit. For example, in upstate New York a salt company is seeking to fill a huge salt mine with incinerator ash, arguing that this will help keep the mine roof from collapsing; if this approach succeeds, the company might avoid the state's exacting regulations on solid waste landfills.[93]

There are many other instances of controversial exemptions, or attempted exemptions, from the definitions of regulated materials in an effort to save money for generators and to reduce the demand on HW/RW disposal facilities.[94] In many cases both the proposals and the outcomes are driven more by politics than by science.

Sometimes major accidents have occurred involving chemicals that had slipped between the regulatory cracks. For example, in 1991 a train derailment sent six tank cars flying off a bridge and into the Sacramento River in northern California. One of the cars released 19,500 gallons of metam sodium, a pesticide, wiping out the fish in a forty-five-mile stretch of the river. To the considerable embarrassment of the EPA, metam sodium did not fall within the definition of "hazardous substance" under CERCLA, and the shipment was not regulated under the Hazardous Materials Transportation Act.[95]

The regulatory ax swings both ways, and legal or political developments will often render a given substance subject to the HW/RW laws. For example, the enactment of the Clean Air Act Amendments of 1990 had the effect of adding forty new chemicals to the list of hazardous substances regulated under CERCLA.[96] The abandonment of commercial reprocessing converted the spent fuel rods kept at nuclear power plants from a resource into a radioactive waste. Current discussions between the United States and the former Soviet republics will determine how much of the fissile material inside old Soviet warheads becomes "waste."

Remote Siting

A common aspiration is to transport waste to some desolate, unpopulated area, where no one will be hurt if there is a leak. As has been amply demonstrated by the Yucca Mountain controversy, this is no guarantee of political acceptance. Moreover, deserts, for all their advantages of remoteness and aridity, have their own special environmental problems. They are subject to strange wind effects, such as tornadoes and "dust devils"; they are often in tectonically active places; and they typically have corrosive salts in their upper layers.[97] The U.S. Bureau of Land Management has supported the construction of a commercial hazardous waste treatment facility and landfill in the Broadwell Dry Lake basin in San Bernardino County, California, an area bounded by several wilderness areas, but the EPA has objected, largely because of impacts on wetlands and air quality (although there are no neighbors for many miles).[98] Moreover, reliance on remote areas also raises serious questions of geographic equity. In the words of a spokesman for the governor of New Mexico, "We always suspect there's an idea the desert Southwest is this great big empty space, and if you have something unpleasant, you can stash it out there and nobody will ever object. But to us it's God's country, and every square inch of it is very important to us and very fragile."[99] Thus deserts are no panacea.

One form of remote siting is prevalent, however: locating a facility near a border so that many of its neighbors will be in some other jurisdiction. An uncanny number of actual or proposed facilities are very close to a state or municipal border. Sometimes this arguably occurs because the borders are formed by bodies of water that attract heavy industry, but often this justification does not apply. Examples include the hazardous waste landfill in Emelle, Alabama, less than five miles from the Mississippi border; the Envirosafe HW landfill in Idaho near the Nevada and Oregon borders; California's proposed LLRW facility in Ward Valley, near the Arizona border;[100] Michigan's proposed LLRW facility in rural Riga Township, across the Ohio border from the suburbs of Toledo;[101] a proposed LLRW facility in North Carolina two miles from the South Carolina border;[102] the new hazardous waste incinerator in East Liverpool, Ohio, on the Ohio River across from West Virginia;

the Model City hazardous waste complex in the extreme northwestern corner of New York State, near Canada;[103] New Jersey's plan to put a hazardous waste incinerator on a narrow body of water called the Arthur Kill, directly across from Staten Island, New York;[104] new resource recovery plants staring at each other across the border of Gloucester and Camden counties, New Jersey;[105] several facilities in Hudspeth County, wedged in the western panhandle of Texas between New Mexico and Mexico;[106] and the effort of Pascagoula, Mississippi, to burn medical waste in an incinerator right across the city line from Moss Point, Mississippi.[107] Additionally, plans are underway to build several hazardous waste facilities in Texas near the Mexican border.[108]

Export

Remote siting refers to locations within the same jurisdiction. Even more attractive to most politicians is export—sending waste to a different jurisdiction. (This might be called SITTOGD—Send It To The Other Guy's District.) This approach includes transporting wastes to other cities or states; shipping it to other nations; disposing of it in the oceans; and shooting it into outer space. The idea is embodied in a *Saturday Night Live* "commercial" for a fantastic new device, the Yard-a-Pult, which allows suburbanites to dispose of their garbage by catapulting it over the back fence into the yards of their neighbors.[109] All these methods impose risks on the unwitting and powerless neighbors of the transport routes, while relieving the burdens on the waste generators.

Other Cities and States

The massive political and legal wars recounted earlier over the interstate transportation of waste all involve efforts of one jurisdiction to transfer its waste to another. Sometimes export of residue is an explicit part of a siting deal. For example, to secure approval by his city council for the construction of a large MSW incinerator, New York City Mayor David Dinkins was forced to cancel plans to dispose of the ash at the city's landfill in Staten Island and instead promise to export it[110]—to Virginia, as it developed.[111] Massachusetts abandoned plans to build a landfill for sewage sludge from Boston Harbor and instead decided to ship it to

Utah.[112] On other occasions, a community will demand that an entire landfill be exhumed and its contents shipped elsewhere,[113] even though the process of exhumation may well release gases into the neighborhood that could cause considerably greater health risks than does leaving the landfill in place.[114]

Other Nations

A huge international trade exists in hazardous waste, with waste travelling mostly from north to south. The West African nations of Benin, Guinea, and Guinea-Bissau have been among the most active importers. Guinea-Bissau once negotiated a contract of $120 million per year—more than the country's annual budget—to store HW from other countries, but public outcry forced the government to rescind the deal.[115] Unscrupulous waste merchants are known or suspected to have dumped numerous loads of HW at sea.[116]

RCRA bars the international export of HW unless the United States and the receiving country have a waste exchange agreement, and the receiving country has agreed to accept the shipment.[117] In August 1992, the U.S. Senate ratified the Basel Convention on the Control of Transboundary Movements of Hazardous Waste and Their Disposal, establishing broad international controls—although not as broad as some would have liked.[118] (Congress has not yet enacted implementing legislation).[119] The net result of these and related international controls is that it is unlikely that U.S. waste generators will lawfully be able to export large quantities of HW. In 1990 only about 134,000 tons of hazardous waste—0.05 percent of the HW generated—was legally exported from the United States, mostly to Canada.[120] However, stringent enforcement will be required to guard against illegal exports.

Oceans

The oceans have long been used for the disposal of HW/RW. Allied forces disposed of approximately 250,000 tons of German chemical munitions at sea after World War II.[121] The United States began dumping radioactive waste at sea in 1946. By the time the practice stopped in 1970, the United States had dumped an estimated 107,000 containers of RW—mostly concrete-capped fifty-five-gallon drums—at twenty-eight sites in the Atlantic and Pacific Oceans and the Gulf of Mexico.[122] Until the early 1970s near-shore areas were commonly used as dumping sites for several types of HWs. The United States also formerly dumped obsolete nerve gas and other chemical weapons in the ocean.[123] Most of this stopped with the passage of the Marine Protection, Research and Sanctuaries Act of 1972,[124] but dumping of acid wastes, contaminated dredge spoils, sewage sludge, and treated municipal and industrial waste waters[125] continued until most of those practices, in turn, were barred by the Ocean Dumping Ban Act of 1988, effective at the end of 1991.[126] (One illegal practice persists off the Atlantic and Gulf coasts—filling barges with HW and abandoning them.)[127] The London Dumping Convention of 1972[128] has also governed waste disposal at sea since 1975, together with a complex of other international agreements.[129] In clear violation of these agreements, Russia dumped hundreds of tons of low-level radioactive waste into the Sea of Japan in October 1993, sparking strong protests from Japan.[130]

Ocean incineration of hazardous waste has received considerable attention. Four sets of research or interim burns occurred under the EPA's authority between 1974 and 1982, all using a ship called the Volcanus I.[131] In 1983, when the EPA held a public hearing concerning ocean incineration in Brownsville, Texas (which was proposed as an embarkation port), more than 6,200 people attended, the overwhelmingly majority of whom were opposed, making it the largest hearing in EPA history.[132] The EPA proposed regulations concerning ocean incineration in 1985[133] but never made them final. Several companies wishing to burn at sea failed to persuade the EPA to grant them permits.[134] Even if the regulatory climate for ocean incineration improved,[135] it could not have much of an impact on the overall HW disposal picture. For technical reasons, only about 8% of all hazardous wastes (mostly liquid chlorinated wastes) are suitable for ocean incineration,[136] and the most ardent proponents foresee an incineration fleet of no more than a few ships, together burning only a tiny fraction of all the nation's HW.[137] Ocean incineration began on a commercial basis in Europe in 1969 and peaked at 108,000 tons per year in 1980. It has been declining ever since, plagued by technical problems (including discovery of dioxin in the exhaust gases), market failures, and political protests.[138]

Proposals occasionally arise to build offshore islands for deep-sea ports, refineries, waste disposal, and the like.[139] The Deepwater Port Act authorizes licenses for artificial ports that handle oil but not for those handling other commodities.[140] The federal government has jurisdiction over the seabed from the three-mile limit to the edge of the continental shelf,[141] and it is likely that an offshore island for any other purpose would require explicit congressional authorization.[142]

Proposals for disposal of HW/RW under the deep-sea floor still arise from time to time.[143] In 1987, when Congress designated Yucca Mountain, a provision was inserted into the NWPA (as a concession to the Nevada Congressional delegation)[144] calling for a study of sub-seabed disposal of HLW.[145] Scientists at the Woods Hole Oceanographic Institution have suggested dropping torpedo-shaped canisters of HLW into the sea floor's sediments,[146] but this proposal, like all the others, does not seem to be going anywhere.

A kindred proposal was made in 1992 by Admiral Stansfield Turner, a former director of the Central Intelligence Agency. He suggested taking the world's old nuclear weapons—especially those from the former Soviet republics—and storing them near America's scientific base at McMurdo Sound in Antarctica.[147]

Outer Space

Disposing of high-level radioactive waste or plutonium by sending it into outer space has been seriously discussed.[148] The concept entails using a space shuttle to carry a waste package to a low-level earth orbit and then transferring the waste to another rocket for insertion into a solar orbit, where it would be expected to remain for at least one million years.[149] This would not be entirely unprecedented; already orbiting the earth is the detritus of thirty years of space flight, ranging from large objects, such as discarded rocket bodies and derelict satellites, to smaller items such as trash bags heaved over the side of spacecraft from previous missions; clouds of urine ice crystals; and a lost Hasselblad camera. Approximately forty nuclear-powered devices are currently in space, carrying about a ton of radioactive material. One of these devices, a Soviet satellite containing Uranium-235, fell back to earth in 1978 and landed in Canada. The USSR paid $3 million to reimburse Canada for the cost of finding and cleaning up after the satellite.[150]

Five international treaties have been cited that arguably would prohibit waste disposal in space, although their effect is not clear.[151] Wholly apart from the troubling ethical issues, outer-space disposal has serious practical problems. It currently costs about $10,000 per kilogram to put materials into space;[152] debris in space pose a very real hazard to spacecraft;[153] rockets sometimes crash; the manufacture of rocket fuel has involved fatal accidents,[154] is highly polluting, and creates its own hazardous wastes;[155] and the burning of rocket fuel is so polluting that it has its own subsection in the Clean Air Act.[156] In fact, the test firing and actual use of rockets is claimed to be a major cause of stratospheric ozone depletion.[157]

Do Nothing

The final alternative that has been seriously offered to building new disposal facilities is simply to stop building them. Several states have stopped even entertaining permit applications.[158] One environmental advocate has called for a "progressive not-in-anybody's-backyard solidarity required for a democratic challenge to a socially unjust and an environmentally unstable production process."[159] This approach is founded on the idea that if no more disposal capacity is provided, companies will be forced to stop generating the waste.[160]

Robert W. Lake, adopting a similar viewpoint, has called for reframing the entire siting question: "Siting hazardous waste incinerators, for example, constitutes a locational solution to an industrial production problem (hazardous waste generation). But the incinerator siting solution is only one of a number of possible strategies for hazardous waste management. The facility siting strategy concentrates costs on host communities, as compared to the alternative strategy of restructuring production so as to produce less waste, which concentrates costs on capital."[161]

In the words of Lois Gibbs of the Citizens Clearinghouse for Hazardous Wastes, grassroots groups "are doing some terrific stuff. People are following the strategy. They're stopping landfills, stopping incinerators, and backing up the wastes. They're plugging up the toilet."[162]

This attitude shows a major difference between the groups seeking to halt landfills, incinerators, and other waste disposal facilities, on the one hand, and those fighting homeless shelters, low-income housing projects, group homes for the mentally ill, and other social service facilities, on the other hand. I am frequently approached by (and always decline to represent) groups fighting proposed social service facilities. Their representatives almost always begin by saying: "We're not opposed to [homeless shelters, or whatever], but this just isn't the right place for it." The pejorative term NIMBY fits them well. However, groups opposing waste management facilities usually start with the premise that the unit should not be built anywhere. Especially when hazardous or radioactive waste is involved (though not as frequently for ordinary solid waste), these groups will often form alliances with people in other localities also faced with the same kind of facility—even though a site elsewhere will often mean that the local site is spared. These groups do not see themselves as shirking their civic responsibility; to the contrary, they believe their efforts to "plug up the toilet," in Lois Gibbs's phrase, serve the higher good of inhibiting the unnecessary production of waste. They believe the problem is one of production, not location, and their cry is NIABY—Not in Anyone's Backyard.

III

A Proposed Solution:
Local Control, State Responsibility,
National Allocation

9

A Description of the Proposal

Reprise: The Lessons of Experience

The reader by now has endured a detailed discussion of the many things that have gone wrong, and the few things that have gone right, in HW/RW facility siting. I will now try to summarize the lessons of this experience, as a prelude to proposing an alternative approach based on those lessons.

The current siting impasse has led to the perpetuation of old, poorly sited, environmentally unsound on-site and off-site disposal facilities, which are disproportionately located in low-income and minority communities. The neighbors endure the externalities of these old facilities without recompense, and thus they effectively subsidize the creation of HW/RW. An even more powerless group, our descendants, provides a further subsidy because the siting impasse has left HW/RW contained in impermanent vessels (such as landfills and capped Superfund sites) that will have to be cleaned up in the more or less distant future. These vessels proliferate around the country because most places do not want to accept waste from any other place.

Partly as a result of the siting impasse, the price of waste disposal has greatly increased. This is both good and bad. It is good in that the creation of hazardous waste and LLRW (although probably not HLW and TRU) is highly price elastic, and the high price of disposal sparks the development of alternative, waste-reducing production techniques. These techniques may drastically reduce creation of hazardous waste; generation of LLRW has already been cut by more than half. The high price is bad because it increases the cost of cleaning up the waste that has already been created, leading to delayed and impermanent remedies.

There is no "out of sight, out of mind" solution to HW/RW. International law and politics prevent reliance on Third World countries, the oceans, and outer space. Domestic law and politics preclude shipping all the waste to remote deserts and wilderness. HW/RW recycling is environmentally problematic. New techniques for treating (as opposed to preventing) HW/RW show promise, but some present their own environmental problems, and it is too early to know if they will make much of a contribution. Thus accomplishing our task—finding the system of HW/RW management that maximizes social welfare, taking full account of the social and economic costs, while achieving fairness—must focus on two methods: minimizing the creation of new HW/RW; and finding a limited number of sites for new disposal facilities to replace the old, poorly sited units and to handle the remedial waste that already exists. Although these new facilities will be far superior environmentally to the old ones they replace, they will not be benign. Accordingly, their numbers should be minimized.

The determinative issues in the success or failure of facility siting attempts seem to be (1) the culture of the local community and (2) the host state's sense of national fairness. Preemption of local control magnifies the sense of incursion and never works in the face of determined opposition backed by the local government. In those communities that fear the facilities will endanger their children's health, offers of compensation and negotiation are ineffective and offensive. On the other hand, there are some communities whose culture of risk perception does not lead them to fear (or at least to loathe) HW/RW facilities; in these communities, compensation and negotiation, as well as some degree of local control, can achieve local acceptance.[1] It appears possible to reduce or eliminate a sense of incursion, though not a preexisting dread. But even if the community is willing to host a facility, the state will often veto the idea, at least partly because the state feels that a disproportionate share of the nation's waste disposal burden is being hoisted upon it. The compartmentalization of disposal programs and laws fosters this sense of geographic inequity: one state may feel it handles an unfair share of the nation's hazardous waste disposal, for instance, while forgetting that other states are taking its radioactive waste, medical waste, and sewage sludge.

The dread that individuals and communities feel about hazardous or radioactive waste may or may not be rational, in the sense of whether the perceived dangers can be confirmed by present-day science. Those who regard the fears as irrational or even superstitious may be offended by the formulation of a system that gives legitimacy to those fears by working around them rather than by confronting them directly. However one feels about the current state of the art of risk assessment—about whether it is the citizens' intuitions or the scientists' calculations that will ultimately prove more accurate—experience over the last two decades is conclusive on one point: attempts to reduce, minimize, or ignore these fears are doomed. In the realpolitik of siting, the only way to succeed is to work around these fears rather than confront them directly.

This understanding of what kills siting attempts should lead to a way out of the thicket. Communities with cultures congenial to HW/RW disposal facilities can be found by offering compensation and asking for volunteers; places that fear those facilities can just say no. To obtain acceptance at the state level, it will be necessary to have a comprehensive national program that considers all the different kinds of wastes and allocates the burdens equitably among the states.

Drawing on these lessons of experience, I will now propose a system that minimizes the creation of HW/RW and, for the waste that still must be disposed of, identifies willing communities and gives the states reason to agree as well. After making this proposal, I will then evaluate it by the same criteria I used to assess the current siting system and will try to forecast who would support and who would oppose such a system. I will close the book with a hypothetical case showing how one particular site might be selected under this scheme.

Centralized Facilities

Chapters 2 and 3 of this volume discussed the numerous sources and types of hazardous and radioactive wastes. These wastes are all regulated separately and are generally disposed of separately in a multitude of different sorts of facilities, even though the same physical characteristics (such as geological setting and transportation access) are desirable for most disposal facilities. The Congressional Office of Technology

Assessment has proposed the establishment of a National Cleanup List to track the cleanup of all chemically contaminated sites, but this has not occurred.[2] Nor is there any coordinated national effort to site disposal facilities for these varying waste streams. This fragmentation fosters a sense of inequity in the siting of each type of facility.

A national program for allocating waste disposal facilities would have several advantages. If every state had at least one facility, and the larger states had the larger facilities, the states would have much less of a sense of regional unfairness. The larger states might have centralized facilities, taking a variety of waste streams and subjecting them to several different kinds of processes; a physical-chemical treatment plant, an incinerator, a landfill, an aqueous treatment plant, and a liquid organics recovery facility might all be located on the same site. Each kind of waste would be more likely to find its ideal treatment process. Sometimes one kind of waste can be used to treat another; a caustic could treat an acid, for example, or the oily waste can be used as fuel to burn other organic waste.[3] Incinerator ash and some other treatment residue might be suitable as landfill cover.

Such a comprehensive approach would afford considerable economies of scale.[4] This is starkly apparent when we look at low-level radioactive waste. Given the number of multistate compacts that have been formed, the country is theoretically heading toward about ten LLRW disposal sites. It takes on the order of $40 million just to perform the studies necessary to determine if a given site is suitable for LLRW, and another $60 million or so to build a facility.[5] Thus ten sites will cost about $1 billion. The volume of LLRW produced in the United States, however, about 36,000 tons a year, is the rough equivalent of what one small municipal landfill might accept. All of this could comfortably be disposed in two or three sites. (It would actually fit readily in one site, but that might involve an undesirable amount of cross-country hauling.) Reducing the number of sites from ten down to two or three would save several hundred million dollars. Moreover, the prospects of actually finding ten geologically and politically suitable sites in the United States for LLRW alone are bleak.

Municipal solid waste is, of course, produced in much larger volumes than LLRW, and many cities and counties have learned the virtues of regionalization and consolidation, both because of the lower costs and because of the more specialized waste processing technology that centralized facilities permit.[6] These advantages would be even more true with hazardous wastes. Among the greatest costs in opening an HW disposal facility are the geological and other studies of the site; the preparation of the permit application; the application process itself; and site improvements such as monitoring wells, roads, laboratories, and administration areas. These costs tend to be affected only marginally by the amount of waste the facility will accept.[7] Thus a few large facilities will tend to be much more economical than many smaller ones.

Several European nations have successfully established centralized hazardous waste disposal facilities. For example, Denmark sends nearly all its hazardous waste to a central facility on the island of Funen in the city of Nyborg, via a network of twenty-one transfer stations. The facility, operated by a public corporation, houses an incinerator, a waste oil recovery plant, a physical/chemical treatment unit, and (twelve miles away) a landfill for treatment residuals. Similar systems are working well in Finland, Sweden, and the German states of Bavaria and Hessen. In most cases, on-site treatment at industrial plants is discouraged.[8] As described by Gary Davis, Joanne Linnerooth, and Bruce Piasecki, all of these European systems "have built technologically advanced, integrated facilities for the storage, treatment and disposal of hazardous waste with significant investment of public funds, placed a high priority on relatively expensive treatment and incineration technologies with little direct land disposal, have required generators to use the publicly owned management facilities in a monopolistic fashion, and have shared the cost of hazardous waste management between industry and the taxpayer."[9] As seen in Chapter 6, several Canadian provinces have adopted a similar approach.

Sites

Any site selected for a waste disposal facility would have to meet minimum technical criteria. Beyond that threshold, however, certain factors would make some sites more appropriate than others.

For the last decade the tendency in siting new disposal facilities, and new industrial operations in general, has been to look for "green fields"—farms or other lands with no prior industrial use. There are two main reasons for this approach: (1) to avoid the possibility of assuming a prior owner's CERCLA liability for contamination of the ground,[10] and (2) to avoid the greater difficulty of monitoring leaks from a facility whose groundwater is already contaminated from prior uses.[11] This approach, however, also increases social conflict, by attempting to impose waste disposal onto pristine areas, and it leads to the spoiling of ever larger swatches of American soil. Because of their low population densities and land costs, rural areas tend to rank very high in the numerical scoring systems used in trying to find disposal sites, but the people living in those areas often have a greater emotional attachment to (and economic dependence on) the land than do their urban cousins. In the Southeast, the "green fields" approach tends to site facilities in areas with mostly black populations.

A growing body of scientific evidence suggests that once the groundwater is contaminated in certain ways, it will never be clean again, at least with current technology.[12] No matter how much money is spent, many badly contaminated sites will always be dirty.[13] Off-site waste migration must be halted, but I believe it makes little sense to spend huge sums in futile attempts to make badly contaminated sites suitable for residential use, while productive farmland is seized for fresh waste disposal sites. Here the perfect is the enemy of the good—by insisting on total cleanup, we pour billions of dollars into a few unproductive holes and allow many other sites to go unaddressed.[14] Moreover, making a site so clean that children can play on it—the standard of cleanliness that is often required—often involves digging up much of the old dirt and trucking it someplace else. This approaches a zero-sum game. Site A is cleaned at the expense of Site B, and in between there is the risk of spillage and the certainty of fuel consumption (with consequent adverse environmental effects in the places the fuel is extracted and processed). There is a good chance that Site B, the destination of Site A's diggings, is an old, grandfathered, poorly sited disposal facility. Alternatively, Site B may be a recently virgin site that has now been converted into a permanent waste repository. Though on-site treatment of Site A's waste at Site A would ordinarily be preferred, that involves risks if Site A is in a populated community. Many on-site treatment methods involve a risk of airborne emissions, and they tend to take months or years to implement, preventing reuse of the site and depressing property values in the neighborhood (since a chain link fence with a "Keep Out—Hazardous Waste Site" sign is hardly a real estate broker's dream).

Instead, to the extent that waste needs to be moved from one place to another, it should go to places that are already contaminated and unlikely ever to be thoroughly clean.

Many sites around the country are available for using this approach. The Pentagon is closing thirty-five major and ninety-five minor domestic military bases,[15] most of which have serious contamination problems.[16] As noted in chapters 2 and 3, the military's inventory includes several thousand smaller contaminated sites. The Nuclear Weapons Complex (NWC) contains seventeen major facilities, all of them contaminated and several of them likely to cease all production.[17] (In 1979 the General Accounting Office recommended that these NWC facilities be investigated for the nation's HLW repository, but the idea was ignored.)[18] The closing facilities (and the surrounding civilian communities that relied on them as their economic foundation) are desperately seeking a new mission. Fort Dix in New Jersey may be converted into a large federal prison,[19] and officials at many military research facilities aspire to convert them to environmental research.[20] Congress has eased the sale of military property for civilian uses.[21]

Some of this is already happening. Waste Management, Inc. has announced plans to build a large HW incineration and treatment complex at Hanford, in the hopes of attracting much of the remedial waste from the cleanup there.[22] The army plans to incinerate hazardous wastes from chemical weapons and pesticide production at the Rocky Mountain Arsenal in Colorado.[23] The Nuclear Regulatory Commission is considering proposals to allow the disposal of LLRW at uranium mill tailings sites.[24] In late 1992, the DOE revealed it was considering using NWC facilities for interim HLW storage.[25] Waste Management's HW complex in Niagara County, New York, was formerly a defense installation, the Lake Ontario Ordinance Works.[26] On a much smaller scale, in 1973 and again in 1979, a private company built small HW disposal facilities on two abandoned Titan missile sites in Idaho.[27]

Many civilian sites might also be suitable. Some former Superfund sites are being reused for such purposes as transportation centers, industrial parks, and shopping centers, but that is the exception rather than the rule.[28] The taint of contamination has killed development on many sites.[29] Banks suffer from what might be called Superfundphobia; they will go to great lengths to avoid lending for activities on contaminated properties, for fear of lender liability. In the 1970s, a few small hazardous waste treatment facilities were built at vacated industrial plants,[30] but this does not seem to have occurred since the enactment of CERCLA (except at sites already used for waste management).

The idea that some places will always be contaminated, and will become "sacrifice zones," has been widely attacked.[31] However, the permanent physical alteration of large areas of land is hardly a novel human activity. Lake Mead, the reservoir created by Hoover Dam on the Colorado River, is 175 square miles, and if the area were drained, the land would still be covered by many feet of sediment.[32] The artificial Lake Volta in Ghana, formed by the Akosombo Dam in 1965, inundated 3,275 square miles of land.[33] A decade of hydrogen bomb tests seems to have made the coral island of Bikini, in the Marshall Islands, permanently uninhabitable, despite massive cleanup efforts.[34] Many sites contaminated by nuclear weapons production and other military and civilian activities will never be completely clean. In fact, the only hope these sites would ever have of complete purity would be to excavate huge quantities of soil (with unknown health and safety impacts) and replace the soil with clean fill. The removed soil would, of course, have to go to some other disposal facility,[35] through the use of massive numbers of trucks.

Although we must regret the initial loss of land, the past cannot be undone. We should focus now on minimizing the risks at existing sites and avoiding the contamination of new sites. Building a limited number of centralized HW/RW treatment and disposal facilities on already contaminated land seems to be the way to achieve this goal.

The prospects of spending billions of dollars cleaning up a small number of badly contaminated sites should also be examined from another perspective. President Bush's proposed federal budget for fiscal year 1993 included $25 billion for hazardous waste site remediation but only $2 billion for cancer research, $1.4 billion for childhood immunization, $1.2 billion for HIV/AIDS research, $700 million for heart disease research, and $400 million for breast and cervical cancer screening. All of these health programs also received a great deal of private funding, but so did hazardous waste remediation. The number of lives potentially saved thanks to all this effort, however, was far lower for hazardous waste remediation than for any of these other health protection measures. Still other programs receiving even less federal money—such as smoking cessation programs, and screening for hypertension or cholesterol— would also have projected health benefits that dwarf those of hazardous waste cleanup.[36] If the prime goal of cleaning up hazardous waste sites is protecting public health, then the intensive remediation of a small number of highly contaminated properties with few neighbors is a very poor investment relative to these other programs. There is no mechanism in society today for ensuring that money not spent on cleaning contaminated sites will instead go to childhood immunization programs or something else with great health benefits, but work is underway to devise such mechanisms.[37]

A different kind of analysis might look at where the money for cleanup comes from, as opposed to where it might alternatively go. Different kinds of taxes, charges, and assessments percolate through the economic system in different ways, and it is very difficult to know exactly who ultimately bears the cost; a $10 million charge against a publicly traded corporation will in the end be paid by some unknowable combination of the company's employees, customers, and stockholders (including large numbers of retirees whose pension funds are invested in the company). These payers will, in turn, have less disposable income for personal expenditures or investments. Especially for people at the lower end of the income scale, a reduction in income in itself has adverse health effects, in the form of poorer diet, more alcoholism and drug use, more heart attacks, and more suicides. Under one crude estimate, every $7.25 million in reduced economic activity induces one additional fatality through this "income effect."[38] On the other hand, the money that goes to cleanup does not vanish into thin air—it is spent somewhere and thus increases the income of other people. The community around the Hanford Reservation is enjoying a great economic boon (at the taxpayers' expense) from the cleanup of radioactive contamination there, just as large natural calamities (such as hurricanes and floods) suck into the affected region massive amounts of insurance money from the rest of the country. Some tasks are also far more labor intensive than others, meaning that (at least at the first layer of analysis) they create different numbers of jobs per million dollars spent. It may never be possible to know the net effect on employment or health of a given large expenditure. One thing is certain, however: it will lead to disruption. It will displace jobs and economic activity. This disruption itself may be stressful and thus, ultimately, unhealthful, unless it is spread so broadly among the population that no one notices the difference.

What is the point of all this to the present discussion? It is that the decision to undertake huge expenditures to clean up a small number of highly contaminated sites is not a trivial or obvious choice. The decision will have significant consequences. Predicting those consequences involves major uncertainties but probably no greater than those involved in the quantitative risk assessments that underlie the desire to remediate a site in the first place. Before spending hundreds of millions of dollars cleaning up a site on the speculation that some day it may thereby be made suitable for residential use, serious thought should be given to whether society as a whole would benefit more if the site were reserved for waste management purposes (or just fenced off and left alone).

Of course, every site should be remediated to the extent that it does not leak pollutants off-site, through the water or the air. Once that occurs, the health benefits of cleaning a remote unpopulated site are less clear. (Of course, wildlife may still be affected.) Hazardous waste poses great dangers to people, but only if there is exposure. It must be borne in mind that none of the major environmental disasters leading to heavy loss of life—Bhopal, Chernobyl, Seveso—involved inactive hazardous waste sites; they all stemmed from the explosion of operating facilities. The health effects of the most famous hazardous waste sites, such as Love Canal and Times Beach, are still disputed.

Finding Volunteer Communities

Under my proposal, a site for a centralized disposal facility would not merely have to be physically suitable—it would also have to be acceptable to the neighboring community. As noted in chapter 6, numerous communities in the United States have volunteered for HW/RW facilities. The

World Health Organization is currently considering a proposed "code of practice" for HW facility site selection that recommends a voluntary siting process. How does one find such communities and secure their consent?

In 1984 a consulting firm hired by the California Waste Management Board said that disposal facilities were more likely to be successfully sited in rural areas where the residents were older, had an educational level of high school or lower, were low-income, Catholic, politically conservative with a free-market orientation, and had occupations such as farmer, rancher, or other jobs the report called "nature exploitative."[39] This characterization generated considerable outrage, particularly in the Catholic press.[40] A few years later, Lawrence Summers, chief economist of the World Bank, suggested in a memo that it made some sense to encourage the migration of dirty industries to Third World countries, where people were more complacent and needed the money. This so annoyed the incoming Clinton administration that the memo may have cost Summers the job of chairman of the Council of Economic Advisers.[41] The lesson here is that it is neither possible nor wise to characterize the communities that might accept HW/RW facilities. Some communities have cultures that make them eager to attract defense-related industries,[42] and other communities (perhaps some of the same) are amenable to HW/RW. Predicting which communities these will be is very difficult, although some have suggested that areas with heavy industry already are likely candidates.[43] A better approach to finding volunteer communities is simply to ask.

Herbert Inhaber has devised a procedure he calls a "reverse Dutch auction," which would presumably be carried out through the newspapers. The auctioneer would propose a compensation amount that would be paid to a volunteer community. Any county that thought it might be willing to accept the facility for that amount would bid. For example, the auctioneer might declare a bid of $10 million and keep it open for a month. If no bids were received, the bid amount would be raised to $20 million the second month, $30 million the third month, and so on until a bid were received. (This is similar to the auction sometimes conducted by airlines seeking volunteers to give up their seats on overbooked flights.) A bid would have to specify a proposed site. Once a bid was

received, the auction would stop until site studies determined acceptability. During this study period, bidding communities would receive
funds from the state to hire their own consultants to do their own studies,
and the communities could withdraw their bids at any time. Communities that did not want the facility under any circumstances would simply
not bid.[44]

This procedure raises several questions:

**What mechanism would ensure that the facility is endorsed by the whole
community, and not just a (biased/unrepresentative/corrupt) governing
body?** The best procedure—and the one actually used in seeking volunteers for several solid waste facilities—is a referendum of the entire
electorate after the detailed studies but before the final decision.

What is the geographic extent of the electorate? Inhaber suggests a
county. The county should consent, but sometimes there will be a politically isolated municipality within the county—for instance, the county
could be dominated by one party but the municipality by another. Thus
the referendum should be required to succeed in *both* the county and the
municipality where the proposed site is located. (In the voluntary process
in Alberta, Canada, the province decided to reject a site which was
supported by the municipality but opposed by the county.)[45] If the
proposed site is near a border, people in the adjoining jurisdiction need
a voice as well. One method to provide this might be to include in the
electorate all voters outside the voting jurisdiction but within a certain
radius of the facility. An alternative method would be to draw a radius
around the facility, and allow only people who live within that radius to
vote in the referendum, regardless of their political jurisdiction. (This
method does, admittedly, create administrative difficulties for the voting
authorities, who typically organize their records by precincts or similar
units.) The approach adopted by Alberta in the one instance with a
proposed site near a municipal border was to allow representatives of
the adjoining municipality a voice on the community liaison group.[46]

**What percentage of the electorate must vote yes for the proposition to
pass?** Agreement to host a major HW/RW facility is a momentous event
in the history of a community, like the adoption of a constitution. It can

be argued that positive votes from more than just 50 percent—perhaps two-thirds—of those voting should be required. On the other hand, this will make it that much harder to secure assent. The Canadian approach has been to allow each municipal council to decide for itself what level of "yes" voters is appropriate.[47]

What consideration should be given to close neighbors of the proposed site who are adamantly opposed to the facility but who lose the referendum?   Several public opinion studies (some done in real situations in the field, others based on hypothetical cases in the laboratory) have shown that opinion on the risks from waste sites and the like tends to be "bipolar": some people believe the risks are very high, some believe they are very low, but few people are in the middle. (Younger people, females, and those who notice odors from a facility are the most likely to believe the risks are high.)[48] Volunteer communities will obviously be those with a predominance of people who believe the risks are low, but there are likely to be some residents at the opposite pole, and they will probably want to move out. People within a close radius of the site should not be trapped; they should be offered the pre-proposal value of their property plus relocation costs. This would follow the experience of several chemical plants that have bought out all the homes around their plants to create buffer zones.[49] One small community of twenty-two homes in Baltimore was voluntarily bought out and relocated to make way for a new HW landfill nearby,[50] and the developers of several MSW landfills in Minnesota and Wisconsin have brought up land and homes near the proposed facilities to reduce conflicts with neighbors.[51] To be sure, forced relocation from a home can impose serious psychic costs above and beyond any purely financial costs, especially when the residents consider themselves part of a community[52] or when the residents have a limited range of job opportunities and housing alternatives.[53] Perhaps some premium above fair market value should be offered.

It should be pointed out here that large-scale relocation is hardly unusual for major public projects. Construction of the Oak Ridge facility displaced about 1,000 families.[54] About 1,500 people were displaced for construction of the Hanford Reservation.[55] Some 4,000 families were evicted to build one approach road to the Triborough Bridge in New

York City.[56] For the Tennessee Valley Authority's Kentucky Reservoir project, 2,609 families and 102 schools, churches, and businesses were relocated from a 300,000-acre area.[57]

What would be done with the compensation money? Those neighbors who want to move should have a priority claim to being bought out. Beyond that, the governing bodies of the county and the municipality should initially determine the intended use of the compensation (for such purposes as tax relief, new schools and hospitals, or more police and teachers) and then include the proposed uses in the referendum question. (Counties and municipalities play differing relative roles in different states; it would be up to each state legislature to decide the decision-making power, and compensation allocation, of its local governments in this process.) Unless the money is spent on capital facilities, it should probably be made available over a period of years, to prevent all the benefits from going to current residents even though the facility will affect future residents for many years to come.

Does every site have neighbors? Some sites may be so remote that there are no residents for miles. At some point, it becomes unreasonable for people to believe that they could be affected by a facility, even in the event of a catastrophic accident. A worst-case analysis may reveal what this radius is for various kinds of facilities. (It would, of course, have to consider proximity to the transport route, and all other paths of exposure to contaminants.) The impact radius would obviously be much larger for an incinerator than for a landfill. Individuals outside the radius should not be entitled to compensation. Moreover, if there are no people in the impact zone, perhaps the municipality's consent should not be required. The same might also hold if there are a few people in the zone, but the government is willing to buy their property for at least what its fair market value would be without the siting proposal.

Needs Assessment

The facility siting process will require specific information on the nature, quantities, and generation patterns of waste. This assessment should address all RCRA hazardous waste, all regulated radioactive waste, and other categories of non-RCRA or RCRA-exempt waste for which a multistate disposal market and significant interstate siting conflicts exist. Examples include incinerator ash, PCBs, medical waste, and asbestos waste. Those waste streams that are usually handled locally would be excluded at this stage, even if there are occasional interstate conflicts—such as municipal solid waste, sewage sludge, and dredge spoil. (Table 3.6 lists the wastes that are usually handled locally.)

For the waste streams included in the process, the next step would be to prepare a disposal needs assessment. The EPA would take the lead in assessing nonradioactive wastes, and the Nuclear Regulatory Commission would assess radioactive wastes. The needs assessment would have these elements:

Current Generation Patterns

The assessment would describe the quantities of wastes generated; where the generation occurs geographically; and the physical form of the wastes when they leave the site of generation. Waste streams disposed of on-site would not be the concern of the federal allocation process *if,* in the view of the state environmental agency, the on-site methods are environmentally satisfactory. Similarly, if the waste is treated on-site before being shipped off-site for disposal, then the form of the material as it leaves the site is what is relevant to the assessment, subject to the same proviso.

Future Generation Projections

The assessment should include projections of how much of this waste will be generated in the future. This will require predictions of future patterns of economic growth and technological development. Although this is a complex undertaking, it is hardly novel. Waste disposal companies engaged in long-term planning, and financial analysts assessing the stock of those companies, perform such analyses routinely. Similar work is also performed in preparing permit applications and environmental impact statements for disposal facilities. Many states have their own hazardous waste planning processes that have addressed these questions. Thus there is a large body of existing research and analysis on which these projections could be based. The projections would include not only

recurrent waste streams (such as waste from ongoing industrial processes) but also remedial waste from the cleanup of past contamination.

Future Waste Generation Targets

The assessment would also establish goals for waste reduction by adjust-
ing future waste generation projections downward to reflect waste minimization, recycling, and other methods to reduce the amount of waste
requiring disposal. An important goal of the process is to provide for the
sound disposal of the waste that *has* to be disposed but not to create so
much disposal capacity that waste generation will be encouraged. Indus-
try-by-industry analyses of opportunities for waste minimization will be
necessary in calculating these targets. (This is similar to the process used
by the EPA in formulating technology-based effluent limitations under
the Clean Water Act, land disposal restrictions under RCRA, and new
source performance standards and air toxics limitations under the Clean
Air Act—processes that have already given the EPA a large database
about the production techniques used in virtually every major industry.)
Congress would have to determine the degree of technological stringency
that will be required for the waste minimization technology to be assumed. The federal pollution control statutes have a large grab bag of
concepts using combinations of "reasonably," "best," or "maximum"
and "available," "achievable," "practicable," or "demonstrated" or
similar words to modify "technology." The planning efforts now underway pursuant to the Pollution Prevention Act will help answer this
question and will provide much of the required industry-specific information. In particular, the EPA's Source Reduction Review Project is
initially focusing its study on seventeen important industrial categories.[58]

Current Disposal Facilities

It will be necessary to inventory the nation's waste disposal facilities, their
present and future capacity, and their regulatory status (for example,
whether they are under orders to close). The EPA's RCRA database and
commercially produced directories already contain most of this information. Facilities operating under RCRA interim status and unable to obtain
full RCRA Part B permits should be excluded from future capacity
projections because they should ordinarily be presumed to be environ-

mentally unsatisfactory. Facilities under construction or in the permit application stage should also be inventoried, with an assessment of the likelihood of their ever coming into operation, and if so, when.

Future Disposal Needs

The inventory of current and proposed disposal facilities will allow a projection of future disposal capacity. A comparison to the future generation targets (prepared in the previous step) will allow a projection of future capacity shortfall: the deficit (if any), per type of waste, of future waste disposal capacity under future waste generation, even after the use of waste minimization. The deficit projections should specify the type of facilities necessary (such as incinerator, landfill, or aqueous treatment). This determination will require knowledge about the treatment technology involved with each type of waste; the EPA has already compiled such information for most RCRA wastes in formulating its land disposal restrictions under HSWA.

The draft needs assessments from the EPA and NRC should be widely distributed and should be subject to public comment. This will ensure that all the affected industries and localities will be able to check the accuracy of the data and assumptions. The final needs assessments—prepared by the EPA and NRC after receiving the public comments—will reveal what new waste disposal capacity needs to be created.

A somewhat similar process, limited to recurrent streams of RCRA hazardous waste, is now underway pursuant to the capacity assurance provision of SARA. In May 1993, the EPA released its *Guidance for Capacity Assurance Planning*,[59] requiring each state to submit base-year (1991) data and projections of commercial hazardous waste capacity and demand for the applicable waste streams generated within that state. This information is due from the states by May 1, 1994. Based on this data, the EPA will project, to the year 2013, whether national shortfalls in disposal capacity will exist. If national shortfalls are projected, then states with disposal demand exceeding supply in the national shortfall categories will be required to proceed to Phase II—the submission of waste minimization plans and information on permitted but not-yet-built facilities. If, after this information is assessed, national shortfalls are still projected, then the states will be required to go to Phase III—submission of plans to eliminate the gap between supply and demand by added waste minimization, new disposal capacity, or interstate agreements. The timing of Phases II and III has not yet been determined. The data developed during this process will go a long way toward providing the information on RCRA hazardous wastes needed for the federal allocation process. Furthermore, the statistical methodologies used will be helpful for non-RCRA wastes as well. A similar assessment has already been prepared for California.[60]

Federal Allocation

The needs assessment prepared by the EPA and the NRC will reveal how much new disposal capacity will be required. This determination will certainly address capacity for high-level and low-level radioactive waste, and perhaps for other categories of RCRA and non-RCRA wastes. Once the capacity needs are known, the process of allocating the satisfaction of those needs among the states should be assigned to an independent federal entity, perhaps called the Federal Waste Disposal Commission (FWDC).[61] The FWDC would have a thankless task: allocating hated facilities among reluctant states. To avoid unending, fruitless debate and rampant political interference, I suggest that the FWDC be a politically independent commission whose recommendations are subject only to the approval or rejection of the entire package by Congress, under the model of the Defense Base Closure and Realignment Act.[62] The Defense Base Closure Commission created by this act has performed admirably in carrying out a similarly unpopular mission. Composed of distinguished people with no future political aspirations, this commission has been able to make base closure decisions on the merits, grounded upon detailed information provided by the Department of Defense, and its recommendations have been accepted by Congress (with, of course, the customary outraged speeches on the floor of the House and Senate by members whose districts lost bases).[63]

The FWDC would have the job of determining what needed capacity should be provided by what states.[64] Although it would announce its decisions all in one package, I believe that it should go about its internal deliberations in a step-by-step fashion. Since Congress has already decided, as national policies, that HLW should go to Yucca Mountain in Nevada and TRU should go to WIPP in New Mexico, this should serve as the starting point for state allocation, and no further capacity should be allocated to either of those states. (If in some future year these facilities are canceled—or they are built and later fill up or otherwise stop accepting waste—this issue would be revisited.) If the NRC believes that a separate repository should be established for the remains of decommissioned nuclear power plants, the FWDC would have to find a state where that repository would go. The NRC would specify the minimum physical conditions necessary for such a repository, and the FWDC would have to allocate the repository to a state that had an ample supply of land meeting those conditions. The NRC could also be asked to specify an optimal general location—that is, the region of the country where the facility should go, determined on the basis of safety and cost, without reference to politics. As will be noted later, sophisticated computer programs have been developed to include transport risk in this kind of calculation. This specification would play heavily in the FWDC's decision. Since most commercial nuclear power plants are located east of the Mississippi, this repository would probably be located in an eastern state that contains such plants.

Next should be repositories for LLRW. The FWDC should also look to the NRC for guidance on the physically optimal number and general location of LLRW facilities; I suspect the answer will be either two or three. If the optimal number is two, then there should be one in the east and one in the west; if three, there should also be one close to the center of the country. As it happens, California, North Carolina, and Texas are all well along in selecting LLRW sites, so the FWDC's task here may be fairly simple.

The process of finding a site for a monitored retrievable storage facility for HLW is already underway. If a site has been selected by the time the FWDC gets to work, that state should be spared further allocations of new facilities. (Once the permanent HLW repository is opened and the MRS facility has been emptied, decades from now, the MRS state might again be eligible for a future allocation.) If no site has been selected, and it appears that the initial volunteer process is not going to work, then the FWDC would allocate this facility to a state, again looking to the NRC for guidance on physically optimal location.

Uranium mill tailings should be allocated next. These wastes are extremely voluminous. Thus, unlike arrangements for HLW and LLRW, long-distance transport is not feasible. Once again, the NRC would designate the optimal number of facilities and their general locations. It is virtually certain that these facilities should be allocated to the western states where the largest mill tailing piles are located. (For this physical reason, a mill tailing repository for in-state waste might be located in a state that already has a HLW, TRU, or MRS facility, as an exception to the rule that those states would be exempt from future allocation.) If a national policy decision has been made by then to develop repositories for naturally occurring radioactive material, such sites would be allocated on a similar basis.

By now all the major radioactive waste streams will have been allocated. The next subject of the FWDC's deliberations would be the non-radioactive waste streams, including RCRA hazardous waste; PCBs, which by historical accident are regulated under TSCA rather than by RCRA; asbestos; medical waste; and any other waste streams under the FWDC's jurisdiction, as well as mixed radioactive/hazardous waste. The EPA will have declared whether there are any nationwide capacity shortages. If there are, the EPA should also reveal whether there are any unique geological or other physical characteristics that must be met by a site for the required facility. It is unlikely that this will disqualify any states altogether; the EPA has already promulgated location standards for RCRA and TSCA disposal facilities, and on their face they do not require, for example, an arid climate that would only be found in the west.

Based on the estimates of future waste generation and future waste disposal capacity calculated earlier by the EPA, the FWDC would determine which of the states still eligible for allocations (i.e., those without a radioactive waste disposal facility) are projected to become net exporters of nonradioactive FWDC waste streams (primarily RCRA hazardous waste). The largest net exporters would receive the first allocations of new disposal capacity. The FWDC would allocate the largest new facility (measured in terms of tons per year) to the largest projected exporter; the second-largest new facility to the second-largest projected exporter; and so on down the list, until every needed facility had been allocated to a state.

Under this allocative method, large centralized facilities would likely go to the largest exporting states; small transfer stations might go to importing states. In the following circumstances, however, the FWDC could vary from this otherwise mechanical process:

• if the EPA advised that special physical characteristics were needed for a particular kind of facility, and the presumptively designated state for that facility lacked those characteristics;

• if most of a particular waste stream is projected to be generated in places so distant from the target state that it would be clearly unsafe or inefficient to send it to that state; or

• if it is unfair to assess exports purely on a tonnage basis because some of the wastes involved have high volume but low toxicity, or vice versa.

It is also possible that the rule against allocation of nonradioactive facilities to states with RW facilities would be breached in one other circumstance. As noted, California and Texas might get LLRW disposal facilities. Both of these are large states in which a great deal of HW is generated, and they are also by far the two largest net exporters of HW. (See table 9.1.) Thus it is fair to allocate to them some HW facilities if doing so posed clear safety or efficiency advantages. (The FWDC should refine table 9.1 by, among other things, adding in the other waste streams that are within the FWDC's mandate.)

Once all the needed facilities had been allocated to states, the FWDC would issue its comprehensive report on where all the RW and HW facilities would go. The report would be submitted to Congress, which would be required to vote yes or no on the entire package. As with the Defense Base Closure and Realignment Act, the statute establishing the FWDC would mandate that Congress consider the package as a whole and may not modify the FWDC's recommendations.

Because every state generates hazardous waste, and every state exports HW to other states,[65] every state should have some disposal obligations; no state should think it can get a free ride. (The possibility of a free ride—the knowledge that only one state in a compact region would

Table 9.1
Imports and exports of RCRA hazardous waste, 1989

Rank	State	Net imports/exports (tons per year)
1	Indiana	181,170 (largest importer)
2	Alabama	176,160
3	South Carolina	113,117
4	Louisiana	88,691
5	Ohio	75,722
6	Oklahoma	66,068
7	Illinois	41,414
8	Utah	40,135
9	Oregon	38,513
10	Idaho	35,945
11	Tennessee	26,206
12	Kentucky	26,173
13	Arkansas	16,338
14	Nevada	13,188
15	Virginia	12,121
16	Nebraska	10,742
17	Rhode Island	8,506
18	Trust Territories	−22
19	Guam	−203
20	South Dakota	−1,250
21	Hawaii	−1,300
22	District of Columbia	−2,409
23	Montana	−2,497
24	Wyoming	−2,954
25	North Dakota	−2,991
26	Alaska	−3,470
27	New Mexico	−5,236
28	North Carolina	−7,363
29	Missouri	−8,210
30	New York	−8,450
31	Puerto Rico	−8,630
32	New Hampshire	−12,889

Table 9.1 (Continued)

Rank	State	Net imports/exports (tons per year)
33	Vermont	−13,740
34	Delaware	−14,083
35	Kansas	−14,653
36	Iowa	−19,071
37	Arizona	−22,637
38	Mississippi	−24,765
39	Massachusetts	−25,444
40	Minnesota	−26,224
41	Michigan	−28,225
42	Colorado	−31,061
43	Washington	−41,975
44	Florida	−44,298
45	New Jersey	−52,082
46	Maine	−52,635
47	Maryland	−61,182
48	West Virginia	−61,993
49	Pennsylvania	−69,411
50	Georgia	−71,272
51	Connecticut	−87,248
52	Wisconsin	−87,803
53	Texas	−108,820
54	California	−176,340 (largest exporter)

Source: Adopted from U.S. Environmental Protection Agency, Office of Solid Waste and Emergency Response Pub. No. OS-312, *National Biennial RCRA Hazardous Waste Report (Based on 1989 Data)* (Feb. 1993), p. 2–67.

probably have to host a facility—is one of the major reasons for the failure of the federal siting efforts for LLRW.)[66] States would be given credit in this allocation process for existing private HW/RW disposal facilities within their borders, such as Emelle and Barnwell, inasmuch as such facilities will tend to make these states importers rather than exporters. It could be argued that states should also be given credit for other undesirable items, such as federal prisons; strip mines; and abandoned, contaminated military installations. This list could be endless, and the line must be drawn somewhere; in trying to site HW/RW disposal facilities, drawing the line to include only other HW/RW disposal facilities seems sensible, or at least defensible.

After Congress has acted, states should then be able to trade allocations among themselves.[67] The National Governors Association or a similar group could establish a trading mechanism.[68] States might also want to trade disposal rights for waste streams not within the FWDC's jurisdiction, such as MSW; if New Jersey, for instance, wanted to export municipal trash to Indiana, then Indiana might agree if it could send some of its hazardous waste back to New Jersey.

Site Selection

Once the state-by-state allocations are established, each state should have the responsibility to find the necessary sites for any newly required facilities. Perhaps the state would look for volunteer communities using Inhaber's bidding process. Land on federal facilities would be made available to the extent that it was physically suitable. (Such land should be sold or leased to the state or the new facility operator at the prevailing price for comparable industrial land, so as not to create a hidden subsidy.) In any state that shirked its responsibility, the FWDC could step in and find sites itself. This resembles the process under the Clean Air Act in which a federal implementation plan can be prepared for any state that fails to submit a satisfactory state implementation plan.[69] Such a role for the FWDC would involve a limited violation of the anti-preemption principle but may be necessary in order to induce states to provide sufficient incentives for volunteer communities to step forward. (Perhaps

FDWC-selected sites would have to be on land already owned by the federal government, if any such land is physically suitable for this purpose.)

Before sites are tentatively identified, it is important that minimum physical characteristics be defined. This should be undertaken by the EPA for nonradioactive waste facilities, and jointly by the EPA and the NRC for radioactive waste facilities. (For many types of disposal units, such standards are already in place.)[70] Fatal characteristics (such as presence in a floodplain) and necessary characteristics (such as depth to bedrock) would be defined in advance; features that are merely desirable or undesirable can wait. With these technical standards in place, an initial screening can be performed to see if a given site has a good chance of surviving the licensing process. The screening can be undertaken by the volunteer community, or by the state once it has received an expression of interest from such a community. This would help avoid a repetition of the disastrous experience in Martinsville, Illinois, where a community volunteered for a LLRW facility, but where after years of litigation and the expenditure of $85 million it was determined that the underlying geology was unsuitable. The state siting board that denied a permit to the Martinsville site wrote, "Politics and the need for local approval were what selected the [site]. Science and engineering were asked to come in after the fact to justify this politically acceptable site."[71]

Once sites were selected, the states would be responsible for overseeing the detailed characterization studies and the permitting and construction of the facilities, all under the applicable guidelines of the EPA or NRC. (An exception would be the HLW and TRU repositories, which would have to be in federal hands because they will contain fissile materials.) Local communities should be given technical assistance grants to participate in the process. Perhaps each facility would have its own board of visitors, with federal, state, and local representation. This board should have full access to the site and its records and be able to conduct inspections at will to ensure that all environmental standards are met. It would also regularly meet with facility management to discuss mutual concerns and could make the discussions public if its recommendations were not followed. The agreement, described in chapter 5, between the communities in Niagara County, New York, and Chemical Waste Management has many of the desirable elements for such a program of local oversight.

Additionally, the FWDC might set caps on how much waste each facility could accept, to avoid the creation of excess capacity that might encourage waste generation.

The sanction for a state's failure to meet its FWDC-set allocation of waste disposal would be that other states could initially tax and ultimately exclude that state's waste from their FWDC-allocated facilities. A similar sanction for states failing to site facilities was upheld under the Low-Level Radioactive Waste Policy Amendments Act.[72]

To advance the goal of closing existing older facilities, private companies should be given special consideration to relocate to FWDC-allocated sites if they (or someone from whom they bought capacity rights) shut down older, environmentally deficient disposal units. (The consent of the receiving state and locality to this move would still have to be obtained.) This idea of disposal capacity trading is based on the emissions offset trading program under the Clean Air Act.[73] Many disposal companies are currently in constant conflict with their neighbors, and I believe many would embrace this option of new, preselected sites. To encourage these moves, perhaps the older facilities with the greatest environmental problems should be *required* to move after a few years (such as when their existing permits expire). Companies moving to FWDC-allocated sites might be able to bypass ordinary site selection processes.

Actual construction and operation of facilities could well be contracted to the private sector. Private companies could also attempt to site and construct new commercial facilities above and beyond those found necessary by the FWDC. In such cases, however, the firms would be on their own.

10

Evaluating the Proposed Siting Process

Chapter 9 described my proposed siting mechanism. Now I will evaluate the mechanism by the same criteria I used in chapter 5 in evaluating the current siting processes.

Needed Disposal Capacity

In evaluating the existing siting processes, I showed that, despite some regional disparities, no major national shortage of hazardous waste disposal capacity exists, though some old facilities should be replaced with new ones. My proposed alternative of local control, state responsibility, and national allocation is designed to eliminate any remaining shortages of capacity, anticipate future ones, and replace aging units. The FWDC would identify shortages and allocate among the states the responsibility to site facilities to meet these needs. Because it ordinarily takes five to eight years to site and build a new waste disposal facility,[1] the task should begin soon.

Without such hidden subsidies as zoning overrides, those wishing to site new facilities will have to negotiate with local governments in volunteer communities for siting approvals. These local governments, to secure support from their constituents, might well require the disposal companies to guarantee neighboring property values and otherwise compensate for any injury, thereby internalizing many of the externalities of such facilities.[2] This is a natural market mechanism, not an artificial one with the attendant transaction costs and uncertainties. The proposed approach will increase the cost of building new facilities. Much of the cost will be passed along to waste generators in the form of higher disposal prices, leading to decreased waste production.

There is a greater societal interest in high disposal prices for production wastes than for remedial wastes (i.e., waste that already exists, often lying dormant in the ground), since the cleanup and disposal of remedial wastes should be encouraged. One possible approach would be to impose a tax on the production of HW and LLRW, and use the proceeds to subsidize the cleanup and disposal of remedial waste. However, this would have high transaction costs and would invite many disputes over whether a given waste stream consists of production or remedial waste. Such a tax is not central to my proposal.

Any threat that higher prices will cause more illegal dumping—a concern that was shown in chapter 5 to be largely misplaced—would be addressed through heightened enforcement. Hidden subsidies from inadequate tort remedies would be reduced, to the extent that the older facilities (presumably the worst nuisances) are shut down and their capacity traded to new FWDC sites. One important uncompensated cost, psychic injury, should decline greatly because the new facilities would be in volunteer communities and any neighbors who still opposed the siting would be fully compensated for moving.

To encourage further the replacement of old facilities with new ones, consideration should be given to creating a tort-like remedy for facility neighbors, so that external costs cannot be so readily ignored. One possible approach is to expand the definition of recoverable "response costs" under CERCLA and allow such costs to be recovered through private actions under RCRA as well.[3]

Protecting Health and the Environment

The proposed system's method of capacity trading, and its acceleration of the closure of interim status facilities, will reduce the number of old facilities, which tend to have the worst health and environmental impacts. To the extent that the new facilities are located on already contaminated land, such as military, NWC, or uranium mill tailings sites, then agricultural and virgin land will be spared degradation. The aesthetic impact of these facilities will also be reduced, as they will be located in settings that are already industrial.

The allocations assigned by the FWDC are likely to lead to the creation of centralized, integrated waste management facilities, on the Canadian, Scandinavian, and German models. (A state given the responsibility to handle waste streams requiring five different kinds of processes will have an easier time siting one big rather than five small facilities.) Such centralized facilities have several advantages for health and the environment over a much larger number of smaller, decentralized facilities:

- the available sites with the best physical characteristics could be used for multiple waste recycling, treatment, and disposal purposes;

- specialized forms of waste recycling and treatment can be economically developed and provided, making it more likely that each load of waste will be optimally handled;

- with the much higher revenues per site, a higher caliber of management and technical staff can be provided;

- employees can enjoy better on-site medical care and receive better training and supervision in safe work practices;

- treatment residues are more likely to find an on-site use, rather than requiring landfilling or off-site transportation;

- each centralized facility can be assigned a full-time staff of governmental monitors, some with specialized functions and some accountable to the local community;

- public exposure to contamination (and adversely affected property values) will be reduced because ordinarily one large facility will have fewer neighbors within a given radius than will multiple small facilities;

- a larger buffer zone around the facility will be economically feasible; and

- only one site, as opposed to several, would be subjected to possible groundwater contamination.

Persons living near existing contamination, such as CERCLA sites, would also benefit, because the proposed system, by ensuring adequate disposal capacity for remedial wastes, should expedite cleanup. Similarly, the centralized facilities might reduce the amount of on-site disposal at factories, thereby benefiting the factory neighbors.

Transportation poses the biggest problem with a centralized system. The risk of transportation accidents is a very controversial issue, both legally[4] and politically.[5] Accidents are an inevitable feature of any system of transportation. On average, trucks are in accidents once every 400,000 miles of travel.[6] Between 1971 and 1985, trucks carrying radioactive materials were in 167 transportation accidents (mostly on highways) involving 2,602 packages of radioactive materials; a total of 67 packages experienced some release of their contents.[7] (The three worst accidents all involved the rupture of fifty-five-gallon drums of yellowcake, the product of uranium mills.)[8] To put this in perspective, of all hazardous materials shipments annually (numerically, not by volume), about 1 percent involve RCRA hazardous waste and about 2 percent contain radioactive materials; the most shipments by far are of gasoline, with chemical products or intermediates a distant second.[9]

Hazardous waste transportation is a big business. The cost is about $0.23 per ton per mile,[10] and this adds up very quickly; more than half of the $6 billion spent on hazardous waste services in 1990 went to transportation.[11] Sophisticated computer models have been developed to determine the least costly and least risky routes to a given set of disposal facilities.[12] Some models go further and also identify ideal facility locations as well as transportation routes.[13] All these models rely on a very large number of assumptions, however; one of the models, for example, in considering the optimal location for an LLRW facility in Pennsylvania, performed three different computer runs, each depending on a different value for human life: $300,000, $18 million, or $300 million.[14] This modeling literature does not lend itself to a generalization about whether, considering transportation risks, centralized or dispersed HW/RW facilities are better. Moreover, the prior work generally focuses on just one type of waste at a time and does not reflect the lower transportation risks that result when a centralized facility eliminates the need for external shipment of wastes and allows them to be handled internally (for example, ash or sludge from on-site treatment units that could go to an on-site landfill).

To a considerable extent, however, the question of many small versus few large facilities is moot. The experience of the past fifteen years shows that, at least under present siting procedures, a large number of new, dispersed facilities is not an option. The country already possesses a large number of old, dispersed facilities, plus a great deal of "temporary" on-site storage and other substandard management techniques; finding new sites will, under any scheme, still be difficult enough that a centralized system is likely to be necessary.

Affording Fairness

The third factor in evaluating the proposed siting system is whether it affords fairness. As discussed in chapter 5, there are two primary ethical systems that have been discussed in the siting context: the utilitarian and the egalitarian. My proposal seeks to achieve utilitarian goals by determining, and then providing, the necessary waste disposal capacity, in locations that meet scientifically established physical criteria. Minimizing the number of sites should also realize significant economies of scale and reduce negative externalities, thus achieving another utilitarian goal.

My proposal also achieves egalitarian goals by rejecting the imposition of facilities on unwilling communities. Instead, facilities would be sited only where they are wanted. Any individual neighbors who were opposed would be compensated to move.

The egalitarian argument has considerable force in the siting context. Many activities that cause a great deal of anguish in large segments of the population are allowed by the law (e.g., flag burning, abortion, marches by hate groups) or even required (e.g., busing to achieve racial integration). But each of these activities invokes important constitutional values, at least as the Supreme Court now construes those values. The forced siting of a waste facility in an unwilling community causes no less anguish, but it serves no fundamental constitutional value. As shown throughout this book, ramming these dread facilities down the throats of unwilling communities is neither necessary nor effective, and the power of government should not be invoked in the attempt.

To address the egalitarian goals, I have also focused on the fairness of the proposal to locations, to classes and races, and to generations.

The national allocation process to be conducted by the FWDC would be designed to provide a high degree of fairness among regions and states. When each waste stream is considered separately, the results are lumpy—

a few states bear the burdens while everyone else gets a free ride. The lumpiness can be eliminated by considering together all of the many kinds of hazardous and radioactive waste. The FWDC would have the task of making sure that each state bears a fair share of the cumulative national burdens; there should be no clear winners or losers. Any state that failed to carry its fair load would then be penalized by being unable to export its HW/RW—a serious penalty, as no state is self-sufficient in waste disposal. (The average state exports waste to twenty other states for treatment and disposal, and imports waste from twenty other states.)[15]

Within states, some localities would inevitably bear a very high burden. However, these would be the volunteers. As Mark Sagoff has written, in the context of the voluntariness of risk, "There is an ethical difference between jumping and being pushed—even if the risks and benefits are the same."[16] No one would be forced to bear a disproportionate risk; dissenting neighbors in volunteer communities could move away at no economic cost. The new facilities would be subject to strict environmental regulation and local as well as state and federal oversight. The facilities will not be without hazard, but they will still be less risky than such unpleasant but necessary neighbors as petrochemical plants, oil refineries, and steel mills, which society has deemed to pose acceptable risks.[17] There is usually little immoral about informed consent to risk, unless the consent is obtained under conditions of coercion.

One form of coercion could be a poverty that forces a community to trade a case of cancer tomorrow for a loaf of bread today. There is concededly some danger of this under my proposal, but it is not at all clear that the volunteer communities will have low-income profiles. As seen in chapter 5, the chief factors in whether communities volunteer appear to be cultural rather than demographic. Many of the municipalities that have offered to receive HW/RW facilities have had well-educated populations, although newly hard times (such as the closure of a major employer) certainly contribute to willingness to accept a facility. To the extent that the suggestions made here are followed, the centralized facilities are likely to be located at old military, nuclear weapons production or uranium mining sites, which—unlike old industrial areas—do not tend to be surrounded by minority communities (except that uranium mines are often located near Native American communities).[18]

Thus, fairness among classes and races would be improved by the proposed system. The system would aim to close down (and certainly prevent expansion of) old grandfathered facilities, which tend to be located in low-income, high-minority areas, and to open new facilities in places with less skewed demographic profiles. More fully internalizing the external costs of HW/RW disposal will tend to reduce the aggregate social costs of disposal (because it will drive up the price of and therefore lower the demand for waste disposal) and shift the remaining social costs away from facility neighbors and toward those who benefit—the shareholders, employees, suppliers, and consumers of hazardous waste generating companies.[19]

Future generations should also benefit significantly from this proposal. Higher disposal prices, or a command-and-control system of waste minimization, would lower HW/RW production. The availability of new disposal facilities would divert waste from temporary storage units to the more permanent new centers and would encourage permanent rather than containment remedies at CERCLA sites and other contaminated locations. Proper entombment of the HLW stored in leaking tanks in Hanford and Savannah River could be brought within reach. The use of already contaminated land would preserve more agricultural and virgin land for posterity.

This is not, of course, a solution for all time. Presumably, after some years, new waste streams will be created that were not accounted for in the initial national allocation. It is hoped that, by then, at least one of the volunteer communities would offer to accept this new waste stream; if not, then the federal government would have to step in again. If, in the intervening years, a state emerged as a large net exporter of waste, that state might be a prime candidate for this new stream.

Political Viability

Success in finding volunteer communities is necessary to the success of my proposal. As shown in chapter 6, many communities have already volunteered, typically with only minor financial inducements. If offers of serious compensation were forthcoming, still more localities with compatible cultures of risk perception should also come forward. To be appropriate, volunteer areas would also have to be physically suitable; land that sits atop a productive aquifer, for example, would have to be rejected for land disposal units. Could enough sites be found where both the sociology and the geology are suitable? I believe so, but I cannot say for sure. It is impossible to know in advance how much local support will be enough, or whether this support will, coincidentally, arise in places with appropriate physical conditions. The best we can do is to provide potential host communities with enough information, cooperation, and technical assistance to allow them to make intelligent choices for themselves, and to avoid the self-made trap into which too many facility developers fall of gratuitously offending communities with displays of arrogance or secrecy.

This proposal requires the consent of the states as well as the localities. Here, I am more confident of success. A major reason for state opposition to locally acceptable waste disposal projects is the fear of being exploited, being stigmatized, becoming the national patsy. A slight variation of a recurring phrase—"nation's dumping ground"—permeates state declarations of opposition to proposed HW/RW facilities.[20] The national allocations provided by the FWDC should go a long way toward eliminating that sentiment. The map the newspapers will print the day after the FWDC announces its proposed allocations will show that every state is bearing a piece, but only a piece, of the national disposal burden.

On several occasions various mayors of New York City have tried to site simultaneously multiple incinerators, homeless shelters, or other unpopular facilities in different communities. The resulting newspaper maps did not achieve community acceptance of these proposals, and in each case the attempt failed politically. However, that experience is much different from the current proposal. In New York City, the mayor named specific sites, leaving no choice to the affected communities. Under my proposal, the FWDC would allocate facilities among *states;* the states would be left to pick sites for the facilities (or to trade allocations with other states). The New York City experience suggests that central allocation is ineffective if not accompanied by decentralized site selection and local control.

Each state will have considerable incentive to allow a facility at a volunteer site to go forward, because if it does not participate in the national allocation scheme it would be subject to the sanction of being unable to send its waste to facilities in other states. A small opposition group can stop a project that has no large constituency; but if an entire state's access to the nation's waste disposal facilities depends on that project, then all the interests dependent on that access will form a constituency that could overcome small pockets of resistance. If the community where the facility would go has knowingly and freely consented, then remaining pockets of resistance to siting in that volunteer community will have diminished legitimacy and power.

The reasons for the failure of most siting approaches were analyzed by two Dutch researchers, Matthijs Hisschemoller and Cees J. H. Midden, who arrayed four primary approaches to siting:[21]

1. The technical approach, under which experts select "optimal" sites based on objective, usually quantitative criteria.
2. The public participation approach, which allows the public to select sites through referenda or other collective methods.
3. The market approach, under which communities are compensated for the costs imposed upon them by unwanted facilities.
4. The distributive justice approach, under which the government decides the best places to put facilities based on fairness.

The researchers explained why each of these methods usually failed. The technical approach ignored public perceptions and met fatal political resistance. The public participation approach led to technically inferior sites or to no sites at all. The market approach failed to overcome public fear of danger to health. The distributive justice approach left the government with no clear or acceptable method of reaching decisions.

Hisschemoller and Midden seemed to despair of reconciling these four approaches. However, the approach presented in this book seeks to embody the best elements of all four. The technical approach is utilized in establishing screening criteria to determine if any given site is physically suitable in terms of geology, access, and the like. The public participation approach is used to secure local consent to the facility. The market approach is used to compensate the community as a whole, and the closest neighbors in particular, for the adverse effects of a facility and to allow states to trade waste disposal allocations among themselves. The distributive justice approach would be applied by the proposed Federal Waste Disposal Commission to set initial allocations among the states. By combining all these elements, an approach of federal allocation, state responsibility, and local control aims toward a siting system that reconciles the apparently disparate needs for fairness, safety, economic soundness, and political viability.

11

Practicalities of Implementation

Obstacles to Success

This proposal attempts to tackle a monumental problem, and policymakers should be fully aware of what might go wrong. The following are what I see as the greatest obstacles to success, together with my thoughts on how to get around each problem. Afterward, I discuss my predictions for how various players in the siting process will react to my proposal.

Insufficient Waste Minimization

One objective of the proposed system is a reduction in the generation of hazardous and radioactive waste. I have proposed technology studies to determine reasonable rates of waste minimization for each generating industry. These studies would be the basis for the projections of future needed capacity, to be used in turn by the FWDC in allocating capacity among the states. This step, in itself, will not necessarily lead to a reduction in the generation of waste. If the new system causes the price of HW disposal to rise, that will help drive down the amount of HW generated. There is no assurance that prices will rise, however.

Two ways to drive down the generation of HW are immediately apparent. The first is a tax on the generation of HW. Its chief advantage is its administrative simplicity. Every generator who has HW transported for disposal off-site must now prepare a manifest that is filed with the regulatory agencies. These manifests already reveal how much HW is generated by each large quantity generator, and a per-ton tax could readily be applied. (Of course, compliance inspections would have to be increased to detect evasion.) A tax's chief disadvantage—other than the political stigma attached to any new tax—is that it does not distinguish between those who can, but have chosen not to, implement vigorous waste minimization programs and those who have done their best but are still left with an irreducible minimum of waste if they are to continue in business. (This problem could be ameliorated by giving tax relief to companies that performed a waste minimization audit and adopted its recommendations, much as auto insurers give discounts to motorists who install antitheft devices.) Additionally, we have no clear information on the price elasticity of waste disposal, so we do not know how much of a tax will be necessary in order to induce significant waste minimization.

A second alternative would be to graft on top of RCRA—one of the most relentlessly command-and-control statutes ever written—a regulatory requirement for waste reduction. Fortunately, the mechanism is already in place for such a requirement. As noted in chapter 8, RCRA manifests contain a rather feeble self-certification of waste minimization efforts. The certification could be made much more specific, and could provide, for example, that the generator is using a technology that generates as little waste as the technology assumed in the EPA's capacity studies. Each certification would have to be signed by the facility manager and, perhaps, by a professional engineer. The EPA should establish an inspection force to audit facilities to ensure that their certifications were accurate. In this manner, there will be much greater assurance that significant waste reduction will be achieved. The advantages and disadvantages of this regulatory mechanism are the opposite of those of the waste generation tax: it would be administratively cumbersome, but it would be sensitive to the waste minimization efforts of particular companies.

Relatively less attention has been paid to minimizing the generation of radioactive waste, but there is no reason to believe that the opportunities here are any less than those with hazardous waste. The NRC should consider establishing a system to explore the availability of waste-minimizing technologies and practices and then to require their use. The state public utilities commissions can also play an important role in requiring their nuclear utilities to reduce the waste they generate.

Another potential problem would arise if the projections prepared by the EPA and NRC overstated the amount of waste minimization that could be achieved, or if other changes in the economy or in technology led to the creation of more waste than anticipated. In that event, the disposal facilities established by the FWDC may be insufficient to handle all the waste created. Of course, there is a natural feedback mechanism: a shortage of disposal supply will drive up the price and thus reduce demand. If for some reason this mechanism does not work, or if the higher price is having other undesirable effects (such as an increase in illegal dumping, or excessive prices for certain goods or services), then it may be desirable to increase disposal capacity at that time. As discussed in chapter 4, it is usually much easier to expand capacity at existing facilities than to create capacity at new sites. Thus the sites allocated by the FWDC might readily become the locations of expanded capacity, especially if they had established good relations with their neighbors.

No Suitable Sites

The proposal depends upon a congruence of willing communities and suitable physical settings. Most states (except for the smallest) have a reasonably broad range of topography, and thus there are usually multiple possible locations for a given type of facility.

The facilities where geology is most critical are land disposal units—landfills and other kinds of underground permanent repositories. That is because waste is designed to be deposited there permanently, so if the waste retains its dangerous characteristics longer than the projected life of the artificial container (such as the landfill liner), the surrounding geology will be depended upon to prevent the waste's spread into the broader environment. Fortunately, as seen in chapter 2, there now seems to be a plentiful supply of landfill capacity, and much of it is located on sites with ample expansion room. Thus the question of finding new landfill sites may simply not arise.

No Volunteers

It is possible that some of the states will be unable to find volunteer communities for the waste streams they have been allocated, at least in physically suitable locations. Several courses of action would then be available:

• Add significantly to the financial inducements offered to communities.

• Make arrangements for another state to take the allocated waste (in return, presumably, for a large payment, or for the acceptance of some alternative waste stream from that other state, or possibly for some other favor).

• Go to an abandoned military installation and buy out all its neighbors, so that there will be no community to object.

• Subsidize (or, within the state, decree) waste minimization programs so that much less of the particular waste is generated, perhaps leaving a small enough residual that alternative disposal methods or sites can be found.

Failure of Yucca Mountain or WIPP

I have assumed that HLW will go to Yucca Mountain and TRU will go to WIPP. Yucca Mountain is still under intensive study, and though WIPP has been constructed, it too is undergoing a reanalysis. If these new studies yield negative results and new sites must be found, the proposed system would be seriously disrupted, especially if this occurs after the FWDC has completed its allocation, and thus all the states have already received all the sites they are expecting.

The obvious first alternative would be to look for new sites in Nevada or New Mexico, as the case may be; these are two states with among the highest probability of containing suitable sites because of their aridity and their low population densities. If that fails, it may be necessary to reopen the FWDC process and determine, at that time, which states remain as significant net exporters, particularly of radioactive waste. Sites in the Nuclear Weapons Complex would be prime candidates for these facilities, if—and this is a big if—they are geologically suitable. (Many of them are located on major bodies of water, which is highly undesirable for such a facility.) In any event, if we proceed with the FWDC system and the Yucca Mountain or WIPP sites fail, we will be in no worse a position than if the FWDC system had not been adopted.

Old Sites Remain Open

One objective of the system is to shut down old, grandfathered facilities. This is one of the principal mechanisms by which greater equity among classes and races is to be achieved. Nothing in the proposed system, so far, would automatically ensure the closure of these facilities. If too many of them remain open, this problem can be addressed readily, however: Congress could amend RCRA to provide that, within x months of enactment, all facilities operating under interim status would have to close.

Siting in Minority Communities

A major objection to the present-day siting system is that many HW facilities are located in low-income and minority communities. Successful closure of many grandfathered facilities will address some of this problem. If it develops that most of the volunteer communities are low-income or minority, then equity concerns will persist. I believe that such an outcome would represent a failure, however, only if the communities invited the facilities on less than full information, after less than full deliberation, or based on unmet expectations; or if the facilities, once approved, were not built and operated so as to afford full protection to the health of their neighbors. In the absence of these conditions—that is, if there was informed consent and if the facilities are not dangerous— then I do not believe that outsiders should second-guess the communities' own judgments on costs and benefits. In the words of an official of Fremont County, Wyoming, one of the communities that has expressed interest in hosting the monitored retrievable storage facility, "To my knowledge no one's ever lost their life because of a MRS facility. But we have people in Fremont County who have blown their brains out because they've lost their jobs, their homes and their life savings. There are some people who are strong advocates of MRS because of economic considerations."[1] A community should be able to make its own decisions about balancing risks and benefits. It should not, and need not, rely on government assurances; in the voluntary process used by several Canadian provinces, for example, the communities that contemplated volunteering spent a great deal of time (with the help of government money) assuring themselves that the proposed facilities were safe. Some of the volunteer communities in Canada were economically depressed, but others were not.[2]

However, there is reason to believe that, at least for certain minority groups, volunteers will not be forthcoming. A considerable amount of public opinion polling has shown that on a statistical basis, blacks, in particular, have a heightened sensitivity to technological risk, and a greater concern over environmental exposure, than do whites or other ethnic or racial groups. This has been found in studies concerning nuclear power, nuclear waste, pesticides, and radon gas. This may be because blacks have less of a sense of control over these exposures (for such reasons as traditional exclusion from the political processes involved in siting, as discussed in chapter 5) or because they are already more affected by environmental hazards.[3] Whatever the reason, it certainly suggests that black communities may be disinclined to volunteer for these facilities. Several Native American tribes, on the other hand, have already volunteered. The issue is clouded if all ethnic minorities are lumped together as "non-white" or "minority"; there is considerable variation in cultures and relevant attitudes among these groups, and even among subgroups—different tribes, for instance, have much different attitudes toward nuclear waste.

Indeterminate Borders

If a facility's neighbors are to have a say in whether it is built, and are to be compensated for the costs it imposes on them, lines will have to be drawn to determine the electorate and the area eligible for compensation. As discussed in chapter 10, the county seems a natural electoral unit, but special provisions will have to be made to protect the municipality and any close neighbors in adjoining counties.

In addressing this kind of problem, a team of researchers at the University of Michigan coined the term *risk perception shadow* for the area in which residents perceive themselves at risk from a project. They then applied this concept to a proposed LLRW facility in Michigan near the Ohio and Indiana borders. The factors in drawing this shadow area are public awareness of the project; directness of impacts; significance of impacts; numbers of impacts; and duration of impacts. Using sophisticated public opinion polling techniques, the researchers drew an irregular circle around the proposed facility, crossing into all three states. The line extended from twenty to thirty miles from the site. There were also three outlying "islands" of concern.[4]

This works out to about 2,000 square miles, a very large area indeed. The radius is also much larger than that found in studies of the impacts of landfills and similar facilities on property values; a circle with a 5-mile radius (encompassing 78 square miles) is a more typical finding there. Part of the problem with opinion polling to set the borders for compensation is that the results can be manipulated—people are more likely to answer that they are worried about a facility, if that answer will increase their chances of being paid.

Since it will be each state's responsibility to produce a site the state finds politically acceptable, it will be up to each state to define the electorate for the referendum; it would be inappropriate for the federal government to dictate this process, unless of course racial discrimination or other practices banned by the voting rights or civil rights laws became apparent. (Federal guidelines may also be necessary if sites are proposed near state borders, so that close neighbors in an adjoining state are not shut out of the consent and compensation programs.) In affording compensation, there should be reliance on standard techniques of property appraisal, which can measure the actual effect of an external event on property values without resort to public opinion polling.

Disposal Monopolies

Centralized facilities for hazardous waste disposal might give great market power to the facility operators. The private trash-hauling business has certainly been plagued by frequent allegations of price fixing and even less-savory methods of ensuring market dominance. It is unlikely, however, that centralized facilities would lead to a full-scale monopoly because for most waste streams there will still be alternative disposal options, although they might be geographically remote. The operators of the nation's three LLRW disposal facilities have similar market power, and the price of LLRW disposal has indeed soared (though it is still a fraction of one percent of the operating costs of nuclear power plants, the facilities' prime market).[5] Rising disposal prices make waste minimization and recycling more economically attractive, and thus have a positive environmental aspect. (High prices could also encourage more on-site disposal, which may or may not be environemntally desirable.) If prices rise high enough, other companies will be attracted to the market, and nothing in my proposal would preclude them from trying to find volunteer sites on their own (though without the benefit of the federal allocation mechanism). Especially since the federally allocated sites will all have been created under the aegis of the state governments, the states should retain the power to regulate prices if necessary, as they do with other public utilities. There is also a built-in control mechanism; if State A allowed its centralized facility to charge outrageous prices (perhaps because the state received a share of the revenues), this would invite the other states to charge retaliatory prices for the disposal of the other waste streams from State A, if the federal statute establishing the siting mechanism allowed for such price discrimination.

The Players

Numerous constituencies would be significantly affected by this proposal. For the proposal to be adopted and to succeed, proponents of the system must politically outweigh opponents. In the following pages I try to predict how different constituencies will react.

Hazardous Waste Generators

I believe that generators will like the aspect of the proposal that would increase the probability of new sites being created but would dislike the mandatory waste minimization provisions. An assurance of a lawful place to dispose of each of their waste streams would be very popular, especially since the system would allow them to enter into long-term transportation and disposal arrangements. Generators will be especially happy about being able to send waste to modern new facilities, since the older, grandfathered facilities carry a particular risk of eventual CERCLA liability for their customers. Generators will not want to be taxed further for creating HW (though, in some states, such a tax already exists), and they will not want the EPA peering over their shoulders to see if they are adequately minimizing their waste. However, most generators have managed so far to comply with RCRA's ever-changing requirements (though not without considerable discomfort), and they would undoubtedly be able to comply with these as well.

Radioactive Waste Generators

These fall into three main categories: electric utilities; medical and industrial users of nuclear materials; and the Department of Energy. I will discuss the DOE separately.

The electric utilities are extremely unhappy about the current inability of the government to site HLW and LLRW disposal facilities. They are forced to store their spent fuel rods at their nuclear power plants, and the efforts of various states to site LLRW repositories are leading to a resurgence of anti-nuclear-power activism all over the country. Thus the utilities should be pleased with any plan that could accelerate the availability of RW disposal capacity.

The medical and industrial users are equally unhappy about the unavailability of LLRW disposal facilities. In fact, some have shut down certain operations because they have no place to send their waste, especially since the LLRW Policy Amendments Act has allowed the states with LLRW repositories to close their borders to states not making adequate efforts to site their own facilities. Thus these users should also be pleased with this plan.

Environmentalists

This constituency must be treated in three different groups—the mainstream national environmental groups, the grassroots-based organizations, and the anti-nuclear groups.

The mainstream groups will probably favor the proposal's emphasis on waste minimization and would work to ensure the vigorous implementation of that aspect of the plan. To the extent that the plan also minimizes the aggregate number of disposal facilities, and encourages each of these facilities to be better constructed and regulated, they will probably favor that aspect as well. They would be certain to keep very close tabs on the siting process, to ensure that environmental concerns are fully addressed at every step.

The grassroots-based groups have been philosophically opposed to the siting of virtually *every* waste disposal facility, on the theory that waste should not be created in the first place. If they are not convinced that literally everything possible has been done to reduce the amount of HW and RW generated, they are likely to oppose this plan. However, if they believe that the plan will speed the remediation of contaminated sites, their opposition might ease somewhat. They are still likely to be very unhappy with the idea of putting waste disposal sites on already contaminated property that will not be thoroughly cleaned up. They will also likely oppose the idea of attracting volunteer communities through offers of compensation; they generally abhor this practice, though if they have the opportunity to meet with people from these volunteer communities and are persuaded that a reasoned choice has been made based on full disclosure, this reaction may ease somewhat. On the other hand, these groups may be happy to know that no community will be forced to take a facility it does not want.

The anti-nuclear groups (some of which are international, some national, and some local) are similar to the grassroots anti-toxics groups in that they have opposed the siting of any facility. The agenda of many of these groups involves not only the blocking of any new nuclear power plants, but also the closure of all existing nuclear plants and the elimination of the entire nuclear industry. Unlike the anti-toxics groups, the anti-nuclear groups have not particularly focused on the cleanup of existing contaminated sites. They hope that the unavailability of waste disposal sites will increase pressure to eliminate the industry. Thus no siting system of any kind is likely to be acceptable to them, to the extent that it facilitates the survival of the existing nuclear plants.

Waste Management Industry

This proposal has some major advantages and major disadvantages for the waste management industry. On the plus side, it should greatly reduce the transaction costs involved in siting new facilities. Many companies have spent tens of millions of dollars attempting to site facilities, only to be defeated, politically or in the courts or regulatory agencies. The proposal should greatly reduce the number of failed siting attempts, because no application will have to be filed until the local community and the regulatory agencies have agreed in principle to the proposal. Moreover, to the extent that the plan actually leads to the siting of new capacity, it is likely that this industry will vigorously work to get pieces of that business and that the states will be happy to allow these companies to operate the facilities (just as most of the facilities in the Nuclear Weapons Complex are operated by private contractors to the federal government).

On the negative side, the plan's emphasis on waste minimization will, to the extent it is successful, cut seriously into this industry's business. The companies will undoubtedly try to become involved as waste minimization consultants and experts, but the most profitable part of the business is actual waste disposal. For many of the businesses, however, the most profitable aspect of waste disposal is MSW, not hazardous or radioactive wastes, and MSW disposal is not included in this proposal. The MSW disposal business might even go more smoothly to the extent that this proposal, by reducing adversarial siting processes, eases some of the serious political and public relations problems faced by the waste management industry.

The Environmental Protection Agency

This scheme would add to the already considerable burdens on the EPA. The agency's reaction may well depend on whether Congress gives it sufficient resources to carry out its new tasks. (The same could be said about another important actor in this scheme, the Nuclear Regulatory Commission.)

One aspect of the proposal will run against the established grain at the EPA: the use of already contaminated sites for waste disposal, without full remediation. However, if the congressional mandate is clear, the EPA is likely to follow.

The Department of Energy

One of the greatest challenges facing the DOE is the cleanup of the Nuclear Weapons Complex facilities. If this plan succeeds in accelerating the opening of RW disposal facilities, in turn advancing the day when RWs from the NWC can be disposed of, the DOE should be delighted. On the other hand, the DOE, and many other organizations involved in facility siting, are accustomed to wielding their governmental power and their expertise to make decisions on their own. A process with state responsibility and local control would require these organizations to yield power in an unfamiliar way. The leadership of these organizations should realize that the manifest failure of the traditional mode of exercising power must give way to a new, more open process.

The Department of Defense

In the post–Cold War era, the military is in search of new missions to justify its appropriations. Use of old military facilities for waste disposal would constitute a new, socially useful mission and is thus likely to meet with favor in the Pentagon. However, the military's environmental restoration program is committed to cleaning up sites more or less completely before turning them over for civilian uses (even if it takes decades to find the necessary resources and perform the work). The military will probably also not be pleased by the prospect of further contaminating the land it now owns. A change in the statutory allocation of liability may be necessary to relieve these concerns.

Congress

Congress has not distinguished itself in recent years in its ability to resolve complex, controversial issues. However, if a reasonable degree of political consensus can be reached on this proposal, then Congress may well put it into law. If major controversy erupts, the odds of enactment would seriously decline.

States

This proposal is designed to foster a sense of fairness among the states. It has the potential to be popular with most governors because it would put a limit on the number of facilities each of them must try to accommodate, and it would allow the governors to point to an outside agency— the FWDC—as the reason they must undertake any siting efforts at all. To be sure, Nevada and New Mexico are likely to be unhappy because the proposal contemplates that the Yucca Mountain and WIPP projects will proceed, but by the same token these may become easier pills to swallow if the two states are convinced the proposal will make all the other states bear an equitable portion of the nation's waste disposal problem. If Nevada and New Mexico subscribe to this logic, then they may even allow the projects within their borders to proceed more rapidly.

The National Governors Association (NGA) has played a constructive role in previous efforts to develop national facility siting policy. It was instrumental in devising the federal legislation for siting LLRW facilities (which, unfortunately, has failed), and it is now a major actor in the implementation of SARA's capacity assurance provisions. The NGA would be looked to for a similar leadership role in the formulation and implementation of the FDWC proposal.

Local Communities

The history of siting efforts shows that local communities usually do not become involved until they are specifically targeted for a particular facility. Though states usually have series of workshops and public forums on overall siting methodologies, local representatives seldom participate at that early stage of the process because they do not know they will be affected. Thus I do not expect that most communities will participate heavily in the formulation of this legislation (though it is possible that some of the national associations of cities and towns might become involved). If the FDWC is created, then the only municipalities likely to become actively involved are those considering offering to host a facility (or are close neighbors of such municipalities). Whether local governments around the country do indeed choose to invite facilities within their borders will be one of the ultimate tests of my proposal.

Epilogue

How a Siting Decision Could Be Made

The task I set at the beginning of this book was to find a system of hazardous and radioactive waste management that maximizes social welfare, taking full account of the social and environmental costs, while achieving fairness. I believe that the proposed system of local control, state responsibility, and national allocation can fulfill this task.

Social welfare would be maximized by reducing waste disposal requirements to a minimum (primarily through price incentives and elimination of hidden subsidies for waste generation, and possibly through regulatory controls) and then determining how much disposal capacity is still required. This capacity would be allocated among the fifty states, based primarily on how much waste they each generate; what disposal facilities they already have; and their geological and other physical attributes. Volunteer communities would be sought in each state to handle that state's allocation. Social and economic costs would be minimized through a sound siting process, and site neighbors would be compensated for any losses. The national allocation process would achieve fairness among states; the search for volunteer communities would achieve fairness within states; the closure of antiquated facilities would reduce the disproportionate burden on the poor and minorities; and the construction of centralized destruction and disposal units, especially on already contaminated federal land, would reduce the number of affected neighbors, preserve now-clean land for posterity, and help relieve future generations of the burden of caring for our waste.

To illustrate how the system would work in finding a particular site, let us take one hypothetical example. Suppose the EPA has determined that there is too little disposal capacity nationwide for the sludge that results from water pollution control devices at petrochemical plants. Ten million tons a year of the sludge is generated, but there are permitted landfills for only six million. (All of these numbers and facts are fictitious.) The EPA, in cooperation with the chemical industry, prepares a study of the technology of water pollution control at these plants. Based on the study, the EPA concludes that it would be feasible to reduce the amount of sludge generated from ten down to seven million tons a year by the use of different chemical processes that reduce the amount of sludge created and by drying devices that reduce the volume. Use of these methods would still leave a deficit of one million tons a year of disposal capacity.

The EPA tells the Federal Waste Disposal Commission about this one-million-ton deficit. The FWDC is at the same time considering all the other deficits the EPA and NRC have calculated in the disposal of waste streams. The FWDC decides that half of this deficit should be made up by New Jersey and the other half by California. This is because most of the plants generating this waste are on the east, Gulf, and west coasts, but Texas already has ample disposal capacity for this sludge; California and New Jersey each have relatively few off-site hazardous waste disposal facilities; the land disposal of the particular sludge in question poses little risk of groundwater contamination, due to the physical properties of the sludge, and thus it is not essential that the disposal facilities be located in arid lands; and because, in the overall national allocation process in which the FWDC is engaged, these two states should get these waste streams.

The FWDC then writes to the governor of New Jersey and says that her state has responsibility to come up with disposal capacity for 500,000 tons of chemical sludge each year but that New Jersey will not be allocated any other waste streams by the FWDC. (For example, all the low-level radioactive waste generated in the state will be handled by a repository in North Carolina.) In consultation with the EPA, New Jersey determines that this capacity should be provided by a landfill at least twenty-five acres in size, and that, with a buffer, a one-hundred-acre site is required, with particular geological conditions. The governor holds a press conference announcing the specifications of the required site, and asking for expressions of interest. Five such expressions are received—

one from the owner of a large Christmas tree farm; one from the chamber of commerce of a town with a recently closed military base; one from a town with a large tract of land adjacent to its closed municipal landfill; one from the trustee in bankruptcy of an abandoned factory now being cleaned up under the Superfund program; and one from the mayor of an old industrial city with a large unbuilt urban renewal tract in the middle of the city.

State environmental officials visit all five sites. They determine based on preliminary inspections that the Christmas tree farm would be unsuitable because much of the site is covered with federal wetlands, and the site in the old industrial city is unsuitable because the nearby population density is high, and sludge disposal facilities cause offensive odors. That leaves three potential sites. The state has a series of workshops in each of the three localities to discuss the nature and impacts of the proposed facility. At the same time, engineering firms are at work doing more detailed site studies. These studies confirm there are no fatal defects in any of the three sites that would preclude the siting of the sludge facility there.

A third set of discussions takes place during this period. The principal customers of the facility would be the chemical companies with plants on the East Coast. The Chemical Manufacturers Association (CMA) convenes a working group of the plant managers to discuss the size of the compensation package they would pay to find a site for their sludge. The plant managers are eager for the disposal facility to be sited, because sludge is building up in tanks at the factories and the tanks are nearing capacity. The managers conclude, based on advice from their engineers, that building and operating the facility will cost about $10 a ton, and that they would be willing to pay an additional $5 a ton in community compensation, or $2.5 million a year. The CMA announces it will offer $2.5 million a year to any municipality that will take this facility.

The $2.5 million offer is coolly received by the leaders of the three communities; they had hoped for considerably more. After consultation with the plant managers, the CMA increases the offer to $5 million. With that, Smithville, the town with the closed military base, comes forward. Smithville's mayor holds discussions with the adjoining municipality, Jonesville, whose residents would also be affected by the facility, and they agree on a 75 percent/25 percent split of the compensation. A referendum is then held; all the residents of Smithville, and the residents of Jonesville who live within a mile of the site, are eligible to vote. The ballot question specifies how the compensation money will be used (a combination of new schools in Smithville, a new library in Jonesville, and property tax relief in both towns—the allocation selected by the two town councils). The referendum carries by a 60 percent majority of the combined electorate.

New Jersey has decided that it does not want to build and operate the sludge facility itself but rather wants the private sector to assume this task. The state puts out a request for proposals for companies interested in building the facility, indicating that the winning company will be expected to put up $1 a ton for a perpetual-care fund and to offer additional compensation to the state, will have to buy any homes within one-quarter mile of the site if their owners want to move out, and will have to provide fair compensation to all people living within a half-mile radius (the distance at which, according to hypothetical appraisals, property values would likely be affected). The state also reveals that the Department of Defense has offered to sell the site for $10 million. Four proposals are received on the due date. The highest amount of compensation offered to the state and the communities is from a national company called Sludge Management, Inc. The state determines that Sludge Management has the necessary qualifications to undertake the project, and the company's offer is accepted.

Sludge Management contracts with the Department of Defense to buy the site for $10 million if all necessary permits are received. It buys the homes of ten immediate neighbors of the site, for what their fair market value would be were it not for the sludge plant proposal, and also pays their relocation expenses. The company then prepares an application for all the environmental permits required by the facility, and writes a detailed environmental impact statement. After public hearings, the state environmental agency awards the permits to Sludge Management. Sludge Management completes the purchase of the site, builds the facility, and starts operations. It pays compensation for lost property values to the remaining neighbors who live within half a mile of the site. Every month,

Smithville, Jonesville, and the state of New Jersey receive compensation checks from the company.

The cost of providing sludge disposal services, per ton, is $10 for construction and operation; $5 for community compensation; $3 for state compensation; and $2 for land acquisition, for a total of $20. Sludge Management decides to charge its customers $25. (Though that is a high profit margin, trucking the sludge to the nearest sludge disposal facility, in Texas, would cost more than $5 a ton.) Because of this high cost, several of the chemical companies modify their production processes to further reduce the volume of sludge produced. Meanwhile, they have been able to empty and dismantle the old tanks in which sludge had been stored pending construction of the new facility.

This is one purely hypothetical scenario. Many variations are possible. Matters went rather smoothly (except for the communities' rejection of the first compensation offer), which is plausible but hardly assured. One thing is certain, however: had the state tried to force an unwilling community to take the facility, the resulting objections would ensure a far rockier process, with a much lower probability of success.

This system would face many practical, political, and economic obstacles. But it is superior to the current regime of impasse, conflict, fragmentation, and futile attempts at coercion.

Abbreviations

ASWMC	Alberta Special Waste Management Corporation
BFI	Browning Ferris Industries
CCHW	Citizens Clearinghouse for Hazardous Waste
CEQ	U.S. Council on Environmental Quality
CERCLA	Comprehensive Environmental Response, Compensation and Liability Act of 1980
CERCLIS	Comprehensive Environmental Response, Compensation and Liability Information System
CWM	Chemical Waste Management, Inc.
DOE	U.S. Department of Energy
EIS	Environmental impact statement
EPA	U.S. Environmental Protection Agency
ESA	Endangered Species Act
FWDC	Federal Waste Disposal Commission [proposed]
HLW	High-level radioactive waste
HSWA	Hazardous and Solid Waste Amendments Act of 1984
HW	Hazardous waste
LLRW	Low-level radioactive waste
LLRWPA	Low-Level Radioactive Waste Policy Act of 1980
LLRWPAA	Low-Level Radioactive Waste Policy Amendments Act of 1985
MHWMC	Manitoba Hazardous Waste Management Corp.
MRS	Monitored retrievable storage

MSW	Municipal solid waste
NEPA	National Environmental Policy Act
NIMBY	Not in my backyard
NORM	Naturally occurring radioactive material
NPL	National Priorities List
NWC	Nuclear Weapons Complex
NWPA	Nuclear Waste Policy Act
OSHA	Occupational Safety and Health Administration
PCBs	Polychlorinated biphenyls
RCRA	Resource Conservation and Recovery Act of 1976
RW	Radioactive waste
SARA	Superfund Amendments and Reauthorization Act of 1986
SMCRA	Surface Mining Control and Reclamation Act
TRU	Transuranic waste
TSCA	Toxic Substances Control Act of 1976
UST	Underground storage tank
WIPP	Waste Isolation Pilot Plant
WMI	Waste Management, Inc.

Notes

Chapter 1

1. New York State Legis. Comm'n on Toxic Substances & Hazardous Wastes, *Hazardous Waste Facility Siting: A National Survey* (June 1987); William Lyons et al., *Public Opinion and Hazardous Waste,* F. for Applied Res. & Pub. Pol'y, Fall 1987, at 89, 90.

2. The federal government has spent $1.26 billion on the effort to site a repository for high-level radioactive waste at Yucca Mountain, Nevada, *Projected HLW Program Budget Down by $200 Million,* Radioactive Exchange, Nov. 16, 1992, at 14; $1.3 billion on the Waste Isolation Pilot Plant for transuranic waste in New Mexico, Keith Schneider, *Wasting Away,* N.Y. Times, Aug. 30, 1992, § 6 (Magazine) at 42, 43; and more than $100 million on its effort (abandoned in 1971) to site a high-level radioactive waste repository in Lyons, Kansas, *id.* at 56. Illinois spent $85 million in an abortive effort to site a low-level radioactive waste (LLRW) facility in Martinsville. *Conference Notes,* Radioactive Exchange, Dec. 14, 1992, at 14, 15. New York State has spent $37 million trying to site an LLRW facility. Herbert Inhaber, *Of LULUs, NIMBYs, and NIMTOOs,* Pub. Interest, Spring 1992, at 52, 62. California, Nebraska, and Pennsylvania have each spent about $30 million trying to site LLRW facilities, and North Carolina has spent $45 million. Jorge Contreras, *In the Village Square: Risk Misperception and Decisionmaking in the Regulation of Low-Level Radioactive Waste,* 19 Ecology L.Q. 481, 528 (1992). It has been reported that the typical cost of obtaining federal and state permits for hazardous waste facilities is $3 million, U.S. Gen. Accounting Office, Pub. No. GAO/RCED-88-95, *Hazardous Waste: Future Availability of and Need for Treatment Capacity Are Uncertain* 22 (Apr. 1988), but some attempts have been much more expensive. CECOS International spent at least $7 million in a failed attempt to expand a hazardous waste facility in Niagara Falls, New York. Paul MacClennan, *The Stakes Are High in the CECOS Hearings,* Buffalo News, Sept. 18, 1988, at H16. Clean Harbors of Braintree, Inc. spent $16 million on its unsuccessful effort to site a hazardous waste incinerator in Massachusetts. Denis J. Brion, *Essential Industry and the NIMBY Phenomenon* 13–14 (1991) [hereinafter Brion 1991]. In Canada, one

ongoing permitting proceeding has already cost $100 million. Mary Lou Garr, *Ontario Waste Management Corporation (OWMC) Proposal for an Industrial Hazardous Waste Treatment and Disposal Facility, in Innovative Approaches to the Siting of Waste Management Facilities: A Guide to Nonconfrontational Siting Procedures* 95, 105 (Resource Futures International ed., 1992).

3. Hunt v. Chemical Waste Management, Inc., 584 So.2d 1367, 1373 (Ala. 1991).

4. *Commercial Hazardous Waste Management: Recent Financial Performance and Outlook for the Future,* Hazardous Waste Consultant, Sept./Oct. 1990, at 4-1; *Hazardous Waste Facility Siting: A National Survey, supra* ch. 1 note 1; John A.S. McGlennon, *A Model Siting Process and the Role of Lawyers,* 15 Envtl. L. Rep. (Envtl. L. Inst.) 10239 (1985) (no new facilities sited in United States between 1976 and 1981); Mario Ristoratore, *Siting Toxic Waste Disposal Facilities in Canada and the United States: Problems and Prospects,* 14 Pol'y Stud. J. 140 (1985) (no new facilities sited in United States between 1979 and 1985).

5. *Public Must Accept Risk in Siting New Waste Facilities, Conference Told,* 12 Env't Rep. (BNA) No. 10, at 314 (July 3, 1981) (remarks of Rep. James J. Florio). *Similarly,* Daniel Mazmanian and David Morell, *Beyond Superfailure: America's Toxic Policy for the 1990s,* at 95–96 (1992) [hereinafter Mazmanian and Morell 1992] (National Conference of State Legislatures survey in early 1980s concluded that at least 125 new HW management facilities were needed). *See also* 132 Cong. Rec. S14924 (daily ed. Oct. 3, 1986) (statement of Sen. Chafee) ("A critical step in the creation of a rational, safe hazardous waste program is the creation of new [hazardous waste disposal] facilities.").

6. *See, e.g.,* Lawrence S. Bacow and James R. Milkey, *Overcoming Local Opposition to Hazardous Waste Facilities: The Massachusetts Approach,* 6 Harv. Envtl. L. Rev. 265, 266 n.9 (1982) (EPA projects major shortfall in disposal capacity.); Susan Caskey, *Alternative Dispute Resolution and Siting of Hazardous Waste Facilities: The Pennsylvania Proposal in Light of the Wisconsin and Massachusetts Statutes,* 5 Temp. Envtl. L. & Tech. J. 58 (1986) ("Few people who are well informed on the subject of hazardous waste generation and disposal will deny that there is an urgent need, through the United States, for the establishment of safe and efficient waste transportation, storage and disposal facilities."); Kenneth A. Dimuzio, *The Siting and Operation of Hazardous Waste Disposal Facilities, in Current Municipal Problems, 1981–1982* 506 (Bryon S. Matthews ed., 1982) ("The most serious political and moral problem facing New Jersey today is the safe disposal of hazardous wastes."); Daniel Mazmanian and David Morell, *The "NIMBY" Syndrome: Facility Siting and the Failure of Democratic Discourse, in Environmental Policy in the 1990s* 125, 126 (Norman J. Vig and Michael E. Kraft eds., 1990) [hereinafter Mazmanian and Morell 1990] ("One of the most important questions of the 1990s is how to move beyond the current gridlock created by NIMBYism. . . . A workable answer is central to

realizing the nation's environmental goals."); Stephen Sussna, *Remedying Hazardous Waste Facility Siting Maladies by Considering Zoning and Other Devices,* 16 Urb. Law. 29, 32 (1984) ("[T]here is a desperate need for soundly designed treatment and disposal facilities.").

7. E.g., Howard Kunreuther et al., *Public Attitudes Toward Siting a High-Level Nuclear Waste Repository in Nevada,* 10 Risk Analysis 469, 469 (1990) (Richard Bryan defeated Sen. Chic Hecht for U.S. Senate from Nevada in 1988 largely as a result of Hecht's allegedly equivocal stand on the siting of a high-level radioactive waste repository in the state); Robert D. Benford et al., *In Whose Backyard?: Concern About Siting a Nuclear Waste Facility,* 63 Sociological Inquiry 30, 31 (1993) (in Nebraska gubernatorial race, incumbent was replaced by a candidate who opposed locating low-level radioactive waste facility in the state).

8. *See* William L. Andreen, *Defusing the "Not in My Back Yard" Syndrome: An Approach to Federal Preemption of State and Local Impediments to the Siting of PCB Disposal Facilities,* 63 N.C. L. Rev. 811, 813 n.8 (1985) (five-hundred-and-twenty-three persons arrested for attempting to block trucks loaded with PCB-contaminated soil from entering disposal facility in Warren County, North Carolina); E. William Colglazier and Mary R. English, *Low-Level Radioactive Waste: Can New Disposal Sites Be Found?* 53 Tenn. L. Rev. 621, 631 (1986) (several cases of suspected arson and sabotage to vehicles of company attempting to build radioactive waste facility in Texas in 1978); Mary R. English, *Siting Low-Level Radioactive Waste Disposal Facilities: The Public Policy Dilemma* 90 (1992) (threats of violence by opponents of radioactive waste facility in New York State in 1990) [hereinafter English 1992]; Robert W. Lake, *Rethinking NIMBY,* 59 J. Am. Planning Ass'n 87 (Winter 1993) [hereinafter Lake 1993] (protestors in Minnesota reportedly dropped explosives into test wells to disrupt the search for a hazardous waste disposal site); *More Than 100 Arrested in Protests Against Start-Up of Ohio Incinerator,* 23 Env't Rep. (BNA) No. 26, at 1910 (Nov. 27, 1992); David Morell and Christopher Magorian, *Siting Hazardous Waste Facilities: Local Opposition and the Myth of Preemption* 96–97 (1982) (citing other examples).

9. *E.g.,* Michael O'Hare et al., *Facility Siting and Public Opposition* 4 (1983) [hereinafter O'Hare 1983]; George F. List et al., *Modeling and Analysis for Hazardous Materials Transportation: Risk Analysis, Routing/Scheduling and Facility Location,* 25 Transp. Sci. 100, 107 (1991); A. Dan Tarlock, *Anywhere But Here: An Introduction to State Control of Hazardous-Waste Facility Location,* 2 J. Envtl. L. 1, 21 (1981) [hereinafter Tarlock 1981].

10. E.g., Brion 1991, *supra* ch. 1 note 2; Greenberg v. Veteran, 889 F.2d 418, 419 (2d Cir. 1989).

11. E.g., Frank J. Popper, *The Environmentalist and the LULU, in Resolving Locational Conflict* 1 (Robert W. Lake ed., 1987) [hereinafter Lake 1987].

12. *Public Opposition to Incinerating Waste Could Seriously Impede Cleanups, Officials Say,* 23 Env't Rep. (BNA) No. 28, at 2028 (Dec. 11, 1992).

Chapter 2

1. Bruce W. Piasecki and Gary A. Davis, *America's Future in Toxic Waste Management: Lessons From Europe* 4 (1987).

2. *Id. See also Task Force Urges EPA to Consider Adding 1,500 to 2,000 Substances to CERCLA List*, 22 Env't Rep. (BNA) No. 52, at 2866 (Apr. 24, 1992).

3. An exception is that some waste generators, when uncertain about the regulatory status of a given waste stream, send the waste to a hazardous waste disposal facility.

4. 42 U.S.C. §§ 6901 *et seq.*

5. The Safe Drinking Water Act controls aspects of underground injection wells. 42 U.S.C. § 300h-3. The Clean Water Act controls the disposal of hazardous waste through publicly owned treatment works. 33 U.S.C. §§ 1251–1387. Disposal of polychlorinated biphenyls (PCBs) is regulated under the Toxic Substances Control Act. 15 U.S.C. § 2605(e). Ocean disposal of hazardous waste, banned as of December 31, 1991, was governed by the Marine Protection, Research and Sanctuaries Act. 33 U.S.C. § 1414b. The Clean Air Act regulates the release of toxics into the air. 42 U.S.C. § 7412.

6. 42 U.S.C. § 6903(5).

7. 42 U.S.C. § 6921. A large body of judicial and regulatory authority has arisen around this system of definitions. *See generally* David R. Case, *Identifying Hazardous Materials and Hazardous Wastes, in Environmental Law Practice Guide* ch. 26 (Michael B. Gerrard ed., 1992) [hereinafter Gerrard 1992]; John-Mark Stensvaag, *Hazardous Waste Law and Practice* (1989).

8. U.S. EPA, Office of Solid Waste and Emergency Response Pub. No. 05-312, *National Biennial RCRA Hazardous Waste Report (Based on 1989 Data)* 2 (Feb. 1993) [hereinafter *National Biennial Report*].

9. U.S. Environmental Protection Agency, Office of Toxic Substances, *The Toxics Release Inventory: A National Perspective, 1987* 226 (June 1989).

10. *National Biennial Report, supra* ch. 2 note 8, at 1, 2–14.

11. U.S. Gen. Accounting Off., Pub. No. GAO/PEMD-90-3, *Hazardous Waste: EPA's Generation and Management Data Need Further Improvement* (Feb. 1990); *Review of EPA's Capacity Assurance Program: Hearings Before the Subcomm. on Environment, Energy and Natural Resources of the House Comm. on Government Operations*, 101st Cong., 1st Sess. 8 (1991) [hereinafter *Review of EPA's Capacity Assurance Program*] (statement of Eleanor Chelimsky, assistant comptroller general).

12. James E. McCarthy and Mark E. Anthony Reisch, Congressional Research Service, *Hazardous Waste Fact Book* 6–7 (Jan. 30, 1987); *similarly, Review of EPA's Capacity Assurance Program, supra* ch. 2 note 11 (statement of Sylvia K. Lowrance, director, Office of Solid Waste, EPA), at 187. Data for 1989 show a

decline in 40 million tons from 1987, but this is largely because of waste being shifted from surface impoundments, which are subject to RCRA regulation, to tanks, which are often exempt from RCRA. U.S. Environmental Protection Agency, Office of Solid Waste, Pub. No. EPA530-R-93-011, *RCRA Environmental Indicators: FY 1992 Progress Report* 4-2 (Aug. 1993) [hereinafter *RCRA Environmental Indicators*]. The reasons for varying estimates are discussed in Office of Tech. Assessment, U.S. Cong., *Ocean Incineration: Its Role in Managing Hazardous Waste* 63 (Aug. 1986) [hereinafter OTA Ocean Incineration], and Harry M. Freeman ed., *Hazardous Waste Minimization* 18 (1990).

13. Susan L. Cutter, *Living With Risk: The Geography of Technological Hazards* 113 (1993).

14. *National Biennial Report, supra* ch. 2 note 8.

15. *RCRA Environmental Indicators, supra* ch. 2 note 12, at 4–6.

16. *National Biennial Report, supra* ch. 2 note 8, at 4; Doug MacMillan, *Interstate Movement of Hazardous Waste,* Waste Age, May 1993, at 290.

17. William Gruber, *Hazardous Waste Incineration: 1990,* EI Digest, Apr. 1990, at 7. This number includes fifteen incinerators owned and operated by the army or the navy, mostly used to destroy expired small-caliber ammunition. Id.

18. Joan Z. Bernstein, *The Siting of Commercial Waste Facilities: An Evolution of Community Land Use Decisions,* 1 Kan. J. L. and Pub. Pol'y 83, 86 (1991).

19. *National Biennial Report, supra* ch. 2 note 8, at 2–46.

20. McCoy and Associates, *1993 Outlook for Commercial Hazardous Waste Management Facilities: A Nationwide Perspective,* Hazardous Waste Consultant, Mar./Apr. 1993, at 4; *Number of Hazardous Waste Facilities Drops, Many Companies Expanding Services,* 23 Env't Rept. (BNA) No. 41, at 2634 (Feb. 5, 1993); *see also Report Says 11 States Are Lacking Hazardous Waste Facilities,* World Wastes, May 1992, at 60.

21. McCarthy and Reisch, *supra* ch. 2 note 12, at 31. *But see Treatment Capacity Shortage Leads EPA to Extend Variance for Hazardous Debris,* 24 Env't Rep. (BNA) No. 3 at 134 (May 21, 1993).

22. *Hazardous Waste Facility Siting: A National Survey, supra* ch. 1 note 1, at 26.

23. *See* 42 U.S.C. § 6924(d).

24. Sharon N. Green, *Planning for Hazardous Waste Capacity: Lessons from the Northeast States* 31 (1990); OTA Ocean Incineration, *supra* ch. 2 note 12, at 74.

25. Karin Schreifels and Lisa Nelowet, *Opportunities and Dilemmas of Transportable Hazardous Waste Management Facilities, in Proceedings, An International Symposium, Hazardous Materials/Wastes: Social Aspects of Facility Planning and Management* 62 (Institute for Social Impact Assessment, 1992) [hereinafter Inst. for Soc. Impact Assessment].

26. Peter Kemezis, *Among the States: Free Trade—High Court Opens the Doors for Waste Imports,* Chemical Wk., Aug. 19, 1992; *William Gruber, Siting Efforts for Hazardous Waste Incinerators,* EI Digest, May 1990, at 18.

27. Paul Harris, *Spotlight on TSD: "Where did the business go?",* Env't Times, Oct. 1992, at 33.

28. Kristi Highum, *The Incineration Picture Based on the Capacity Assurance Plans,* EI Digest, Apr. 1990, at 16 (shows total annual capacity at commercial hazardous waste incinerators to be over 1,000,000 tons/year, while demand is 670,000 tons; but California has no commercial hazardous waste incinerators, and nation has oversupply of liquid incineration capacity and slight shortfall in solid incineration capacity); U.S. Environmental Protection Agency, Office of Policy Analysis, *1986–1987 Survey of Selected Firms in the Commercial Hazardous Waste Management Industry: Final Report* iv (1988) (concludes that commercial incineration capacity is adequate); Frederic M. Iannazzi and Christine A. O'Shaughnessy, *Hazardous Waste Incineration, Part 2: Leading Participants* (1991) (projects incineration market to become increasingly competitive).

29. Mary Melody, *Hazwaste Treatment Services Weather Challenges,* Hazmat World, June 1993, at 30, 32. *Similarly,* Ray Pospisil, *Radical Change for Hazardous Waste Services,* Chemical Wk., Aug. 18, 1993, at 26 ("[I]ncinerator overcapacity could be exacerbated by the debut of several new units. . . .").

30. Richard Ringer, *Toxic-Waste Concern Sets Big Cutbacks,* N.Y. Times, Oct. 1, 1993, at D3.

31. EPA Draft Strategy for Combustion of Hazardous Waste in Incinerators and Boilers, 24 Env't Rep. (BNA) No. 3 at 157 (May 21, 1993).

32. *See* Jeff Bailey, *WMX Technologies Ends Plan to Build California Incinerator; Demand Shrinks,* Wall St. J., Sept. 8, 1993, at C15; Elisabeth Kirschner, *DuPont Drops Incineration, But Ecoservices Still Growing,* Chemical Wk., Sept. 8, 1993, at 9; *Hazardous Waste: Company Withdraws Application to Build Incinerator, Cites EPA Review of Permits,* 24 Env't Rep. (BNA) No. 20 at 881 (Sept. 17, 1993); *Hazardous Waste: Waste Company Blames North Carolina After Abandoning Bid to Build Incinerator,* 24 Env't Rep. (BNA) No. 25 at 909 (Sept. 24, 1993); *Hazardous Waste: Colorado Firm Abandons Application to Build Waste Incinerator in Florida,* 24 Env't Rep. (BNA) No. 5 at 1174 (Oct. 22, 1993).

33. National Solid Wastes Management Ass'n v. Alabama Dep't of Envtl. Management, 910 F.2d 713, 715 (11th Cir. 1990), *modified,* 924 F.2d 1001 (11th Cir.), *cert. denied,* 111 S.Ct. 2800, 115 L.Ed.2d 973 (1991).

34. Hunt v. Chemical Waste Management, Inc., 584 So.2d 1367, 1373 (Ala. 1991).

35. Testimony of John Ianotti, N.Y. St. Dep't of Envtl. Conservation, before hearing of N.Y. St. Industrial Hazardous Waste Facility Siting Board, In the Matter of CECOS International, Inc., Dec. 12, 1989, at 7237–38.

36. Jeffrey D. Smith, *Growth Potential of the Hazardous Waste Landfill Business,* EI Digest, Apr. 1990, at 45.

37. *Landfills: Recession Cuts Into Volumes,* Env't Times, Oct. 1992, at 36; *Cyclicity Rears Its Ugly Head,* Chemical Wk., Aug. 21, 1991, at 44; *Recession Shocks Hazwaste,* Chemical Wk., Jan. 8, 1992, at 23.

38. Bernstein, *supra* ch. 2 note 18, at 86. Similar sentiments were expressed in Ross and Assocs., *Observations on Capacity Assurance: The Western States' Experience in Implementing CERCLA 104(c)(9)* 3 (W. Governors Ass'n, Denver, Colo., June 18, 1990).

39. Paul Kemezis, *Interstate Waste Transport is Still a Fair Game,* Chemical Wk., Aug. 18, 1993, at 30.

40. Alex. Brown and Sons, Environmental Research: Chemical Waste Management, Inc., Rollins Environmental Services, Inc., at 17 (Apr. 5, 1993).

41. *Id.*

42. *Id.* at 15, 20. *Similarly, see Haz Waste Market Growth Fueled by Remediation, Disposal and Services Flat,* Envtl. Business J., June 1993, at 1; Neil Springer, *Plenty of Room at the Inn,* Chemical Marketing Reporter, Nov. 16, 1992, at SR8. A differing view is taken in Freedonia Group, Inc., *Commercially Managed Wastes to Grow at Twice the Rate of Total Generation* (April 1993).

43. *S&P Places Debt of Waste Management Firms Under Review,* Wall St. J., Sept. 7, 1993, at A5.

44. Jeff Bailey, *WMX Slates Big Write-Down, Job Cuts as Hazardous-Waste Industry Struggles,* Wall St. J., Oct. 1, 1993, at A3; *Safety-Kleen to Take 4th-Period Charge, Cut Jobs Amid Slump in Waste Industry,* Wall St. J., Oct. 5, 1993, at B4.

45. United States v. Hooker Chemicals and Plastics Co., 680 F. Supp. 546 (W.D.N.Y. 1988); Adeline Gordon Levine, *Love Canal: Science, Politics, and People* (1982); Michael Brown, *Laying Waste: The Poisoning of America by Toxic Chemicals* (1981).

46. Marc K. Landy et al., *The Environmental Protection Agency: Asking the Wrong Questions* 133–71 (1990).

47. 42 U.S.C. §§ 9601 *et seq.*

48. Technically, these are hazardous substance sites, since CERCLA regulates hazardous substances—a broad category that incorporates by reference all RCRA hazardous wastes, as well as several other lists of materials. 42 U.S.C. § 9601(14).

49. 42 U.S.C. § 9605.

50. This process is set forth in the National Contingency Plan, 40 C.F.R. Part 300. *See also* Thomas W. Church and Robert T. Nakamura, *Cleaning Up the Mess: Implementation Strategies in Superfund* (1993).

51. Office of Tech. Assessment, U.S. Cong., Pub. No. OTA-ITE-433, *Coming Clean: Superfund Problems Can Be Solved* . . . 6 (1989).

52. *Id.* at 11.

53. *Id.* at 179.

54. Mazmanian and Morell 1992, *supra* ch. 1 note 5 at 13, 43, 47.

55. U.S. Environmental Protection Agency, Pub. No. EPA 542-R-92-012, *Cleaning Up the Nation's Waste Sites: Markets and Technology Trends* 4 (Apr. 1993).

56. *Id.* at 7. See also Congressional Budget Office, *The Total Costs of Cleaning Up Nonfederal Superfund Sites* (1994).

57. Keith Schneider, *Debate on Burning Dioxin Divides Arkansas Town,* N.Y. Times, Oct. 28, 1992, at A14; Keith Schneider, *Judge to Oversee Burning in Arkansas,* N.Y. Times, Oct. 30, 1992, at A16; Keith Schneider, *In Arkansas Toxic Waste Cleanup, Highlights of New Environmental Debate,* N.Y. Times, Nov. 2, 1992, at B11; Liane Clorfene Casten, *Toxic Burn: Agent Orange's Forgotten Victims,* The Nation, Nov. 4, 1991, at 550; James Ridgeway, *Toxic Waste Syndrome: Is It Business as Usual at EPA?* Village Voice, Mar. 23, 1993, at 17.

58. Arkansas Peace Center v. Arkansas Dep't of Pollution Control and Ecology, 992 F.2d 145 (8th Cir. 1993).

59. *Pollution Remedy is Hotly Debated,* N.Y. Times, Oct. 10, 1993, at 28.

60. Keith Schneider, *As Superfund Ages, Resistance to Cleanups Grows,* N.Y. Times, Sept. 6, 1993, at 7.

61. 42 U.S.C. § 6924(u), (v).

62. *Cleaning Up the Nation's Waste Sites, supra* ch. 2 note 55, at 5.

63. Alex. Brown and Sons, *supra* ch. 2 note 40, at 8.

64. Michael Satchell, *Uncle Sam's Toxic Folly,* U.S. News and World Rep., Mar. 27, 1989, at 20.

65. Bill Richards and Andy Pasztor, *Why Pollution Costs of Defense Contractors Get Paid by Taxpayers,* Wall St. J., Aug. 31, 1992, at A1.

66. Exec. Order No. 12,088 (1978), *reprinted in* 42 U.S.C. § 4321 nt. (Supp. 1992).

67. Federal Facilities Compliance Act, Pub. L. No. 102-386, 106 Stat. 1505.

68. Katie Hickox, *Swords Into Bankshares: How the Defense Industry Cleans Up on the Nuclear Build-Down,* Wash. Monthly, Mar. 1992, at 31; Cheryl L. McAdams, *Targeting Closing Military Bases: The Clean Up of Hazardous Wastes on Closing Military Bases Could Represent a Waste Management Business Bonanza,* Waste Age, Oct. 1992, at 53.

69. U.S. Dep't of Defense, *Defense Environmental Restoration Program: Annual Report to Congress for Fiscal Year 1991* 4, 6 (Feb. 1992).

70. *Id.* at 10; Seth Shulman, *The Threat at Home: Confronting the Toxic Legacy of the U.S. Military* 105–06 (1992).

71. U.S. House of Representatives, Committee on Natural Resources, Subcommittee on Oversight and Investigations, Majority Staff Report, *Deep Pockets: Taxpayer Liability for Environmental Contamination* 10–11 (July 1993).

72. *Id.* at 107; Luis H. Francia, *Bases Loaded: Pentagon Strikes Out at Subic and Clark,* Village Voice, June 15, 1993, at 17.

73. Michael Satchell, *Uncle Sam's Toxic Folly,* U.S. News and World Rep., Mar. 27, 1989, at 20, 21.

74. *Cleaning Up the Nation's Waste Sites, supra* ch. 2 note 55 at 7; *see also* U.S. General Accounting Office, Pub. No. GAO/RCED-93-119, *Superfund: Backlog of Unevaluated Federal Facilities Slows Cleanup Efforts* (July 1993).

75. Unless otherwise noted, this subsection is based on John W. Birks, *Weapons Forsworn: Chemical and Biological Weapons, in Hidden Dangers: Environmental Consequences of Preparing For War* 161 (Anne E. Ehrlich and John W. Birks eds., 1990); Vicki Kemper, *Deadly Debris: How the Army Plans to Rid the World of Chemical Weapons,* Common Cause Mag., July/Aug. 1990, at 20; and Eugene L. Meyer, *Toxic Fallout: Citizens Are Up in Arms Over Plans to Incinerate the Nation's Chemical Weapons,* Audubon, Sept.–Oct. 1992, at 16.

76. Seymour M. Hersh, *Chemical and Biological Warfare: America's Hidden Arsenal* (1968).

77. Pub. L. 99-145, § 1412, 99 Stat. 747.

78. Pub. L. 102-484, § 171, 106 Stat. 2341.

79. *Nervous About Nerve Gas,* Time, Aug. 6, 1990, at 28. *See* Greenpeace USA v. Stone, 748 F. Supp. 749 (D. Haw. 1990).

80. Paul Kemezis, *Congress Kills Funds for Alabama Chemical Weapons Incinerator,* Chemical Wk., Oct. 14, 1992, at 14; Ronald Smothers, *Plan to Destroy Toxic Weapons Polarizes a City,* N.Y. Times, Sept. 24, 1992, at A16; *60 Minutes: Time Bomb* (transcript of CBS News broadcast, July 19, 1992).

81. Jacki Jones, *Is the Nerve Gas Burning?* Village Voice, Oct. 6, 1992, at 16.

82. Pub. L. 102-484, § 173, 106 Stat. 2342. *See also* Office of Tech. Assessment, U.S. Congress, *Disposal of Chemical Weapons: Alternative Technologies* (June 1992); National Research Council, *Alternative Technologies for the Destruction of Chemical Agents and Munitions* (1993) (finds that numerous technologies other than incinerators are available; their economics and feasibility are now under study).

83. U.S. General Accounting Office, Pub. GAO/NSIAD-93–50, *Chemical Weapons Destruction: Issues Affecting Program Cost, Schedule, and Performance* 2 (Jan. 1993).

84. U.S. Department of the Army, Memorandum for Correspondents No. 353-M (Nov. 23, 1993); *Army Announces Program to Manage Old Chemical Sites, Non-Stockpile Material,* 23 Env't Rep. (BNA) No. 52, at 3191 (Apr. 23, 1993); *Army Reports on Sites That May Contain 'Non-Stockpile' Chemical Material Facilities,* Environmental Update (U.S. Army Environmental Center), July 1993, at 5.

85. Michael R. Gordon, *Negotiators Agree on Accord to Ban Chemical Weapons,* N.Y. Times, Sept. 2, 1992, at A1; Alan Riding, *Signing of Chemical-Arms*

Pact Begins, N.Y. Times, Jan. 14, 1993, at A16. *See also Making a Chemical Warfare Treaty Work,* 261 Sci. 826 (Aug. 13, 1993) (implementation of treaty will require extensive disclosure of data by chemical manufacturers that make, use, or buy treaty-controlled chemicals).

86. Michael R. Gordon, *Moscow Is Making Little Progress In Disposal of Chemical Weapons,* N.Y. Times, Dec. 1, 1993, at A1; John J. Fialka, *Russia Seeks to Include a Recycling Plan In Its Chemical-Weapons Dismantling,* Wall St. J., Sept. 2, 1992, at A7; Karen Elliott House and Philip Revzin, *Toxic Dump: Arsenal of Poison Gas Languishes as Russia Is Unable to Destroy It,* Wall St. J., Feb. 25, 1993, at A1; David N. Clark, *Chemical Weapons Disposal,* 257 Sci. 12 (1992). *See also* Tariq Rauf, *Soviet Union: Cleaning Up With a Bang,* Bull. Atomic Scientists, Jan./Feb. 1992, at 9 (proposal by Moscow company to destroy or encapsulate chemical and nuclear waste with underground nuclear explosions).

87. Barry Meier, *Breaking Down an Arms Buildup,* N.Y. Times, Oct. 15, 1993, at D1.

88. Asbestos Hazard Emergency Response Act, 15 U.S.C. §§ 2641 *et seq.*

89. *Coming Clean: Superfund Problems Can Be Solved . . . , supra* ch. 2 note 51, at 101.

90. *Abatement Industry Sees Decline in Removal, Trend Toward In-Place Management Strategy,* 7 Toxics L. Rep. (BNA) No. 26, at 758 (Nov. 25, 1992).

91. The Lead-Based Paint Poisoning Prevention Act, 42 U.S.C. §§ 4801–4846, authorizes regulations prohibiting the use of lead-based paint in cooking, eating, and drinking utensils and in toys, furniture, and house paint. It further authorizes the U.S. Department of Housing and Urban Development to establish procedures to eliminate lead-based paint hazards. The Lead Contamination Control Act of 1988, 42 U.S.C. § 300j-21, is aimed at reducing the hazard of lead in drinking water at schools and day-care centers.

92. E.g., Steven Lee Myers, *Sandblasting Halted Again on Bridge,* N.Y. Times, Oct. 27, 1992, at B1 (Williamsburg Bridge in New York City); Marcia Willhite, *Sandblasting of Lead Paint Contaminates Residential Area,* NATICH (National Air Toxics Information Clearinghouse) Newsletter, Sept. 1991, at 5 (water tower in Cedar Park, Texas).

93. 40 C.F.R. § 261.24.

94. *See* Suzette Brooks, *Legal Considerations of Disposal of Lead-Contaminated Construction Debris,* N.Y. L.J., July 9, 1993, at 1.

95. *But see* Metal Trades, Inc. v. United States, 810 F. Supp. 689 (D.S.C. 1992).

96. *See* United States v. Nicolet, Inc., 712 F. Supp. 1205 (E.D. Pa. 1989).

97. 15 U.S.C. § 2605(e).

98. 40 C.F.R. pt. 761.

99. *Cf. State, Los Angeles Officials Grapple With Cleanup of Ruins Caused by Rioting,* 23 Env't Rep. (BNA) No. 2, at 284 (May 8, 1992) (concern over

presence of asbestos and other hazardous materials in many of the buildings gutted by fire during the disturbances that followed the acquittal of police officers accused of beating Rodney King).

100. *Nonhazardous Industrial Waste: A Question of 'Orphans'*, Smith Barney Pollution Control Monthly, June 1990, at 1. *See also* Richard M. Schlauder and Robert H. Brickner, *Setting Up For Recovery of Construction and Demolition Waste*, Solid Waste and Power, Jan./Feb. 1993, at 28.

101. *See also Bridgeport's Plan to Burn Mountain of Trash Draws Threat of Lawsuit*, N.Y. Times, Nov. 12, 1992, at B7 (controversy over plan to incinerate contents of illegal construction and demolition debris landfill, including lead, creosote, and asbestos).

102. 40 C.F.R. § 261.4.

103. *Incineration of Hazardous Waste at Sea: Hearings Before the House Comm. on Merchant Marine and Fisheries*, 101st Cong., 1st Sess. 213 (1987) [hereinafter *Incineration of Waste at Sea*].

104. *Nonhazardous Industrial Waste: A Question of 'Orphans,'* Smith Barney Pollution Control Monthly, June 1990, at 1.

105. John C. Dernbach, *The Other Ninety-Six Percent*, Envtl. F., Jan./Feb. 1993, at 10.

106. *Incineration of Hazardous Waste at Sea, supra* ch. 2 note 103, at 213.

107. Beneficiation is the process during which the ore is crushed and ground and minerals are recovered by physical or chemical techniques. T.S. Ary, *The Importance of Waste Management Regulations to the Minerals Industry, in Proceedings of the First International Conference on Environmental Issues and Waste Management in Energy and Materials Production* 6 (T.M. Yegulalp and Kunsoo Kim eds., 1992).

108. *Incineration of Hazardous Waste at Sea, supra* ch. 2 note 103, at 213.

109. *Coming Clean: Superfund Problems Can Be Solved . . . , supra* ch. 2 note 51, at 100, 194. *Similarly, see* Mineral Policy Center, *The Burden of Gilt* (July 20, 1993) (estimates that there are 557,650 hardrock mines in the United States, of which 14,950 contaminate ground or surface waters); Keith Schneider, *New Approach to Old Peril: Abandoned Mines in West*, N.Y. Times, Apr. 27, 1993, at A1.

110. 42 U.S.C. § 6921(b), enacted as part of the Solid Waste Disposal Amendments of 1990, Pub. L. No. 96-482, 94 Stat. 2334 (1980).

111. E.g., Environmental Defense Fund v. EPA, 852 F.2d 1309 (D.C. Cir. 1988); Environmental Defense Fund v. EPA, 852 F.2d 1316 (D.C. Cir. 1988), *cert. denied*, 489 U.S. 1011 (1989); Solite Corp. v. EPA, 952 F.2d 473 (D.C. Cir. 1991).

112. *See* John R. Jacus and Thomas E. Root, *RCRA Regulation of Mine Waste: An Overview*, 5 Nat. Resources and Env't No. 3, at 26 (1991); Glenn C. Van Bever, *Mining Waste and the Resource Conservation and Recovery Act: An Overview*, 7 J. Min. L. and Pol'y 249 (1991–92); Constantine Sidamon-Eristoff,

Mining in the Environmentally Conscious '90s, Yegulalp and Kim, *supra* ch. 2 note 107, at 19.

113. Steve Hwang, *An Assessment of Health and Environmental Impact of Contaminant Releases from a Mine Tailings Pile,* Yegulalp and Kim, *supra* ch. 2 note 107, at 174.

114. *Incineration of Hazardous Waste at Sea, supra* ch. 2 note 103, at 213.

115. 42 U.S.C. § 6921(b)(2).

116. *Deep Pockets: Taxpayer Liability for Environmental Contamination, supra* ch. 2 note 71, at 13.

117. 42 U.S.C. § 6901(b)(3).

118. Of this, 41.9% was land applied (used, after processing, as a soil conditioner or fertilizer, or as a fill material in land reclamation projects); 22.2% was landfilled; 13.6% was incinerated; 5.8% was processed, distributed, and marketed for compost and other uses; 9.2% was spread on vacant land; 4.8% was dumped in the ocean; and 2.5% was disposed by other means. George A. Ravenscroft, *Managing a Special Waste: Sewage Sludge,* Solid Waste and Power, Dec. 1992, at 50.

119. *Final EPA Sludge Management Standards Promote Beneficial Use, Official Says,* 23 Env't Rep. (BNA) No. 32, at 1932 (Dec. 4, 1992). *See also* 33 U.S.C. § 1345(d)(1).

120. 33 U.S.C. § 1414b; *see* United States v. City of New York, 972 F.2d 464 (2d Cir. 1992); City of New York v. EPA, 543 F. Supp. 1084 (S.D.N.Y. 1981).

121. Michael Specter, *Ocean Dumping is Ending, But Not Problems,* N.Y. Times, June 29, 1992, at B1.

122. Bureau of Impact Assessment and Meteorology, N.Y. St. Dep't of Envt'l Conservation, *Incineration 2000: A Joint Study of Impacts of a Sludge Disposal Alternative to Ocean Dumping in the New York–New Jersey Metropolitan Region* 35–36 (Aug. 1992).

123. Stephen Breyer, *Breaking the Vicious Circle: Toward Effective Risk Regulation* 22 (1993).

124. *See* Jonathan Rabinovitz, *Nassau's New Sludge Plan Is To Dump It Elsewhere,* N.Y. Times, Nov. 24, 1992, at B5; Jonathan Rabinovitz, *Nassau Given Time to Explore Sites for Sludge,* N.Y. Times, Dec. 5, 1992, at 25 (One week before it was to award a $200-million contract to build two plants to process sewage sludge, Nassau County, New York, decided instead to ship the sludge elsewhere.); Jonathan Rabinovitz, *U.S. Judge Lets Nassau Sludge Be Shipped Out, Not Shaped,* N.Y. Times, Mar. 20, 1993, at 26.

125. Reginald Patrick, *Island Won't Be Getting Sludge Plant,* Staten Island Advance, Dec. 16, 1993, at A1.

126. *See* Frank Edward Allen, *Western Farmers Love New York Sludge,* Wall St. J., Nov. 24, 1992, at B1; *Illinois: Christian County May Receive N.Y. City Sludge,* Solid Waste Dig. (Midwest ed.), Aug. 1992, at 6; Michael Specter,

Ultimate Alchemy: Sludge to Gold—Big New York Export May Make Desert, and Budget, Bloom, N.Y. Times, Jan. 25, 1993, at B1.

127. *See* Stephen Lester, *Sewage Sludge . . . A Dangerous Fertilizer,* Everyone's Backyard, Oct. 1992, at 8; *Nassau Sewerage Sludge Not Wanted in West Virginia,* Solid Waste Digest (Northeast ed.), May 1993, at 7.

128. *See Immediate Stop to Spreading New York Sludge, Permit Revocation Advised by Attorney General,* 23 Env't Rep. (BNA) No. 25, at 1601 (Oct. 16, 1992); *Out-of-State Sludge Disposal May be Illegal, Texas Attorney General Tells Water Commission,* 23 Env't Rep. (BNA) No. 21, at 1434 (Sept. 18, 1992); *Texas Sues EPA for Failing to Study Risks of Spreading New York Sewage Sludge on Ranch,* 23 Env't Rep. (BNA) No. 11, at 770 (July 10, 1992).

129. Louis Blumberg and Robert Gottlieb, *War on Waste: Can America Win Its Battle With Garbage?* 110 (1989).

130. Office of Tech. Assessment, U.S. Cong., *Facing America's Trash: What Next for Municipal Solid Waste?* 247 (Oct. 1989) [hereinafter *Facing America's Trash*].

131. Id. at 250.

132. 42 U.S.C. § 6921(i). *Compare* Environmental Defense Fund, Inc. v. Wheelabrator Technologies, Inc., 725 F. Supp. 758 (S.D.N.Y. 1989), *aff'd,* 931 F.2d 211 (2d Cir.) (holding exemption does apply), *cert. denied,* 112 S. Ct. 453 (1991), *with* Environmental Defense Fund v. City of Chicago, 948 F.2d 345 (7th Cir. 1991) (holding exemption does not apply), *vacated,* 113 S.Ct. 486, 121 L.Ed.2d 426 (1992) (remanded for reconsideration in view of EPA policy statement that MSW ash should not be treated as hazardous waste), *on remand,* 985 F.2d 303 (7th Cir. 1993) (adhering to prior decision), *aff'd,* 128 L.Ed.2nd 302 (1994). *See also* Benjamin Hershkowitz, *Analysis of the Household Waste Exclusion for Municipal Solid Waste Incinerator Ash—42 U.S.C. § 6921(i),* 2 N.Y.U. Envt'l L.J. 84 (1993); David C. Wartinbee, *Comment: Incinerator Ash May Not Be a Hazardous Waste, But the Story Doesn't End There!* 9 T.M. Cooley L. Rev. 115 (1992).

133. Chicago v. Environmental Defense Fund, 128 L.Ed.2nd 302 (1994).

134. International Trade Admin., U.S. Dep't of Commerce, *Overview on the Use and Storage of Coal Combustion Ash in the United States* 1 (Nov. 1988).

135. U.S. E.P.A., Pub. No. EPA/530-SW-99-002, *Wastes From the Combustion of Coal by Electric Utility Power Plants: Report to Congress* (Feb. 1988).

136. *Regulation of Fossil-Fuel Combustion Wastes As Hazardous Under RCRA Inappropriate, EPA Says,* 24 Env't Rep. (BNA) No. 14, at 588 (Aug. 6, 1993).

137. For a listing of some of the items that fit within this term, *see* 42 U.S.C. § 6992a(a).

138. Agency for Toxic Substances and Disease Registry, U.S. Public Health Serv., *The Public Health Implications of Medical Waste: A Report to Congress* 3.6 (1990).

139. Id. at E.9; Sue Darcy, *State Laws Boost Medical Waste Handling,* World Wastes, Nov. 1992, at 44.

140. Milt Freudenheim, *A Re-usable Solution to Health Costs,* N.Y. Times, Nov. 30, 1993, at D1.

141. 42 U.S.C. §§ 6992 *et seq.*

142. Ortwin Renn and Vincent Covello, *Medical Waste: Risk Perception and Communication,* in *Perspectives on Medical Waste* (Nelson A. Rockefeller Institute of Government, State University of New York, June 1989) at VII.1.

143. E.g., Frances Frank Marcus, *Medical Waste Divides Mississippi Cities,* N.Y. Times, June 24, 1992, at A16; *Oklahoma Court Revokes Incinerator Permit,* 22 Env't Rep. (BNA) No. 10, at 571 (July 5, 1991); Ian Fisher, *Builders and Foes Using Bronx Incinerator as Test,* N.Y. Times, Sept. 8, 1992, at B3; BFI Medical Waste Systems v. Whatcam County, 983 F.2d 911 (9th Cir. 1993).

144. E. Malone Steverson, *Provoking a Firestorm: Waste Incineration,* 25 Envtl. Sci. Tech. 1808, 1810–11 (1991).

145. Id. at 1808. The overcapacity of medical waste incinerators in some markets is also discussed in Jeff Bailey, *How Two Garbage Giants Fought Over Medical Waste,* Wall St. J., Nov. 17, 1992, at B6.

146. Shawna Vogel, *Biotech Wastes,* Tech. Rev., Feb./Mar. 1988, at 13.

147. Solid Waste and Emergency Response Office, U.S. E.P.A., Pub. No. EPA/530-SW-90-042A, *Characterization of Municipal Solid Waste in the United States: 1990 Update, Executive Summary* ES-3 (June 1990).

148. Id. at ES-6.

149. National Solid Wastes Management Ass'n, *Landfill Capacity in North America—1991 Update (1992).*

150. *Facing America's Trash, supra* ch. 2 note 130, at 273; *similarly, Review of EPA's Capacity Assurance Program, supra* ch. 2 note 11 (statement of Sylvia Lowrance, director, Office of Solid Waste, EPA), at 195.

151. The most important of these is 40 C.F.R. Parts 257, 258, promulgated at 56 Fed.Reg. 50,978 (1991). *See* Kathleen Farrelly, *The New Federal Standards for Municipal Solid Waste Landfills: Adding Fuel to the Regulatory Fire,* 3 Villanova Envt'l L.J. 383 (1992).

152. *Cost of Compliance With RCRA Subtitle D Will Lead to "Mega-Landfills,"* *Consultant Says,* 23 Env't Rep. (BNA) No. 16, at 1204 (Aug. 14, 1992).

153. *Subtitle D Enforcement Or The Emperor's New Clothes,* Solid Waste Digest (Northeast ed.), Nov. 1993, at 12.

154. U.S. Gen. Accounting Office, Pub. No. GAO/RCED-89-165BR, *Nonhazardous Waste: State Management of Municipal Landfills and Landfill Expansions* 2 (June 1989).

155. B.F. Goodrich Co. v. Murtha, 958 F.2d 1192 (2d Cir. 1992).

156. EPA, *Solid Waste Disposal Facility Criteria*, 56 Fed.Reg. 50,978, 50,982 (Oct. 9, 1991); Kirsten Engel, *Environmental Standards as Regulatory Common Law: Toward Consistency in Solid Waste Regulation*, 21 N.M. L. Rev. 13 (1990); G. Fred Lee and R. Anne Jones, *Groundwater Pollution by Municipal Landfills: Leachate Composition, Detection and its Water Quality Significance*, Presentation to National Water Well Ass'n, Fifth National Outdoor Action Conference, Las Vegas, Nev. (May 1991).

157. *Landfill Capacity in North America—1991 Update, supra* ch. 2 note 149, at 4.

158. Id. at 5; Edward W. Repa and Susan K. Sheets, *Landfill Capacity in North America*, Waste Age, May 1992, at 18; Jeff Bailey, *Space Available: Economics of Trash Shift as Cities Learn Dumps Aren't So Full*, Wall St. J., June 2, 1992, at 1.

159. Barnaby J. Feder, *The Saga of Lonetree Landfill*, N.Y. Times, Dec. 22, 1992, at D1.

160. *Review of EPA's Capacity Assurance Program, supra* ch. 2 note 11 (statement of Sylvia Lowrance, director, Office of Solid Waste, EPA), at 195.

161. *See, e.g.*, Richard W. Tome, *Regional Note—The Siting of a Trash-to-Energy Plant: A Tale of Three Connecticut Towns*, 9 Temp. Envtl. L. and Tech. J. 181 (1990); Steven A. Broiles, *A Suggested Approach to Overcome California's Inability to Permit Urban Resource Recovery Facilities*, 8 Risk Analysis 357 (1988); Newsday, *Rush to Burn: Solving America's Garbage Crisis?* (1989); Michael Specter, *Incinerators: Unwanted and Politically Dangerous*, N.Y. Times, Dec. 12, 1991, at B1.

162. *Facing America's Trash: What Next for Municipal Solid Waste? supra* ch. 2 note 130, at 343.

163. Randy Woods, *Fighting NIMBY With Fire*, Waste Age, Sept. 1992, at 19. Four such plants opened in 1992. Tom Arrandale, *Waste-to-Energy: Promises and Problems*, Governing, Feb. 1993, at 51.

164. Jonathan V.L. Kiser, *Municipal Waste Combustion in North America: 1992 Update*, Waste Age, Nov. 1992, at 26, 28. Projections of future growth are found in Curt Holman, *Gains Expected in Incineration and Gas Recovery*, World Wastes, Aug. 1993, at 8.

165. Steverson, *supra* ch. 2 note 144.

166. Jonathan Rabinovitz, *Costs of Long Island Incinerators Rise With Trash Shortage*, N.Y. Times, Oct. 10, 1992, at 26. *See also* Jeff Bailey, *Up in Smoke: Fading Garbage Crisis Leaves Incinerators Competing For Trash*, Wall St. J., Aug. 11, 1993, at A1; Jeff Bailey, *Waste Project Seeks Court Protection In Dispute With WMX Technologies*, Wall St. J., Sept. 20, 1993, at A9C; Frank Edward Allen, *Some Incinerators Have Capacity to Burn*, Wall St. J., Oct. 2, 1991, at B1.

167. C and A Carbone, Inc. v. Town of Clarkstown, 62 U.S.L.W. 4315 (May 16, 1994); *Flow Control: When Keeping Waste In Is As Important as Keeping*

It Out, Solid Waste Dig. (Midwest ed.), Aug. 1992, at 10; Sue Darcy, *Flow Control: A Tug of Waste,* World Wastes, July 1992, at 58; Eric Peterson, *Whose Waste Is It, Anyway?* World Wastes, Apr. 1993, at 48; Sarah Lyall, *Suddenly, Towns Fight to Keep Their Garbage,* N.Y. Times, Jan. 5, 1992, at E14.

168. Sue Darcey, *Environmental Worries Spark State Burn Bans,* World Wastes, June 1993, at 12.

169. Eileen Berenyi and Robert N. Gould, . . . *Despite Temporary Development Slowdown,* World Wastes, June 1991, at 31, 32.

170. *See* Anne G. Seel, *Regulation of Contaminated Dredged Material From New York Harbor,* 3 Envt'l L. N.Y. 113 (1992).

171. *See* Washington County Cease, Inc. v. Persico, 99 A.D.2d 321, 473 N.Y.S.2d 610 (3d Dep't 1984), *aff'd,* 64 N.Y.2d 923, 477 N.E.2d 1084, 488 N.Y.S.2d 630 (1985); Elsa Brenner, *New Campaign Against PCBs in the Hudson,* N.Y. Times, Sept. 27, 1992, § 13WC, at 1.

172. Pub. L. No. 98-616, 98 Stat. 3221 (1984); 42 U.S.C. § 6991–6991i.

173. *Cleaning Up The Nation's Waste Sites, supra* ch. 2 note 55, at 5. The cost of this cleanup is estimated at $30 billion. *Id.* at 6. *See also Cost of Cleaning Up Leaking UST Sites Could Exceed $41 Billion, Report Says,* 23 Env't Rep. (BNA) No. 30, at 2091 (Dec. 25, 1992).

174. *See* Green, *supra* ch. 2 note 24, at 69–70; Christopher Harris and Gary Fremerman, *Used Oil Management,* and Dale E. Hermeling and Joseph B. Pereles, *Storage Tanks, in* Gerrard 1992, *supra* ch. 2 note 7.

175. *Cost of Cleaning Up Leaking UST Sites Could Exceed $41 Billion, Report Says,* 23 Env't Rep. (BNA) No. 30, at 2091 (Dec. 25, 1992). Another estimate is $67 billion. *See* Milton Russell et al., *Hazardous Waste Remediation: The Task Ahead* (Waste Management Research and Education Institute, Univ. of Tennessee, 1991).

Chapter 3

1. Nicholas Lenssen and Christopher Flavin, *Closing Out Nuclear Power,* World Watch, Sept.–Oct. 1992, at 35.

2. Id.; Komanoff Energy Associates, *Fiscal Fission: The Economic Failure of Nuclear Power* (Dec. 1992); Howard W. French, *Cuba Cancels Atom Plant, Blaming Costs and Russians,* N.Y. Times, Sept. 7, 1992, at 5.

3. Michael Renner, *Finishing the Job,* World Watch, Nov.–Dec. 1992, at 10.

4. Calculated from Robert L. Goble, *Time Scales and the Problem of Radioactive Waste, in Equity Issues in Radioactive Waste Management* 139, 164 (Roger E. Kasperson ed., 1983).

5. Sidamon-Eristoff, *supra* ch. 2 note 112, at 19, 67.

6. Keith Schneider, *The Soviets Show Scars From Nuclear Arms Production,* N.Y. Times, July 16, 1989, § 2, at 2.

7. Nella Banerjee, *Explosion Causes Radiation Leak at Siberian Nuclear Fuel Facility,* Wall St. J., Apr. 8, 1993, at A10; James Ridgeway, *Russian Roulette: The Plutonium Trade is Booming,* Village Voice, Apr. 20, 1993, at 18.

8. U.S. Congress, Office of Technology Assessment, Pub. No. OTA-0-484, *Complex Cleanup: The Environmental Legacy of Nuclear Weapons Production* 43–45 (1991) [hereinafter *Complex Cleanup*]; Y.S. Tang and James H. Saling, *Radioactive Waste Management* 91 (1990); U.S. Senate, Committee on Governmental Affairs, Majority Staff, *On the Risk of Explosions at DOE Radioactive Waste Facilities* (Dec. 9, 1993); *Mishap Casts Pall Over Big Atomic Waste Site,* N.Y. Times, Aug. 15, 1993, at 18; Matthew L. Wald, *At an Old Atomic-Waste Site, The Only Sure Thing is Peril,* N.Y. Times, June 21, 1993, at A1; Matthew L. Wald, *Hazards at Nuclear Plant Fester 8 Years After Warnings,* N.Y. Times, Dec. 24, 1992, at A11; Matthew L. Wald, *Deadly Nuclear Waste Seems to Have Leaked in Washington State,* N.Y. Times, Feb. 28, 1993, at 31.

9. Marvin Resnikoff, *Living Without Landfills: Confronting the "Low-Level" Radioactive Waste Crisis* 10 (1987).

10. Donald L. Bartlett and James B. Steele, *Forevermore: Nuclear Waste in America* 74 (1985).

11. Id. at 75–87. For a much more sanguine view of the hazards at West Valley, *see* John M. Matuszek, *Safer Than Sleeping With Your Spouse—The West Valley Experience, in Low-Level Radioactive Waste Regulation: Science, Politics and Fear* (Michael E. Burns ed., 1988), at 261.

12. Bartlett and Steele, *supra* ch. 3 note 10, at 90–99.

13. Tang and Saling, *supra* ch. 3 note 8, at 103. Great Britain and France still do commercial reprocessing. Glen Zorpette and Gary Stix, *Nuclear Waste: The Challenge is Global,* IEEE Spectrum, July 1990, at 18. The Soviet Union formerly took spent fuel from nuclear power plants in Czechoslovakia for reprocessing; that stopped with the collapse of the Soviet Union, and the new Czech and Slovak republics have been forced to store the spent fuel from their nuclear plants until a solution is devised. Malcolm W. Browne, *Post-Czechoslovak Problem: Spent Atom Fuel,* N.Y. Times, Nov. 22, 1992, at 11.

14. Arjun Makhijani and Scott Saleska, *High-Level Dollars, Low-Level Sense: A Critique of Present Policy for the Management of Long-Lived Radioactive Wastes and Discussion of an Alternative Approach* 10, 14 (1992).

15. Bartlett and Steele, *supra* ch. 3 note 10, at 124.

16. U.S. Gen. Accounting Office, Pub. No. GAO/RCED-91-194, *Nuclear Waste: Operation of Monitored Retrievable Storage Facility is Unlikely By 1998* 4 (Sept. 1991).

17. Board on Radioactive Waste Management, National Academy of Science, *Rethinking High Level Radioactive Waste Disposal* 7 (July 1990).

18. U.S. Dep't of Energy, Energy Information Administration, Pub. No. SR/CNEAF/93-01, *Spent Fuel Nuclear Fuel Discharges From U.S. Reactors 1991* (Feb. 1993) at 7.

19. 42 U.S.C. §§ 10131–10145. *See* James H. Davenport, *The Law of High-Level Nuclear Waste,* 53 Tenn. L. Rev. 481 (1986).

20. 42 U.S.C. § 10132(b)(1)(B).

21. Natural Resources Defense Council v. EPA, 824 F.2d 1258, 1262 (1st Cir. 1987).

22. Nevada v. Herrington, 827 F.2d 1394, 1397 (9th Cir. 1987); Robert D. Hershey, Jr., *U.S. Suspends Plan for Nuclear Dump in East or Midwest,* N.Y. Times, May 29, 1986, at A1.

23. Omnibus Budget Reconciliation Act of 1987, Pub. L. 100–203, 101 Stat. 1330, § 5011; 42 U.S.C. § 10172(a).

24. Fred C. Shapiro, *Yucca Mountain,* New Yorker, May 23, 1988, at 61.

25. E.g., Nevada v. Watkins, 939 F.2d 710 (9th Cir. 1991); Nevada v. Burford, 918 F.2d 854 (9th Cir.), *cert. denied,* 500 U.S. 932 (1990); Nevada v. Watkins, 914 F.2d 1545 (9th Cir. 1990), *cert. denied,* 499 U.S. 906 (1991); Nevada v. Herrington, 827 F.2d 1394 (9th Cir. 1987).

26. Gerald Jacob, *Site Unseen: The Politics of Siting a Nuclear Waste Repository* 172–74 (1990); William J. Broad, *A Mountain of Trouble,* N.Y. Times, Sept. 18, 1990, § 6 (Magazine), at 36. A technical justification for the selection of Yucca Mountain is provided in Katrin Borcherding et al., *Comparison of Weighting Judgments in Multiattribute Utility Measurement,* 37 Mgmt. Sci. 1603 (1991).

27. U.S. Gen. Accounting Office, Pub. No. GAO/RCED-92-73, *Nuclear Waste: DOE's Repository Site Investigations, a Long and Difficult Task* 4 (May 1992). *See also* Energy Policy Act of 1992, Pub. L. 102-486, § 801 (requires EPA to repromulgate 40 C.F.R. pt. 191, concerning licensing standards for Yucca Mountain facility); Matthew L. Wald, *Rules Rewritten on Nuclear Waste,* N.Y. Times, Oct. 11, 1992, at 31, and *Congress Approves Energy Bill With Yucca Mountain HLW Standard,* Radioactive Exchange, Oct. 21, 1992, at 15 (describing origins of this legislation in challenges brought by Nevada).

28. Matthew L. Wald, *U.S. Will Start Over on Planning for Nevada Nuclear Waste Dump,* N.Y. Times, Nov. 29, 1989, at A1.

29. U.S. General Accounting Office, Pub. No. GAO/RCED-93-124, *Nuclear Waste: Yucca Mountain Project Behind Schedule and Facing Major Scientific Uncertainties* 4, 45 (May 1993).

30. Richard A. Kerr, *Nuclear Waste: Another Panel Rejects Nevada Disaster Theory,* 256 Science 434 (Apr. 24, 1992).

31. Jon Jefferson, *Yucca Mountain: How Long Can It Last? How Long Can It Take?* 8 Forum for Applied Research and Public Policy 110 (1993).

32. Jacob, *supra* ch. 3 note 26, at 169.

33. Nicholas Lenssen, *Confronting Nuclear Waste, in State of the World 1992* 46 (Lester R. Brown ed., 1992).

34. *See* Allen v. United States, 588 F. Supp. 247 (D. Utah 1984).

35. 42 U.S.C. §§ 10161–69.

36. Nicholas K. Brown, *Monitored Retrievable Storage Within the Context of the Nuclear Waste Policy Act of 1982,* 52 Tenn. L. Rev. 739 (1985); Tennessee v. Herrington, 806 F.2d 642 (6th Cir. 1986), *cert. denied,* 480 U.S. 946 (1987).

37. U.S. Congress, Office of Technology Assessment, Pub. No. OTA-E-575, *Aging Nuclear Power Plants: Managing Plant Life and Decommissioning* 97 (1993) [hereinafter *Aging Nuclear Power Plants*].

38. Richard P. Mauro, *Tennessee v. Herrington: An End Run Around State Participation in Nuclear Waste Siting Decisions,* 9 J. Energy L. and Pol'y 113 (1988).

39. 42 U.S.C. § 10168(d)(1).

40. Glen Zerpette and Gary Stix, *Nuclear Waste: The Challenge is Global,* IEEE Spectrum, July 1990, at 18, 20.

41. Matthew L. Wald, *A New U.S. Stance on Atomic Wastes,* N.Y. Times, Dec. 3, 1993, at A27.

42. Matthew L. Wald, *Nuclear Industry Seeks Interim Site to Receive Waste,* N.Y. Times, Aug. 27, 1993, at 1; *Mescalero Apache Tribe Renews Efforts to Site MRS on New Mexico Reservation,* 24 Env't Rep. (BNA) No. 6 at 274 (June 11, 1993). *See also* James P. Miller, *Northern States Utility Nears Pact With Apache Tribe,* Wall St. J., Feb. 4, 1994 at B2 (tentative agreement between Minnesota utility and Mescalero Apache tribe for storage of utility's spent fuel at Mescalero reservation).

43. 42 U.S.C. § 2014(ee).

44. Alan Burdick, *The Last Cold-War Monument,* Harper's Mag., Aug. 1992, at 62.

45. John E. Seley, *The Politics of Public-Facility Planning* 94 (1983).

46. U.S. Dep't of Energy National Security and Military Application of Nuclear Energy Act of 1980, Pub. L. 96-164, 93 Stat. 1265, § 213(a), codified at 42 U.S.C. § 7271.

47. Keith Schneider, *Wasting Away,* N.Y. Times Mag., Aug. 30, 1992, at 42, 58.

48. New Mexico v. Watkins, 969 F.2d 1122 (D.C. Cir. 1992).

49. Charles C. Reith and N. Timothy Fischer, *Transuranic Waste Disposal: The WIPP Project, in Deserts as Dumps? The Disposal of Hazardous Materials in Arid Ecosystems* 303, 316 (Charles C. Reith and Bruce M. Thompson eds., 1992); *see also* Keith Schneider, *Idaho Says No,* N.Y. Times, Mar. 11, 1990, § 6 (Magazine), at 50.

50. Pub. L. 102-579; *see also Congress Approves WIPP Bill With EPA Standard Requirements,* Radioactive Exchange, Oct. 21, 1992, at 17.

51. John H. Cushman, Jr., *U.S. Drops Test Plan at Bomb Waste Site,* N.Y. Times, Oct. 22, 1993, at A16. For other discussions of WIPP, *see* Luther J. Carter, *Nuclear Imperatives and Public Trust: Dealing with Radioactive Waste* 176-93 (1987); Allen V. Kneese et al., *Economic Issues in the Legacy Problem, in* Kasperson ed., *supra* ch. 3 note 4, at 203 (cost/benefit analysis of WIPP); Nicholas Lenssen, *WIPP-Lashed: Nuclear Burial Plan Assailed,* World Watch, Nov.–Dec. 1991, at 5.

52. *Safety, Modernization, and Environmental Cleanup of the U.S. Nuclear Weapons Complex: Hearings Before the Senate Comm. on Armed Services,* 101st Cong., 2d Sess. 339 (1989) [hereinafter *Safety, Modernization, and Environmental Clean-up of the U.S. Nuclear Weapons Complex*] (statement of Keith O. Fulz, General Accounting Office).

53. Office of Tech. Assessment, U.S. Cong., *Partnerships Under Pressure: Managing Commercial Low-Level Radioactive Waste—Summary* 5 (1989) [hereinafter *Partnerships Under Pressure*].

54. Resnikoff, *supra* ch. 3 note 9, at 4, 11; Tang and Saling, *supra* ch. 3 note 8, at 197.

55. *See Medical Waste Needs Should Not Be Scare Tactic to Open Low-Level Disposal Facilities, Group Says,* 23 Env't Rep. (BNA) No. 40, at 2584 (Jan. 29, 1993).

56. *Partnerships Under Pressure: Managing Commercial Low-Level Radioactive Waste—Summary, supra* ch. 3 note 53, at 1, 7. More detailed information is contained in U.S. Dep't of Energy, Office of Environmental Restoration and Waste Management, Pub. No. DOE/EM-0091P, *1991 Annual Report on Low-Level Radioactive Waste Management Progress* (Nov. 1992).

57. In 1980, 251,116 million kwh of electricity were generated by nuclear power; in 1989, the figure was 529,355; and in 1991, 612,565. *Nuclear Power Plant Operations,* Monthly Energy Rev., Jan. 1993, at 101.

58. U.S. Dep't of Energy, Office of Civilian Radioactive Waste Management, Pub. No. DOE/RW-0006, Rev. 8, *Integrated Data Base for 1992: U.S. Spent Fuel and Radioactive Waste Inventories, Projections, and Characteristics* (Oct. 1992) at 119.

59. *Complex Cleanup, supra* ch. 3 note 8, at 24, 45.

60. Colglazier and English, *supra* ch. 1 note 8, at 627.

61. The account in this paragraph is drawn from id., 622–24; Timothy J. Peckinpaugh, *The Politics of Low-Level Radioactive Waste Disposal, in* Burns ed., *supra* ch. 3 note 12, at 45; and New York v. United States, 112 S.Ct. 2408, 2415–16 (1992).

62. Pub. L. 96-573, 94 Stat. 3347, codified at 42 U.S.C. § 2021b–2021d.

63. Pub. L. 99-240, 99 Stat. 1843.

64. 42 U.S.C. § 2021e(d)(2)(C).

65. New York v. United States, 112 S.Ct. 2408 (1992).

66. *See* Paul Furiga, *Hot Stuff,* Governing, Nov. 1989, at 50.

67. English 1992, *supra* ch. 1 note 8, at 13–15; U.S. Gen. Accounting Office, Pub. No. GAO/RCED-92-61, *Nuclear Waste: Slow Progress Developing Low-Level Radioactive Waste Disposal Facilities* 4 (Jan. 1992).

68. *Martinsville Rejected for Central Midwest LLRW Facility,* Radioactive Exchange, Oct. 21, 1992, at 1.

69. Desert Tortoise v. Interior Department, No. C93-0114 (N.D. Calif. Jan. 19, 1993); *TRO Against Ward Valley Land Transfer Extended,* 23 Env't Rep. (BNA) No. 40, at 2594 (Jan. 29, 1993); *Burying the Future,* Wall St. J., Jan. 22, 1993, at A10. *See also* Jon Cohen, *Radioactive Waste: California's Disposal Plan Goes Nowhere Fast,* 263 Sci. 912 (Feb. 18, 1994); California Radioactive Materials Management Forum v. Dep't of Health Services, 15 Cal. App. 4th 841, Cal. Rptr.2d 357 (Cal. App. 3d Dist., 1993) (holding that no adjudicatory hearing is necessary on U.S. Ecology's permit application for Ward Valley facility); *but see Radioactive Waste: California Approves License Application; Governor Agrees on Hearing For Disposal Site,* 24 Env't Rep. (BNA) No. 21 at 916 (Sept. 24, 1993).

70. Letter, Randolph Wood, Nebraska Department of Environmental Quality and Mark B. Horton, Nebraska Department of Health, to Richard F. Paton, U.S. Ecology, Inc., Jan. 22, 1993. *See also* Concerned Citizens of Nebraska v. Nuclear Regulatory Comm'n, 970 F.2d 421 (8th Cir. 1992) (affirming dismissal of challenge to development of regional LLRW facility on ground that siting violated unenumerated Ninth Amendment rights to freedom from environmental releases of nonnatural radiation).

71. County of El Paso v. Texas Low-Level Radioactive Waste Disposal Authority, No. 2588-34 (Dist. Ct. Hudspeth Co., Texas, Apr. 25, 1991). The efforts to site a LLRW facility in Hudspeth County are described in Rita R. Hamm, *Coalition Formation in Radioactive Waste Disposal Facility Siting, in* Inst. for Soc. Impact Assessment, *supra* ch. 2 note 25, at 176.

72. *Texas Agrees to Host Disposal Site in Low-Level Compact With Vermont, Maine,* 24 Env't Rep. (BNA) No. 7 at 323 (June 18, 1993).

73. *Radioactive Waste: North Carolina Authority Selects Site for Southeast's Next Low-Level Facility,* 24 Env't Rep. (BNA) No. 32 at 1497 (Dec. 10, 1993); *Radioactive Waste: State Pledges Thorough Investigation of Proposed Regional Low-Level Facility,* 24 Env't Rep. (BNA) No. 33 at 1524 (Dec. 17, 1993).

74. *Efforts to Reopen Beatty Dead After Legislator Conflict Aired,* Radioactive Exchange, May 19, 1993, at 7.

75. Richard R. Zuercher, *The Operator of the Licensed Low-Level Radioactive Waste (LLW) Disposal,* Inside N.R.C., Nov. 30, 1992, at 10; Richard R. Zuercher, *US Ecology Sues for Level Playing Field to Stop Diversion of LLW to Utah,* Inside N.R.C., Apr. 20, 1992, at 5; *Kerr-McGee Sets Funding for W. Chicago Cleanup,* Nuclear News, March 1993, at 72.

76. Robert Reinhold, *States, Failing to Cooperate, Face a Nuclear-Waste Crisis,* N.Y. Times, Dec. 28, 1992, at 1; Sarah Lyall, *Failing to Build a Dump, New York Faces Shutout,* N.Y. Times, May 21, 1993, at A1.

77. *Research, Medical Institutions Ask NRC to Amend Rules on Radionuclide Disposal,* 24 Env't Rep. (BNA) No. 13 at 555 (June 30, 1993).

78. New York State Energy Research and Development Authority, *Low-Level Radioactive Waste Storage Study, Vol. II: Centralized Storage Facility* (Oct. 1993) at 5-5.

79. *Complex Cleanup, supra* ch. 3 note 8, at 15; Thomas B. Cochran et al., *The U.S. Nuclear Warhead Production Complex, in* Ehrlich and Birks eds., *supra* ch. 2 note 75, at 3.

80. Id. at 5. All the facilities are listed, and their status given, in *Oversight of Cleanup and Modernization Proposals for DOE's Weapons Production Complex: Hearings before the Senate Comm. on Governmental Affairs,* 101st Cong., 1st Sess. (1989) at 390–431.

81. Keith Schneider, *U.S. Drops a Plan to Build a Reactor,* N.Y. Times, Sept. 12, 1992, at 5 (government decides not to build a new nuclear reactor to produce tritium at Savannah River plant).

82. Michael Renner, *Finishing the Job,* World Watch, Nov.–Dec. 1992, at 10.

83. *Complex Cleanup, supra* ch. 3 note 8 at 15.

84. *Id.* at 4, 29; Commission on Physical Sciences, Mathematics, and Resources, National Research Council, *The Nuclear Weapons Complex: Management for Health, Safety, and the Environment* 37–38 (1989).

85. Leo P. Duffy, *Environmental and Waste Management Issues, Causes, Characteristics, and Cures, in* Yegulalp and Kim, *supra* ch. 2 note 107, at 15.

86. U.S. General Accounting Office, Pub. No. GAO/RCED-93–71, *Nuclear Waste: Hanford's Well-Drilling Costs Can Be Reduced* 2 (Mar. 1993). *See generally* Michael D'Antonio, *Atomic Harvest: Hanford and the Lethal Toll of America's Nuclear Arsenal* (1993).

87. U.S. Dep't of Energy, Office of Inspector General, Pub. No. DOE/IG-0308, *Report on Packaging, Transporting, and Burying Low-Level Waste* (May 1992) at 5.

88. U.S. Gen. Accounting Office, Pub. No. GAO-T-RCED-92-82, *Federal Facilities: Issues Involved in Cleaning Up Hazardous Waste* 5 (July 28, 1992).

89. *Oversight of Cleanup and Modernization Proposals for DOE's Weapons Production Complex: Hearings before the Senate Comm. on Governmental Affairs,* 101st Cong., 1st Sess. 10 (1989) [hereinafter *Oversight of Cleanup and Modernization Proposals for DOE's Weapons Production Complex*].

90. Russell, *supra* ch. 2 note 175. *See also Energy Department Lacks Valid Cost Figures for Cleanup of Weapons Complex,* GAO Reports, 24 Env't Rep. (BNA) No. 13 at 553 (July 30, 1993).

91. U.S. Gen. Accounting Office, Pub. No. GAO/RCED-92-183, *Nuclear Waste: Defense Waste Processing Facility—Cost, Schedule, and Technical Issues* (June 1992).

92. *Safety, Modernization, and Environmental Cleanup of the U.S. Nuclear Weapons Complex, supra* ch. 3 note 52, at 331 (statement of Raymond P. Berube, Dep't of Energy); *id.* at 417 (statement of Dan W. Reicher and James D. Werner, Natural Resources Defense Council). The GAO has recommended reconsideration of the plan to build a vitrification plant at Hanford. U.S. General Accounting Office, Pub. No. GAO/RCED-93-99, *Nuclear Waste: Hanford Tank Waste Program Needs Cost, Schedule, and Management Changes* (Mar. 1993). *See also Washington State, U.S. Agencies Agree to Changes in Hanford Site Cleanup Plan,* 23 Env't Rep. (BNA) No. 50 at 3121 (Apr. 9, 1993).

93. Matthew L. Wald, *U.S. Reaches Pact on Plant Cleanup,* N.Y. Times, Oct. 3, 1993, at 32.

94. William J. Broad, *Nuclear Accords Bring New Fears on Arms Disposal,* N.Y. Times, July 6, 1992, at 1. However, some Russian officials have alleged that the Soviet arsenal was 45,000 warheads, not 33,000. William J. Broad, *Russian Says Soviet Atom Arsenal Was Larger Than West Estimated,* N.Y. Times, Sept. 26, 1993, at 1.

95. *Nuclear Notebook: Where the Bombs Are,* Bull. Atomic Scientists, Sept. 1992, at 48; William M. Arkin and Robert S. Morris, *Taking Stock: U.S. Nuclear Deployments at the End of the Cold War* (1992).

96. Michael Renner, *Finishing the Job,* World Watch, Nov.–Dec. 1992, at 10, 13; Frans Berkout et al., *Plutonium: True Separation Anxiety,* Bull. Atomic Sci., Nov. 1992, at 28. A very detailed accounting is presented in David Albright et al., *World Inventory of Plutonium and Highly Enriched Uranium, 1992* (1993).

97. U.S. Congress, Office of Technology Assessment, Pub. No. OTA-0-572, *Dismantling the Bomb and Managing the Nuclear Materials,* Sept. 1993.

98. *Id.* The warheads also contain other valuable materials. *See* John J. Fialka, *Alchemists Today Recycle A-Bombs; They Conjure Gold, and That's Not All,* Wall St. J., July 7, 1993, at A5A.

99. Matthew L. Wald, *Study Faults U.S. Program To Dismantle Atomic Arms,* N.Y. Times, Dec. 1, 1993, at B8; Matthew L. Wald, *Nation Considers Means to Dispose of Its Plutonium,* N.Y. Times, Nov. 15, 1993, at A1.

100. Soviet Nuclear Threat Reduction Act, Pub. L. No. 102-228, Dec. 12, 1991; Former Soviet Union Demilitarization Act of 1992, Title XIV of Pub. L. No. 102-484, Oct. 23, 1992; Title V of the Freedom Support Act, Pub. L. No. 102-511, Oct. 24, 1992. *See* U.S. General Accounting Office, Pub. No. GAO/T-NSIAD-93-5, *Soviet Nuclear Weapons: U.S. Efforts to Help Former Soviet Republics Secure and Destroy Weapons* (Mar. 9, 1993).

101. William J. Broad, *From Soviet Warheads to U.S. Reactor Fuel,* N.Y. Times, Sept. 6, 1992, at E4; William J. Broad, *A Plutonium Pact Will Aid Disposal,* N.Y. Times, Apr. 6, 1993, at A11. *See also* Thomas L. Neff, *A Grand Uranium Bargain,* N.Y. Times, Oct. 24, 1991, at A25.

102. Craig R. Whitney, *Illicit Atom-Material Trade Worries Germans,* N.Y. Times, Oct. 20, 1992, at A8.

103. Eric Hoskins, *Making the Desert Glow,* N.Y. Times, Jan. 21, 1993, at A25. *But see* Russell Seitz, *No Uranium Peril in Iraqi Desert,* N.Y. Times, Feb. 9, 1993, at A20.

104. U.S. Gen. Accounting Office, Pub. No. GAO/RCED-89–119, *Nuclear Regulation: NRC's Decommissioning Procedures and Criteria Need to Be Strengthened* 8 (May 1989).

105. Matthew L. Wald, *Nuclear Power Plants Take Early Retirement,* N.Y. Times, Aug. 16, 1992, at E7; Matthew L. Wald, *Cheap and Abundant Power May Shutter Some Reactors,* N.Y. Times, Apr. 4, 1992, at A1.

106. English 1992, *supra* ch. 1 note 8, at 4.

107. Matthew L. Wald, *As Nuclear Plants Close, Costs Don't Shut Down,* N.Y. Times, Sept. 20, 1992, at E18.

108. Pace University Center for Envtl. Legal Studies, *Environmental Costs of Electricity* 389 (1990); Robert Johnson and Ann de Rouffignac, *Closing Costs: Nuclear Utilities Face Immense Expenses in Dismantling Plants,* Wall St. J., Jan. 25, 1993, at A1.

109. Resnikoff, *supra* ch. 3 note 9, at 14.

110. U.S. General Accounting Office, Pub. No. GAO/RCED-93-149, *Department of Energy: Cleaning Up Inactive Facilities Will Be Difficult* (June 1993) at 8.

111. *Full Cleanup of Nuclear Power Plants Urged By Public Citizen Group in Report,* 23 Env't Rep. (BNA) No. 49, at 3081 (Apr. 2, 1993).

112. U.S. Gen. Accounting Office, Pub. No. GAO/RCED-89-119, *Nuclear Regulation: NRC's Decommissioning Procedures and Criteria Need to Be Strengthened* 4 (May 1989); *The Decommissioning and Decontamination Requirements for Closing Nuclear Facilities: Hearings Before the House Comm. on Gov't Operations,* 101st Cong., 1st Sess. 126–30 (1989) (written submission of Kenneth M. Carr, chairman, NRC).

113. *See* 10 C.F.R. § 20.302.

114. *Nuclear Regulation: NRC's Decommissioning Procedures and Criteria Need to Be Strengthened, supra* ch. 3 note 104, at 4, 44–45.

115. William J. Broad, *Russians Describe Extensive Dumping of Nuclear Waste,* N.Y. Times, Apr. 27, 1993, at A1; Eliot Marshall, *A Scramble for Data on Arctic Radioactive Dumping,* 257 Sci. 608 (July 31, 1992); Walter Sullivan, *Soviet Nuclear Dumps Disclosed,* N.Y. Times, Nov. 24, 1992, at C9; Patrick E. Tyler, *Soviets' Secret Nuclear Dumping Raises Fears for Arctic Waters,* N.Y. Times, May 4, 1992, at A1.

116. W. Jackson Davis and Jon M. Van Dyke, *Dumping of Decommissioned Nuclear Submarines at Sea,* Marine Pol'y, Nov. 1990, at 467.

117. *Deep Pockets: Taxpayer Liability for Environmental Contamination, supra* ch. 2 note 71, at 22.

118. Goble, *supra* ch. 3 note 4, at 139, 157.

119. Makhijani and Saleska, *supra* ch. 3 note 14, at 22.

120. *See* Wm. Paul Robinson, *Uranium Production and Its Effects on Navajo Communities Along the Rio Puerco in Western New Mexico, in Race and the Incidence of Environmental Hazards* (Bunyan Bryant and Paul Mohai eds. 1992) at 153; Keith Schneider, *Valley of Death: Late Rewards for Navajo Miners,* N.Y. Times, May 3, 1993, at A1; Jessica Pearson, *Hazard Visibility and Occupational Health Problem Solving: The Case of the Uranium Industry,* 6 J. of Community Health 136 (1980).

121. Roger E. Kasperson, *Social Issues in Radioactive Waste Management: The National Experience, in* Kasperson ed., *supra* ch. 3 note 4, at 24, 30.

122. E.g., *see* Brafford v. Susquehanna Corp., 586 F. Supp. 14 (D. Colo. 1984).

123. Tang and Saling, *supra* ch. 3 note 8, at 277–78.

124. William E. Blundell, *Nuclear Mess: Uranium Mill Wastes, Piled High in West, Pose Cleanup Issues,* Wall St. J., Feb. 25, 1986, at 1.

125. Pub. L. 95-604, codified at 42 U.S.C. § 7901 et seq. *See also* 40 C.F.R. pt. 190 (EPA radiation standards for uranium mill sites); 40 C.F.R. pt. 440 Subpart C (EPA water pollution standards for uranium, radium, and vanadium ores); 40 C.F.R. pt. 192 (EPA standards for remedial actions at inactive uranium processing sites); 40 C.F.R. pt. 61 Subparts T and W (air pollution standards for radon emissions from disposal of uranium mill tailings and from operating mills).

126. *See* Raoul S. Portillo, *Mill Tailings Remediation: The UMTRA Project, in* Reith and Thompson eds., *supra* ch. 3 note 49, at 281.

127. John L. Russell, *Health Risks From Uranium Mill Tailings, in* Yegulalp and Kim eds., *supra* ch. 2 note 107, at 236.

128. Portillo, *supra* ch. 3 note 126. *See also Colorado NPL Site To Be Uranium Waste Dump,* Superfund Week, Mar. 26, 1993, at 4 (remediated uranium mill site may be utilized for off-site disposal of waste from other UMTRA sites).

129. This discussion is drawn from Joseph R. Egan and John F. Seymour, *Disposing of Naturally Occurring Radioactive Material Wastes: A Legal Strategy,* 22 Envtl. L. Rep. (Envtl. L. Inst.) 10433 (1992), and Anthony J. Thompson and Michael L. Goo, *Naturally Occurring Radioactive Material: Regulators Should Look Before They Leap,* 22 Envtl. L. Rep. (Envtl. L. Inst.) 10052 (1992). *See also* James R. Cox, *Comment: Naturally Occurring Radioactive Materials in the Oilfield: Changing the NORM,* 67 Tul. L. Rev. 1197 (1993). Egan and Seymour assume roughly 50 billion metric tons of NORM waste is produced annually. 22 Envtl. L. Rep. at 10439 n.66.

130. 55 Fed.Reg. 16850 (1990).

131. La. Admin. Code tit. 33, § 1401.

132. *Nukewaste,* UPI, May 8, 1987, *available in* LEXIS, Nexis Library, UPI file.

133. *See* Barry Siegel, *A Perfect Place for a Waste Dump,* Los Angeles Times Mag., Dec. 22, 1991, at 25. A different attempt in Texas has also been defeated. *See Texas Denies Low-Level Waste Landfill Permit,* 24 Env't Rep. (BNA) No. 10, at 440 (July 9, 1993).

134. *Review of EPA's Capacity Assurance Program, supra* ch. 2 note 11, at 184, 263–64 (statement of Sylvia Lowrance, director, Office of Solid Waste, EPA); *Integrated Data Base for 1992: U.S. Spent Fuel and Radioactive Waste Inventories, Projections, and Characteristics, supra* ch. 3 note 58, at 212. *See also Study Says Most Mixed Waste in 1990 Consisted of Liquid Scintillation Fluids,* 23 Env't Rep. (BNA) No. 42, at 2682 (Feb. 12, 1993).

135. Legal Envtl. Assistance Found. v. Hodel, 586 F. Supp. 1163 (E.D. Tenn. 1984); 52 Fed.Reg. 15938 (1987) (codified at 10 C.F.R. § 962). *See also* Anthony J. Thompson and Michael L. Goo, *Mixed Waste: A Way to Solve the Quandary,* 23 Envtl. L. Rep. (Envtl. L. Inst.) 10705 (1993).

136. *Partnerships Under Pressure: Managing Commercial Low-Level Radioactive Waste—Summary, supra* ch. 3 note 53, at 17.

137. Id. at 15.

138. *DOE Mixed Waste Inventory Report First Step in Cleanup, Official Says,* 24 Env't Rep. (BNA) No. 3, at 146 (May 21, 1993).

139. 55 Fed.Reg. 22520, 22645 (1990); 57 Fed.Reg. 22024 (1992).

140. Edison Electric Institute v. EPA, 996 F.2d 326, 337 (D.C. Cir. 1993).

141. House Comm. on Interior and Insular Affairs, *The Department of Energy's Failure to Police Standards for Radioactivity in Hazardous Waste* (draft Sept. 30, 1992).

142. U.S. Dept. of Energy, Pub. No. DOE/S-000097P Vol. 2, *Environmental Restoration and Waste Management Five-Year Plan Fiscal Years 1994–1998: Installation Summaries* II-148 (Jan. 1993).

143. Id. at 19, 22.

144. Another set of estimates, using different methodologies, is reported in *Coming Clean: Superfund Problems Can Be Solved . . . , supra* ch. 2 note 51, at 194.

Chapter 4

1. John V. Winter and David A. Conner, *Power Plant Siting* 61–62 (1978); William G. Murray, Jr., and Carl J. Seneker II, *Industrial Siting: Allocating the Burden of Pollution,* 30 Hastings L.J. 301, 304 (1978); Suren S. Singhvi, *A Quantitative Approach to Site Selection,* 76 Mgmt. Rev. No. 4, at 47 (Apr. 1987).

2. Howard A. Stafford, *Environmental Protection and Industrial Location,* 75 Annals Ass'n Am. Geographers 227 (1985); *see also* Craig N. Oren, *Prevention of Significant Deterioration: Control-Compelling Versus Site-Shifting,* 74 Iowa

L. Rev. 1, 36 (1988) (prevention of significant deterioration requirements under Clean Air Act have had little impact on facility siting; underlying economics of projects are far more important in siting decisions).

3. R. Nils Olsen, Jr., *The Concentration of Commercial Hazardous Waste Facilities in the Western New York Community,* 39 Buff. L. Rev. 473, 482–83 (1991); Michael R. Greenberg and Richard F. Anderson, *Hazardous Waste Sites: The Credibility Gap* 389 (1984).

4. Richard Rhodes, *The Making of the Atomic Bomb* 449–51 (1986); *see also* Tad Bartimus and Scott McCartney, *Trinity's Children: Living Along America's Nuclear Highway* 67–90 (1991).

5. Rhodes, *supra* ch. 4 note 4, at 496–97; Shulman, *supra* ch. 2 note 70, at 96; Michele Stenehjem Gerber, *On the Home Front: The Cold War Legacy of the Hanford Nuclear Site* 11–12, 22–23 (1992).

6. J. Newell Stannard, *Radioactivity and Health: A History* 757, 763–64 (1988).

7. Charles W. Johnson and Charles O. Jackson, *City Behind a Fence: Oak Ridge, Tennessee, 1942–1946* 3–8 (1981).

8. Board on Radioactive Waste Management, *supra* ch. 3 note 17, at 7–11; *Oversight of Cleanup and Modernization Proposals for DOE's Weapons Production Complex, supra* ch. 3 note 89, at 420; Charles Piller, *The Fail-Safe Society: Community Defiance and the End of American Technological Optimism* 77 (1991).

9. Roger Rapoport, *The Great American Bomb Machine* 33 (1971).

10. This account is based on Bartlett and Steele, *supra* ch. 3 note 10, at 250–96, and Carol Bradley, *"Environmentalist" is Father of Low-Level,* Gannett News Service, Nov. 25, 1990, *available in* LEXIS, Nexis Library, GNS File.

11. Kathryn Visocki and Sherol S. Breman, *Regional Compacts and Waste Disposal,* 8 Forum for Applied Research and Public Policy 86, 87 (1993).

12. *See* Margaret Kriz, *Slow Burn,* National J., April 3, 1993, at 811, 813.

13. Conner Bailey and Charles E. Faupel, *Movers and Shakers and PCB Takers: Hazardous Waste and Community Power,* 13 Sociological Spectrum 89 (1993).

14. Landy et al., *supra* ch. 2 note 46, at 91.

15. Pub. L. 91-512, 84 Stat. 1227 § 212.

16. Charles E. Davis, *The Politics of Hazardous Waste* 19 (1993); Greenberg and Anderson, *supra* ch. 4 note 3, at 169–70; Bernstein, *supra* ch. 2 note 18, at 83, 86.

17. Alvin L. Alm, *Standing Committee Symposium on Risk Assessment: Opening Address,* 15 Env. L. Rep. (Env. L. Inst.) 10233 (1985)

18. Pub. L. 99-499, 100 Stat. 1617.

19. *Superfund Improvement Act of 1985,* S. Rep. No. 11, 99th Cong., 1st ses., at 22 (1985). *See also* 132 Cong. Rec. S14924 (daily ed. Oct. 3, 1986) (statement of Sen. Chafee).

20. Hazardous Waste Treatment Council v. South Carolina, 945 F.2d 781, 784 (4th Cir. 1991).

21. 42 U.S.C. § 9604(c)(9).

22. U.S. Environmental Protection Agency, Office of Solid Waste and Emergency Response, OSWER Directive No. 9010.00, *Assurance of Hazardous Waste Capacity: Guidance to State Officials* (1988).

23. Jean H. Peretz, *Equity Under and State Responses to the Superfund Amendments and Reauthorization Act of 1986,* 25 Policy Sciences 191, 203 (1992).

24. *Review of EPA's Capacity Assurance Program, supra* ch. 2 note 11, at 281 (statement of Sylvia K. Lowrance, director, Office of Solid Waste, EPA) (only plans not approved were those from Georgia, Mississippi, Arizona, Missouri, and District of Columbia). *See also Capacity Assurance Plan Nullified; Environmental Council Ruled Unconstitutional,* 24 Env't Rep. (BNA) No. 7 at 331 (June 18, 1993) (state judge rules that Mississippi submission violates state Constitution because it requires state legislators to serve on an administrative board).

25. Id. at 153 (statement of Richard C. Fortuna, executive director, Hazardous Waste Treatment Council).

26. State of New York v. Reilly, 35 Env't Rep. Cas. (BNA) 1959 (N.D.N.Y. 1992). The author's law firm represents the town of Porter, the town of Lewiston, and Niagara County, New York, which are plaintiff/intervenors in this action. No decision on the merits of the action has been rendered. *See also* Hazardous Waste Treatment Council v. South Carolina, 945 F.2d 781 (4th Cir. 1991).

27. U.S. Environmental Protection Agency, Office of Solid Waste and Emergency Response, OSWER Directive No. 9010.02, *Guidance for Capacity Assurance Planning: Capacity Planning Pursuant to CERCLA § 104(c)(9)* (May 1993).

28. Green, *supra* ch. 2 note 24, at 75, 78; Thomas D. McKewen and Anne C. Sloan, *A Successful Hazardous Waste Landfill Siting—Maryland's Experience, in Proceedings of the National Conference on Hazardous Wastes and Hazardous Materials* (Hazardous Materials Control Research Institute, 1987); Anne Sprightley Ryan, Massachusetts Hazardous Waste Facility Site Safety Council, *Approaches to Hazardous Waste Facility Siting in the United States* A11 (Sept. 1984).

29. Ariz. Rev. Stat. Ann. § 36-2802(A).

30. *Arizona is First State to Own Treatment Plant,* Engineering News-Record, Apr. 19, 1990, at 12; Richard N.L. Andrews, *Hazardous Waste Facility Siting: State Approaches, in Dimensions of Hazardous Waste Politics and Policy* 117, 121 (Charles E. Davis and James P. Lester eds., 1988); Ryan, *supra* ch. 4 note 28, at A1.

31. *Governor Approves $44 Million Buyout of Planned Hazardous Waste Incinerator,* 22 Env't Rep. (BNA) No. 10, at 567 (July 5, 1991); *Hearing on Hazardous Waste Facility Draws 2,500 to Hear Debate on Permits,* 21 Env't Rep. (BNA) No. 9, at 424 (June 29, 1990).

32. Sidney M. Wolf, *Public Opposition to Hazardous Waste Sites: The Self-Defeating Approach to National Hazardous Waste Control Under Subtitle C of the Resource Conservation and Recovery Act of 1976,* 8 Envtl. Aff. 463, 486 (1980).

33. Hazardous Waste Disposal Advisory Committee and Environmental Facilities Corp., *A Comprehensive Program for Hazardous Waste Disposal in New York State: Report to Governor Hugh L. Cary and the New York State Legislature* (Mar. 1980).

34. Michael K. Heiman, *The Quiet Evolution: Power, Planning, and Profits in New York State* 169 (1988); New York State Joint Legis. Comm'n on Toxic Substances and Hazardous Wastes, *Hazardous Waste Facility Siting in New York State: The Evolution of a Promising Public Policy* 7 (Mar. 1989) [hereinafter *Hazardous Waste Facility Siting in New York State*].

35. *Hazardous Waste Facility Siting in New York State, supra* ch. 4 note 34, at 7–8.

36. Washington County Cease Inc. v. Persico, 99 A.D.2d 321, 473 N.Y.S.2d 610 (3d Dep't 1984), *aff'd,* 64 N.Y.2d 923, 477 N.E.2d 1084, 488 N.Y.S.2d 630 (1985).

37. In the Matter of the New York State Dep't of Environmental Conservation Division of Water Hudson River PCB Project, Decision of the Industrial Hazardous Waste Facility Siting Board, Jan. 5, 1989.

38. *Hazardous Waste Authority Finds No Need for State-Sponsored Management Facility,* 23 Env't Rep. (BNA) No. 26, at 1923 (Nov. 27, 1992).

39. Mary English and Gary Davis, *American Siting Initiatives: Recent State Developments, in* Piasecki and Davis eds., *supra* ch. 2 note 1, at 279, 280–81.

40. Green, *supra* ch. 2 note 24, at 75–79.

41. Mazmanian and Morell, *supra* ch. 1 note 5, at 192, 202; Daniel A. Mazmanian and Michael Stanley-Jones, *Reconceiving LULUs: Changing the Nature and Scope of Locally Unwanted Land Uses, in Confronting Regional Challenges: Approaches to LULUs, Growth, and Other Vexing Governance Problems* 55 (Joseph F. DiMento and LeRoy Graymer eds., 1991); Sarah Crim, *The NIMBY Syndrome in the 1990s: Where Do You Go After Getting to 'NO'?,* 21 Env't Rep. (BNA) No. 1, at 132, 135 (May 4, 1990).

42. In the 1970s the Gulf Coast Waste Disposal Authority, a quasi-governmental nonprofit corporation in the Houston, Texas, area, did succeed in siting an industrial waste landfill in Texas City. Alan L. Farkas, *Overcoming Public Opposition to the Establishment of New Hazardous Waste Disposal Sites,* 9 Cap. U. L. Rev. 451, 461 (1980); Greenberg and Anderson, *supra* ch. 4 note 3, at 166–67.

43. Christopher J. Duerksen, *Environmental Regulation of Industrial Plant Siting* 109–23 (1983); A. Dan Tarlock, *Siting New or Expanded Treatment, Storage, or Disposal Facilities: The Pigs in the Parlors of the 1980s,* 17 Nat. Resources Law. 429, 459 (1984) [hereinafter Tarlock 1984]; O'Hare et al., *supra* ch. 1 note

9, at 54, 60; Mickale Carter, *The Montana Major Facility Siting Act*, 45 Mont. L. Rev. 113 (1984).

44. EPA developed a Model State Hazardous Waste Management Act and urged the states to assume primary responsibility for assuring the availability of hazardous waste disposal capacity. Bram Canter, *Hazardous Waste Disposal and the New State Siting Programs*, 14 Nat. Resources Law. 421, 432–36 (1982).

45. *Hazardous Waste Facility Siting: A National Survey, supra* ch. 1 note 1, at 9.

46. *See* Green, *supra* ch. 2 note 24, at 76–77; Greenberg and Anderson, *supra* ch. 4 note 3, at 179–86; *Hazardous Waste Facility Siting: A National Survey, supra* ch. 1 note 1; Richard N.L. Andrews, *Hazardous Waste Facility Siting: State Approaches, in Dimensions of Hazardous Waste Politics and Policy* 117 (Charles E. Davis and James P. Lester eds., 1988); Canter, *supra* ch. 4 note 44; Isabelle R. Davidson, *An Analysis of Existing Requirements for Siting and Permitting Hazardous Waste Disposal Facilities and a Proposal for a More Workable System*, 34 Admin. L. Rev. 533 (1982); Celeste P. Duffy, *State Hazardous Waste Facility Siting: Easing the Process Through Local Cooperation and Preemption*, 11 Envtl. Aff. 755 (1984); English and Davis, *supra* ch. 4 note 39, at 279; Susan G. Hadden et al., *State Roles in Siting Hazardous Waste Disposal Facilities: From State Preemption to Local Veto, in The Politics of Hazardous Waste Management* 196 (James P. Lester and Ann O'M. Bowman eds., 1983); Albert R. Matheny and Bruce A. Williams, *Knowledge vs. NIMBY: Assessing Florida's Strategy for Siting Hazardous Waste Disposal Facilities*, 14 Pol'y Stud. J. 70 (1985); Ryan, *supra* ch. 4 note 28; Tarlock 1984, *supra* ch. 4 note 43.

47. E.g., Morell and Magorian, *supra* ch. 1 note 8, at 160; O'Hare 1983, *supra* ch. 1 note 9, at 4; Douglas Easterling, *Fair Rules for Siting a High-Level Nuclear Waste Repository*, 11 J. Pol'y Analysis and Mgmt. 442 (1992); Matheny and Williams, *supra* ch. 4 note 46, at 71; Barry D. Solomon and Diane M. Cameron, *Nuclear Waste Repository Siting: An Alternative Approach*, 13 Energy Pol'y 564 (1985).

48. William M. Sloan, *Site Selection for New Hazardous Waste Management Facilities* 73 (World Health Organization Regional Publications, European Series, No. 46, 1993); Temple, Barker and Sloane, EPA Office of Solid Waste, *Study of State Hazardous Waste Facility Siting Criteria* (1987); Thomas P. Ballestero and Mark D. Kelley, *Where Can New Landfills Be Sited?* Waste Age, Oct. 1990, at 145. Certain federal statutes also require the development of siting criteria. E.g., NWPA, 42 U.S.C. § 10132(a); RCRA, 42 U.S.C. § 6924(a)(4). *See Guidelines for Hazardous Waste Treatment Sites: Hearings before Subcomm. on Commerce, Transportation and Tourism of House Comm. on Energy and Commerce*, 99th Cong., 2d Sess. 32 (1986) [hereinafter *Guidelines for Hazardous Waste Treatment Sites*] (statement of Marcia E. Williams, director, EPA Office of Solid Waste).

49. Greenberg and Anderson, *supra* ch. 4 note 3, at 191–234; New York State Dept. of Envt'l Conservation, *Solid Waste Management Facility Siting* (Apr. 1990); George Noble, *Siting Landfills and Other LULUs* (1992); Richard F.

Anderson and Michael R. Greenberg, *Hazardous Waste Facility Siting: A Role for Planners*, APA J., Spring 1982, at 204; Charles D. Hollister and Harry W. Smedes, *Selecting Sites for Radioactive Waste Repositories, in Hazardous Waste Management: In Whose Backyard?* 63 (Michalann Harthill ed., 1981); Stephen K. Swallow et al., *Siting Noxious Facilities: An Approach That Integrates Technical, Economic, and Political Considerations,* 68 Land Econ. 283 (1992).

50. O'Hare 1983, *supra* ch. 1 note 9, at 56; *similarly,* Frank J. Dodd, *Siting Hazardous Waste Facilities in New Jersey: Keeping the Debate Open,* 9 Seton Hall Legis. J. 423, 433 (1985); Tarlock 1984, *supra* ch. 4 note 43, at 457–58.

51. *See* Joseph B. Rose, *Planning for "Fairness": Wrestling With Criteria for the Location of City Facilities,* The Assessor (Citizens Housing and Planning Council), Feb. 1991, at 1 (shows how seemingly objective siting criteria can be readily manipulated).

52. Carter, *supra* ch. 3 note 51, at 410.

53. E.g., serious technical flaws were established in the studies leading to the selection of an LLRW site in Martinsville, Illinois. English, *supra* ch. 1 note 8, at 64; *Martinsville Rejected for Central Midwest LLRW Facility,* Radioactive Exchange, Oct. 21, 1992, at 1. Major questions have been raised as well about the application of siting criteria in locating a LLRW facility in New York State, U.S. Gen. Accounting Office, Pub. No. GAO/RCED-92-172, *Nuclear Waste: New York's Adherence to Site Selection Procedures is Unclear* (Aug. 1992), and in siting an industrial liquid waste disposal facility in Ontario, Edward J. Farkas, *The Nimby Syndrome,* 10 Alternatives No. 2, at 47 (1981), as well as the selection of Yucca Mountain, Jacob, *supra* ch. 3 note 26.

54. *Guidelines for Hazardous Waste Treatment Sites, supra* ch. 4 note 48, at 48 (statement of Marcia E. Williams, director, EPA Office of Solid Waste). Within this sample, on-site facilities tended to be in somewhat worse locations than commercial facilities. Id.

55. U.S. Gen. Accounting Office, Pub. No. GAO/RCED-91-79, *Hazardous Waste: Limited Progress in Closing and Cleaning Up Contaminated Facilities* 44 (May 1991).

56. Hameline v. New York St. Dep't of Envtl. Conservation, 175 A.D.2d 206, 572 N.Y.S.2d 347 (2d Dep't 1991).

57. RECRA Environmental, Inc., *Pine Avenue Expansion: Draft Supplemental Environmental Impact Statement for CECOS International, Inc. Proposed Secure Chemical Residual Facility* 22 (1987).

58. In the Matter of the Application of CECOS Int'l, Inc. (N.Y. St. Industrial Hazardous Waste Facility Siting Bd., Mar. 13, 1990).

59. *See,* e.g., Regina Austin and Michael Schill, *Black, Brown, Poor and Poisoned: Minority Grassroots Environmentalism and the Quest for Eco-Justice,* 1 Kan. J. L. and Pub. Pol'y 69, 70 (1991) (proposed construction of hazardous waste incinerator in Kettleman City, California, at site of existing hazardous

waste landfill); Reg Lang, *Fair Siting in Waste Management, in* Inst. for Soc. Impact Assessment, *supra* ch. 2 note 25, at 237 (concerning expansion of landfill sites in Ontario); Pat Medige, *No More Waste, CWM Told,* Niagara Gazette, June 15, 1990, at 1 (proposed construction of hazardous waste incinerators in Porter, New York, at site of existing hazardous waste landfills) (note: plan has since been abandoned—see pp. 113–17).

60. Gretchen D. Monti, *"All Politics is Local": Integrating Local Concerns Into Facility Site Selection, in* Inst. for Soc. Impact Assessment, *supra* ch. 2 note 25, at 36, 38 (Study of experience in Illinois shows "[i]t is much easier to expand a waste facility that a community has become accustomed to."); Jacob, *supra* ch. 3 note 26, at 161 ("[A] legacy of externalities already evident in the environments of Hanford and Yucca Mountain lowered political opposition to additional increments of environmental degradation."); McKewen and Sloan, *supra* ch. 4 note 28 (successful siting of hazardous waste landfill in area of Baltimore surrounded by existing landfills).

61. Barry G. Rabe, *Low-level Radioactive Waste Disposal and the Revival of Environmental Regionalism in the United States,* 7 Envtl. and Plan. L.J. 171, 177 (1990) ("Survey research has demonstrated that public trust of power generation and waste disposal facilities increases with greater proximity to and familiarity with such facilities." [citing evidence from areas of Beatty, Richland, and Barnwell LLRW facilities]); Michael R. Edelstein, *Contaminated Communities: The Social and Psychological Impacts of Residential Toxic Exposure* 17 (1988) [hereinafter Edelstein 1988].

62. G. Stephen Mason, Jr., *Closure and Rejection of Waste Facilities: What Effect Has Public Pressure?* Hazardous Materials Control, July/Aug. 1989, at 54, 55.

63. Peter Huber, *The Old-New Division in Risk Regulation,* 69 Va. L. Rev. 1025 (1983). For an economic analysis of aspects of this phenomenon, *see* Charles D. Kolstad, et al., Ex Post *Liability for Harm vs.* Ex Ante *Safety Regulation: Substitutes or Complements?* 80 Am. Econ. Rev. 888 (1990).

64. Constance Perin, *Everything in Its Place: Social Order and Land Use in America* 144–45 (1977); Patrick J. Rohan, *Zoning and Land Use Controls* § 41.

65. Niagara Recycling, Inc. v. Town of Niagara, 83 A.D.2d 316, 443 N.Y.S.2d 939 (4th Dep't 1981). A contrary result occurred in Clean Harbors of Braintree Inc. v. Braintree Bd. of Health, 409 Mass. 834, 570 N.E.2d 987 (1991), but in response to this decision, state legislation was enacted exempting preexisting HW incinerators from municipal approval requirements. Clean Harbors of Braintree v. Braintree Bd. of Health, 415 Mass. 876, 616 N.E.2d 78 (1993). *See also* CECOS Int'l, Inc. v. Jorling, 895 F.2d 66 (2d Cir. 1990) (upholds a state statute repealing exemption of expansion of existing landfills from state siting laws).

66. *E.g.,* Township of Chartiers v. William H. Martin, Inc., 518 Pa. 181, 542 A.2d 985 (1988); Speedway Grading Corp. v. Barrow County Bd. of Comm'rs, 258 Ga. 693, 373 S.E.2d 205 (1988); Sturgis v. Winnebago County Bd. of

Adjustment, 141 Wis.2d 149, 413 N.W.2d 642 (Ct. App. 1981). *See generally* Bruce J. Parker and John H. Turner, *Overcoming Obstacles to the Siting of Solid Waste Management Facilities* 21 N.M. L. Rev. 92, 111 (1990).

67. Landy et al., *supra* ch. 2 note 46, at 109. Such choices are common in environmental law. Richard B. Stewart, *Regulation, Innovation, and Administrative Law: A Conceptual Framework,* 69 Cal. L. Rev. 1256, 1314 (1981).

68. 42 U.S.C. § 6925(e).

69. Alex S. Karlin, *Hazardous Waste Management, in* Gerrard 1992, *supra* ch. 2 note 7, § 29.09[2].

70. John E. Bonine and Thomas O. McGarity, *The Law of Environmental Protection* 771 (2d ed. 1992).

71. 42 U.S.C. § 6925(e)(2).

72. Bonine and McGarity, *supra* ch. 4 note 70, at 771; U.S. Gen. Accounting Office, Pub. No. GAO/RCED-92-84, *Hazardous Waste: Impediments Delay Timely Closing and Cleanup of Facilities* 2 (Apr. 1992).

73. *Complex Cleanup: The Environmental Legacy of Nuclear Weapons Production, supra* ch. 3 note 8, at 28.

74. 40 C.F.R. pt. 265.

75. 40 C.F.R. pt. 264.

76. U.S. Gen. Accounting Office, Pub. No. GAO/RCED-91-79, *Hazardous Waste: Limited Progress in Closing and Cleaning Up Contaminated Facilities* 20 (May 1991); Christopher Harris et al., *Hazardous Waste: Confronting the Challenge* 111 (1987).

77. U.S. General Accounting Office, Pub. No. GAO/RCED-92-21, *Hazardous Waste: Incinerator Operating Regulations and Related Air Emission Standards* (Oct. 1991).

78. U.S. Gen. Accounting Office, Pub. No. GAO/RCED-88-95, *Hazardous Waste: Future Availability of and Need for Treatment Capacity Are Uncertain* 24 (Apr. 1988); N.J. Stat. Ann. § 13–1E-87. This partial or total "grandfathering" of existing facilities extends beyond RCRA to many of the other relevant legal requirements, *see generally* Duerksen, *supra* ch. 4 note 43, at 162–65, such as the Clean Air Act's stationary source standards, 42 U.S.C. § 7411(d), prevention of significant deterioration permits, 42 U.S.C. § 7475(b), and solid waste combustion rules, 42 U.S.C. § 7429(b); the Clean Water Act's rules for publicly owned treatment works, 33 U.S.C. § 1311(b)(1)(B); the National Environmental Policy Act's environmental impact statement requirements, *see* Daniel R. Mandelker, *NEPA Law and Litigation* § 5.02 (2d ed. 1992); restrictions on trade in endangered species, 16 U.S.C. § 1539(b)(1); the designation of areas unsuitable for mining under the Surface Mining Control and Reclamation Act (SMCRA), 30 U.S.C. § 1272(a)(6); the prohibition on commercial activities in designated wilderness areas under the Wilderness Act, 16 U.S.C. § 1133(c); and EPA's regulations on MSW landfills, *Solid Waste Disposal Facility Criteria,* 56 Fed.Reg. 50978, 51042 (Oct. 9, 1991); 40 C.F.R. § 258.16(a).

79. A plethora of federal and state permit requirements have been enacted, mostly in the 1970s. Duerksen, *supra* ch. 4 note 43, at 79–88; Fred Bosselman et al., *The Permit Explosion: Coordination of the Proliferation* (1976); Katharine J. Teter et al., *Long Arm of Uncle Sam: Federal Environmental Issues in Siting Decisions,* 7 Natural Resources and Envt. No. 3, at 9 (Winter 1993).

80. Benjamin Walter and Malcolm Getz, *Social and Economic Effects of Toxic Waste Disposal, in Controversies in Environmental Policy* 223, 240 (Sheldon Kamieniecki et al. eds. 1986).

81. Kenski and Ingram have used a much different formulation to list various policy tools for pollution control from the least coercive or regulatory to the most. Henry C. Kenski and Helen M. Ingram, *The Reagan Administration and Environmental Regulation: The Constraint of the Political Market, in* Kamieniecki, *supra* ch. 4 note 80, at 275, 280–81. For other typologies of environmental policies, *see* William J. Baumol and Wallace E. Oates, *Economics, Environmental Policy, and the Quality of Life* 218 (1979), and Rae Zimmerman, *Governmental Management of Chemical Risk: Regulatory Processes for Environmental Health* 41–43 (1990) [hereinafter Zimmerman 1990].

82. Tennessee Valley Authority v. Hill, 437 U.S. 153 (1978).

83. *See* Sequoyah v. Tennessee Valley Authority, 480 F. Supp. 608 (E.D. Tenn. 1979), *aff'd,* 620 F.2d 1159 (6th Cir.), *cert. denied,* 449 U.S. 953 (1980).

84. Pub. L. 90-495, § 23.

85. D.C. Fed'n of Civic Ass'ns v. Airis, 391 F.2d 478 (D.C. Cir. 1968).

86. D.C. Fed. of Civic Ass'ns v. Volpe, 434 F.2d 436 (D.C. Cir. 1970). These events are discussed in Zygmunt J.B. Plater et al., *Environmental Law and Policy: Nature, Law, and Society* 593–95 (1992).

87. Douglas B. Feaver, *Three Sisters Highway Project is Killed—Again,* Washington Post, May 13, 1977, at B1.

88. 42 U.S.C. § 10172.

89. Ariz. Rev. Stat. Ann. § 36-2802(A).

90. Tex. Health and Safety Code Ann. § 402.0921 (West 1992).

91. 33 U.S.C. § 1344(a).

92. 33 U.S.C. § 1344(e)(1).

93. 33 C.F.R. § 330.1.

94. 33 C.F.R. § 320.4(j)(4).

95. 40 C.F.R. § 230.1(c). This paradox is discussed in Oliver A. Houck, *Hard Choices: The Analysis of Alternatives Under § 404 of the Clean Water Act,* 60 U. Colo. L. Rev. 773, 779 (1989).

96. 33 U.S.C. §§ 21–59ff. These statutes provide exemptions from the requirement of § 10 of the Rivers and Harbors Appropriation Act of 1899, 33 U.S.C. § 403, and not from the dredge and fill permit requirement of 33 U.S.C. § 1344.

97. *See,* e.g. Stop H-3 Ass'n v. Dole, 870 F.2d 1419 (9th Cir. 1989) (upholding a statute exempting a controversial Hawaii highway project from § 4(f) of the Department of Transportation Act, 49 U.S.C. § 403, after litigation blocking the project on § 4(f) grounds); Earth Resources Co. of Alaska v. Federal Energy Regulatory Comm'n, 617 F.2d 775 (D.C. Cir. 1980) (upholding statute exempting Alaskan Pipeline from many requirements of NEPA); Intermodal Surface Transportation Efficiency Act, Pub. L. 102–240, 105 Stat. 1914, § 1097 (providing that reconstruction of sections of West Side Highway in Manhattan can be considered separate projects under NEPA); 1990 N.Y. Laws ch. 774 (exempting from the cumulative impact review requirement any widening of the Long Island Expressway between Exits 49 and 57, enacted in reaction to Village of Westbury v. New York St. Dep't of Transp., 75 N.Y.2d 62, 549 N.E.2d 1175, 550 N.Y.S.2d 604 (1989)); Cal. Pub. Res. Code §§ 21080(b), 21080.01–.09 (West 1992) (exempting numerous specified actions from California Environmental Quality Act).

98. Addendum to Annex 1 of the 1972 Convention on the Prevention of Marine Pollution by Dumping Wastes and Other Matter. *See* Christopher B. Kende, *Oceans and Coasts, in* Gerrard 1992, *supra* ch. 2 note 7, at § 23.06[1][a].

99. New York City Planning Commission, *Criteria for the Location of City Facilities* (1990) (implementing N. Y. City Charter § 203(a)). *See* William Valletta, *Fair Share Criteria,* N.Y.L.J., Dec. 18, 1990, at 1; Vicki Been, *What's Fairness Got To Do With It? Environmental Equity in the Siting of Locally Undesirable Land Uses,* 78 Cornell L. Rev. 1001 (1993); Naikang Tsao, *Ameliorating Environmental Racism: A Citizens' Guide to Combatting the Discriminatory Siting of Toxic Waste Dumps,* 67 N.Y.U. L. Rev. 366, 375–78 (1992); Stephen L. Kass and Michael B. Gerrard, *"Fair Share" Siting of City Facilities,* N.Y.L.J., June 21, 1990, at 3.

100. Several commentaries have been written on such laws. *See* James A. Henderson and Richard N. Pearson, *Implementing Federal Environmental Policies: The Limits of Aspirational Commands,* 78 Colum. L. Rev. 1429 (1978); Gerald M. Levine, *The Rhetoric of Public Expectations: An Enquiry Into the Concepts of Responsiveness and Responsibility Under the Environmental Laws,* 8 Pace Envt'l L. Rev. 389 (1991); Baumol and Oates, *supra* ch. 4 note 81, at 282–306.

101. Harold P. Green, *Legal Aspects of Intergenerational Equity Issues, in* Kasperson, *supra* ch. 3 note 4, at 189, 192.

102. 42 U.S.C. §§ 4321, 4331(a).

103. 33 U.S.C. § 1251(a).

104. 42 U.S.C. § 7401(b).

105. 42 U.S.C. § 6902.

106. 42 U.S.C. § 10131.

107. Moira Hayes Waligory, *Radioactive Marine Pollution: International Law and State Liability,* 1992 Suffolk Transnat'l L.J. 674, 682 (1992).

108. E.g., 42 U.S.C. § 6902(b); N.Y. Envtl. Conserv. Law §§ 27-0105, 27-1102(1)(d) (McKinney Supp. 1992).

109. In the Matter of the Application for Combined Air and Solid Waste Permit No. 2211-91-OT-1 for the Dakota Co. Mixed Solid Waste Incinerator, 483 N.W.2d 105 (Minn. Ct. App. 1992).

110. Clean Water Act Sec. 402, 33 U.S.C. § 1342. Note that Sec. 301(a) of the Clean Water Act, 33 U.S.C. § 1311, provides that "the discharge of any pollutant by any person shall be unlawful" unless a permit is obtained under some other section, such as Sec. 402. Thus Sec. 301(a) reads, on its face, like a "Must Be Avoided Unless" law. In reality, however, Sec. 301(a) is only a loophole closer—a clause that sweeps all water pollution discharges within the net of the Clean Water Act so that specific permit provisions, such as Sec. 402, can come into play.

111. 42 U.S.C. §§ 7661–7661f.

112. *See* Hazardous Waste Treatment Council v. Reilly, 938 F.2d 1390 (D.C. Cir. 1991).

113. 42 U.S.C. § 10172a(c).

114. Shapiro, *supra* ch. 3 note 24, at 61, 64.

115. 42 U.S.C. §4332(C). Environmental impact statements prepared under this statute can become binding documents if they commit to mitigation measures and are written for agencies, such as the Federal Highway Administration, with regulations providing that such commitments are legally enforceable. *See* 23 C.F.R. § 771.109(b).

116. 42 U.S.C. § 11023.

117. Alex. Brown and Sons, *supra* ch. 2 note 40, at 10.

118. *See* Stephen Breyer, *Regulation and its Reform* 161–63 (1982).

119. 16 U.S.C. § 1456(c).

120. 30 U.S.C. § 1272(a)(3).

121. 16 U.S.C. § 470a.

122. 33 U.S.C. § 1344(c).

123. 33 U.S.C. § 1412(c)(2).

124. 42 U.S.C. §§ 4001–4128.

125. *See* Kamhi v. Town of Yorktown, 74 N.Y.2d 423, 547 N.E.2d 346, 548 N.Y.S.2d 144 (1989); Golden v. Town of Ramapo, 30 N.Y.2d 359, 285 N.E.2d 291, 334 N.Y.S.2d 138, app. dism., 409 U.S. 1003 (1972); Cherry Hills Farms v. City of Cherry Hills Village, 670 P.2d 779 (Colo. 1983).

126. *See* Murray and Seneker, *supra* ch. 4 note 1, at 1077 (citing Kaiparowits power project in Utah, Dow project in San Francisco Bay area, and Sohio project in southern California); *Formosa Cancels Louisiana Project,* Chemical Wk., Oct. 21, 1992, at 13 (large rayon plant canceled due to protracted delays); Sarah Lyall,

State Court Permits Development in the Pine Barrens of Long Island, N.Y. Times, Nov. 25, 1992, at A1 (Protracted litigation concerning cumulative impact assessment in Long Island Pine Barrens area contributed to cancelation of several development projects). Other examples are cited in Philip Weinberg, *SEQRA's Too Valuable to Trash: A Reply to Stewart Sterk,* 14 Cardozo L. Rev. 1959 (1993).

127. Telephone conversation with Benjamin Conlon, N.Y. St. Dep't of Envtl. Conservation, June 23, 1992. *See also Environmental Preview?* Wall St. J., Dec. 30, 1992, at 6 (similar design change at Ohio incinerator for similar reasons).

128. 16 U.S.C. §§ 470 *et seq.*

129. 49 U.S.C. § 303; *similarly,* 23 U.S.C. § 138, 49 C.F.R. § 1653(f), 23 C.F.R. § 771.135. *See also* Citizens to Preserve Overton Park, Inc. v. Volpe, 401 U.S. 402 (1971).

130. 42 U.S.C. §§ 7470–7479. *See* Oren, *supra* ch. 4 note 2.

131. 30 U.S.C. § 1272(e)(2).

132. 42 U.S.C. § 7475(d)(2)(D).

133. 42 U.S.C. § 300h-3(e).

134. 40 C.F.R. § 264.18(b)(1). EPA's RCRA siting regulations, as promulgated by the Reagan administration in 1981, are considerably weaker than those originally proposed by the Carter administration in 1978. Tarlock 1981, *supra* ch. 1 note 9, at 10 n.26.

135. 40 C.F.R. § 761.75(b).

136. 40 C.F.R. pt. 258.

137. 40 C.F.R. § 227.6(c).

138. 40 C.F.R. § 227.3(c).

139. 10 C.F.R. § 61.50(b)(7).

140. *See* Myrl L. Duncan, *Agriculture as a Resource: Statewide Land Use Programs for the Preservation of Farmland,* 14 Ecology L.Q. 401 (1987).

141. *See* Phillips Petroleum Co. v. Mississippi, 484 U.S. 469 (1988); Illinois Central R.R. Co. v. Illinois, 146 U.S. 387 (1892).

142. Tennessee Valley Auth. v. Hill, 437 U.S. 153, 173 (1978) ("One would be hard pressed to find a statutory provision whose terms were any plainer than those . . . of the [ESA] . . . The language admits of no exception.").

143. Pub. L. § 95-631, § 3, 92 Stat. 3751, 3752, codified at 16 U.S.C. § 1536. *See* Zygmunt J.B. Plater, *In the Wake of the Snail Darter: An Environmental Law Paradigm and its Consequences,* 19 J.L. Reform 805 (1986).

144. 16 U.S.C. § 1133(c). The New York State Constitution affords similar protection to the Forest Preserve within the Adirondack Park. N.Y. Const. art. XIV, § 1. *See* William H. Kissel, *Permissible Uses of New York State Forest Preserve Under "Forever Wild,"* 19 Syr. L. Rev. 969 (1967–68). It is of note that in 1973 opponents of the proposed Rye-Oyster Bay Bridge over Long Island

Sound succeeded in having 3,117 acres of town-owned wetlands, needed for approach ramps, transferred to the U.S. Department of the Interior as a wildlife preserve; this helped to kill the project. Robert H. Connery and Gerald Benjamin, *Rockefeller of New York: Executive Power in the Statehouse* 360–62 (1979).

145. 30 U.S.C. § 1272(e)(1).

146. 42 U.S.C. § 10172a(a).

147. 33 U.S.C. § 1414c(a).

148. 1979 Mass. Acts chs. 574, 742, *cited in* O'Hare 1983, *supra* ch. 1 note 9, at 34 n.*.

149. N.Y. Comp. Codes R. and Regs. tit. 6, § 382.21(a)(1) (1992). This regulation may be repealed. *See* text accompanying ch. 6 note 77 *infra*.

150. 40 C.F.R. § 264.18(a).

151. 40 C.F.R. § 258.14.

152. 40 C.F.R. § 264.18(c).

153. 10 C.F.R. § 61.50(a)(5).

154. 5 U.S.C. § 702; *see* Cass R. Sunstein, *After the Rights Revolution: Reconceiving the Regulatory State* 218–20 (1990).

155. 42 U.S.C. § 9613(h). *See* Alfred R. Light and M. David McGee, *Preenforcement, Preimplementation, and Postcompletion Preclusion of Judicial Review Under CERCLA,* 22 Envtl. L. Rep. (Envtl. L. Inst.) 10397 (1992); Victor M. Sher and Carol Sue Hunting, *Eroding the Landscape, Eroding the Laws: Congressional Exemptions From Judicial Review of Environmental Laws,* 15 Harv. Envtl. L. Rev. 435 (1991).

156. 42 U.S.C. § 10132(d). *See* State of Nevada v. Watkins, 939 F.2d 710 (9th Cir. 1991). Similarly, concerning a hazardous waste facility, *see* Town of Warren v. Hazardous Waste Facility Site Safety Council, 392 Mass. 107, 466 N.E.2d 102 (1984).

157. 42 U.S.C. § 7475(d)(2)(D)(ii).

158. *See* Dalton v. Specter, 62 U.S.L.W. 4340 (May 23, 1994).

159. Arthur J. Harrington, *The Right to a Decent Burial: Hazardous Waste and its Regulation in Wisconsin,* 66 Marq. L. Rev. 223, 275 (1983); Michael O'Hare, *"Not on My Block You Don't": Facility Siting and the Strategic Importance of Compensation,* 25 Pub. Pol'y 407, 441 n.11 (1977) [hereinafter O'Hare 1977]. *See also* Orlando E. Delogu, *"NIMBY" Is a National Environmental Problem,* 35 S.D. Dakota L. Rev. 198, 214 n.48 (1990) (Utah has enacted such a limitation).

160. *E.g.,* Lujan v. Defenders of Wildlife, 112 S.Ct. 2130 (1992).

161. *E.g.,* New Scotland Ave. Neighborhood Ass'n v. Planning Bd. of City of Albany, 142 A.D.2d 257, 535 N.Y.S.2d 645 (3d Dep't 1988).

162. E.g., 42 U.S.C. § 10139(c). *See* Michael B. Gerrard et al., *Environmental Impact Review in New York* § 7.02[4] (1990 and 1993 Supp.).

163. *E.g.,* Marsh v. Oregon Natural Resources Council, 490 U.S. 360 (1989).

164. E.g., Weinberger v. Catholic Action of Hawaii, 454 U.S. 139 (1981) (Navy did not prepare environmental impact statement on risk of nuclear accident and refused to disclose whether proposed ammunition and weapons storage facilities would store nuclear weapons; Court ruled that this information was exempt under the Freedom of Information Act, rendering the action essentially unreviewable).

165. 42 U.S.C. § 9621(e)(1).

166. 42 U.S.C. § 10107; Reith and Fischer, *supra* ch. 3 note 49, at 303, 316.

167. *See* Portland Cement Ass'n v. Ruckelshaus, 486 F.2d 375 (D.C. Cir. 1973), *cert. denied,* 417 U.S. 921 (1974); Comment, *To Police the Police: Functional Equivalence to the EIS Requirement and EPA Remedial Actions Under Superfund,* 33 Cath. U. L. Rev. 863 (1984).

168. E.g., New York Pub. Interest Research Group v. Town of Islip, 71 N.Y.2d 292, 520 N.E.2d 517, 525 N.Y.S.2d 798 (1988) (holding that the expansion of a landfill did not require an EIS because expansion was required by consent decree).

169. 42 U.S.C. § 7503(a)(1)(B). Such exemptions have become quite widespread in the United Kingdom. *See* C. Wood and P. Hooper, *The Effects of the Relaxation of Planning Controls in Enterprise Zones on Industrial Pollution,* 21 Envt. and Planning A 1157 (1989).

170. *E.g.,* Daniel A. Farber, *Playing the Baseline: Civil Rights, Environmental Law, and Statutory Interpretation,* 91 Colum. L. Rev. 676, 685, 697–98 (1991); Zygmunt J.B. Plater, *Statutory Violations and Equitable Discretion,* 70 Cal. L. Rev. 524, 528 (1982).

171. *E.g.,* T.J. Hiles, *Comment: Civil Forfeiture of Property for Drug Offenders Under Illinois and Federal Statute: Zero Tolerance, Zero Exceptions,* 25 J. Marshall L. Rev. 389 (1992); Elizabeth Poliner, *The Regulation of Carcinogenic Pesticide Residues in Food: The Need to Reevaluate the Delaney Clause,* 7 Va. J. Nat. Resources L. 111 (1987).

172. *See* Daniel A. Farber, *From Plastic Trees to Arrow's Theorum,* 1986 U. Ill. L. Rev. 337, 344–46, 348–49.

173. Sierra Club v. United States Army Corps of Eng'rs, 772 F.2d 1043 (2d Cir. 1985).

174. Fox Butterfield, *Idaho Firm on Barring Atomic Waste,* N.Y. Times, Oct. 23, 1988, at 32.

Chapter 5

1. *See* Peter S. Menell, *Beyond the Throwaway Society: An Incentive Approach to Regulating Municipal Solid Waste,* 16 Ecology L.Q. 655, 695 (1990); *see also* Mark Sagoff, *The Economy of the Earth* (1988).

2. Noble, *supra* ch. 4 note 49, at 17; Audrey M. Armour, *The Siting of Locally Unwanted Land Uses: Towards a Cooperative Approach*, 35 Progress Plan. 1, 7 (1991) [hereinafter Armour 1991]; Emilie Schmeidler and Peter M. Sandman, *Getting to Maybe: Decisions on the Road to Negotiation in Hazardous Waste Facility Siting* 22 (Envtl. Educ. Fund, N.J., Jan. 1988); *see also* ch. 1 note 6 *supra*.

3. Harvey Alter, *The Myths of Municipal Solid Waste*, Solid Waste and Power, July/Aug. 1992, at 46 (Consistently since the turn of the century, government officials have proclaimed an imminent crisis in garbage disposal.); 42 U.S.C. § 6901(b)(8) (preamble to RCRA, enacted in 1976, says, "[M]any of the cities in the United States will be running out of suitable solid waste disposal sites within five years unless immediate action is taken.").

4. For example, Western LNG Terminal Co. waged a ten-year battle to site a liquified natural gas terminal in California, claiming dire need. However, the company withdrew the application when deregulation of domestic natural gas prices in 1978 destroyed the market for imported LNG. Howard Kunreuther, et al., *A Decision-Process Perspective on Risk and Policy Analysis, in* Lake 1987, *supra* ch. 1 note 11, at 260; Lawrence E. Susskind and Stephen R. Cassella, *The Dangers of Preemptive Legislation: The Case of LNG Facility Siting in California, in id.*, at 408. After at least twenty-four attempts between 1970 and 1980 to site a new oil refinery somewhere along the eastern seaboard, all failed, Piller, *supra* ch. 4 note 8, at 161, but it is not clear that there remains any crying need for these facilities.

5. In the Matter of the Application of CECOS Int'l, Inc., slip op. at 7–10 (N.Y. St. Industrial Hazardous Waste Facility Siting Bd. Mar. 13, 1990). The board denied the landfill permit because it found the site to be geologically unsuitable.

6. *See* Industrial Fuels and Resources/Illinois, Inc. v. Illinois Pollution Control Board, 227 Ill.App.3d 533, 592 N.E.2d 148 (1992); Green, *supra* ch. 2 note 24, at 74; Gary Davis and William Colglazier, *Siting Hazardous Waste Facilities: Asking the Right Questions, in* Piasecki and Davis, *supra* ch. 2 note 1, at 167, 175–76; Robert W. Craig and Terry R. Lash, *Siting Nonradioactive Hazardous Waste Facilities, in* Harthill, *supra* ch. 4 note 49, at 99, 101–03. Approaches used by state agencies in determining the need for MSW facilities are discussed in Mary Beth Arnett, *Down in the Dumps and Wasted: The Need Determination in the Wisconsin Landfill Siting Process*, 1987 Wis. L. Rev. 543; Joseph C. Gergits, *Enhancing the Community's Role in Landfill Siting in Illinois*, 1987 U. Ill. L. Rev. 97, 122. Approaches used in determining the need for HW facilities are discussed in Dennis M. Toft, *Site Selection of Hazardous Waste Facilities*, 7 Natural Resources and Envt. No. 3 at 6 (Winter 1993).

7. The Bureau of Labor Statistics' Producer Price Index for all industrial commodities rose from 58.4 in 1976 to 116.5 in 1991.

8. D. Kofi Asante-Duah et al., *The Hazardous Waste Trade: Can it Be Controlled?* 26 Envtl. Sci. Tech. 1684, 1685 (1992). *Similarly, see* Ellen Goodbaum and David Rotman, *Hazardous Waste: Faced With Dwindling Choices, Compa-*

nies Must Seek New Ways to Manage It, Chemical Wk., Aug. 23, 1989, at 18; Hsin-Neng Hsieh and Haydar Erdogan, *Cost Estimates for Several Hazardous Waste Disposal Options,* 5 Hazardous Waste and Hazardous Materials 329 (1988); Mazmanian and Morell 1992, *supra* ch. 1 note 5, at 128–29. Disposal fees for solid waste landfills have also escalated. Menell, *supra* ch. 5 note 1 at 665–66. For a theoretical discussion of how disposal prices should be set, *see* Jay R. Lund, *Pricing Solid- and Hazardous-Waste Landfill Capacity,* 116 J. Urb. Plan. and Dev. 17 (1990).

9. Jeff Bailey, *Hazardous-Waste Incinerators, Facing Overcapacity, Seek Regulators' Help,* Wall St. J., Dec. 13, 1993, at A9C; *No relief in hazwaste,* Chemical Wk., Jan. 5, 1994, at 43.

10. 42 U.S.C. §§ 10131(a)(1)–(2). Note that the NWPA also provides that the EIS for Yucca Mountain need not consider the need for the facility, 42 U.S.C. § 10134(f)(6) and that NRC may not consider the need for the monitored retrievable storage facility in licensing it, leaving that decision to DOE and Congress. 42 U.S.C. §§ 10161(d), 10168(c).

11. Makhijani and Saleska, *supra* ch. 3 note 14, at 41; *similarly,* Jacob, *supra* ch. 3 note 26, at 40, 182; Roger E. Kasperson et al., *Confronting Equity in Radioactive Waste Management: Modest Proposals for a Socially Just and Acceptable Program, in* Kasperson ed., *supra* ch. 3 note 4, at 331, 352; Eliot Marshall, *Thirty Ways to Temporize on Waste,* 237 Sci. 591 (1987); Carol Polsgrove, *Where Will We Dump the Nuclear Trash?* Progressive, Mar. 1983, at 22, 25–26.

12. U.S. NRC, *Waste Confidence Decision Review,* 54 Fed.Reg. 39767, 39790 (1989).

13. *Rethinking High Level Radioactive Waste Disposal, supra* ch. 3 note 17, at 7.

14. *Partnerships Under Pressure, supra* ch. 3 note 53, at 1, 8; Robert Reinhold, *States, Failing to Cooperate, Face a Nuclear-Waste Crisis,* N.Y. Times, Dec. 28, 1992, at A1; *States Outside Southeast Compact Face Sharply Higher Fees to Use Barnwell,* 23 Env't Rep. (BNA) No. 19, at 1313 (Sept. 4, 1992). The average cost of disposing of one cubic foot of Class A LLRW (the least radioactive) rose from $1 in 1975 to $42 in 1988. Contreras, *supra* ch. 1 note 2, at 529.

15. Contreras, *supra* ch. 1 note 2, at 520–21.

16. Ronald Coase, *The Problem of Social Cost,* 3 J. Law and Econ. 1 (1960).

17. A. Myrick Freeman III et al. eds., *The Economics of Environmental Policy* 27 (1973); Talbot Page, *Conservation and the Gospel of Economic Efficiency* 83–84 (1977); Kenneth S. Sewall, *The Tradeoff Between Cost and Risk in Hazardous Waste Management* 209 (1990); Roberta S. Gordon, *Legal Incentives for Reduction, Reuse, and Recycling: A New Approach to Hazardous Waste Management,* 95 Yale L.J. 810, 815–16 (1986). *See also* Congressional Budget Office, *Federal Options for Reducing Waste Disposal* 4 (1991) (states that optimal level of waste disposal is the point where the demand for waste disposal

services equals the marginal social cost); *similarly,* C. Miller, *Efficiency, Equity and Pollution: The Case of Radioactive Waste,* 19 Env't and Plan. A 913, 914 (1987).

18. For a more theoretical discussion of the interaction of zoning law and nuisance law, *see* Guido Calabresi and A. Douglas Melamed, *Property Rules, Liability Rules, and Inalienability: One View of the Cathedral,* 85 Harv. L. Rev. 1089 (1972).

19. *See* Norman Karlin, *Zoning and Other Land Use Controls: From the Supply Side,* 12 Sw. U. L. Rev. 561 (1981).

20. Michael F. Sheehan, *Economism, Democracy, and Hazardous Wastes: Some Policy Considerations, in* Kamieniecki et al., *supra* ch. 4 note 80, at 108, 122.

21. For one attempt, *see* John Schall, *Does the Solid Waste Management Hierarchy Make Sense? A Technical, Economic and Environmental Justification for the Priority of Source Reduction and Recycling* (School of Forestry and Envtl. Studies, Yale U., Oct. 1992). Much pertinent information (relevant especially to the externalities of incinerators) is found in Pace U. Center for Envtl. Legal Studies, *supra* ch. 3 note 108.

22. These studies follow extensive earlier work on the effects of air pollution on property values. *See* Lester B. Lave, *Air Pollution Damage: Some Difficulties in Estimating the Value of Abatement, in Environmental Quality Analysis: Theory and Method in the Social Sciences* 213 (Allen V. Kneese and Blair T. Bower eds., 1972); Daniel L. Rubinfeld, *Market Approaches to the Measurement of the Benefits of Air Pollution Abatement, in Approaches to Controlling Air Pollution* 240 (Ann F. Friedlaender ed. 1978). Other studies looked at the effects of nuclear power plants on property values. *See* Hays B. Gamble and Roger H. Downing, *Effects of Nuclear Power Plants on Residential Property Values,* 22 J. Reg Sci. 457 (1982); Jon P. Nelson, *Three Mile Island and Residential Property Values: Empirical Analysis and Policy Implications,* 57 Land Econ. No. 3, at 363 (Aug. 1981).

23. Kusum Ketkar, *Hazardous Waste Sites and Property Values in the State of New Jersey,* 24 Applied Econ. 647 (1992); Gary H. McClelland et al., *The Effects of Risk Beliefs on Property Values: A Case Study of a Hazardous Waste Site,* 10 Risk Analysis 485 (1990); R. Gregory Michaels and V. Kerry Smith, *Market Segmentation and Valuing Amenities With Hedonic Models: The Case of Hazardous Waste Sites,* 28 J. Urb. Econ. 223 (1990); Arthur C. Nelson et al., *Price Effects of Landfills on House Values,* 68 Land Econ. 359 (1992); V. Kerry Smith and William H. Desvousges, *The Value of Avoiding a LULU: Hazardous Waste Disposal Sites,* 68 Rev. Econ. and Stat. 293 (1986); Gerald E. Smolen et al., *Hazardous Waste Landfill Impacts on Local Property Values,* Real Est. Appraiser, Apr. 1992, at 4. *See also* Dana Milbank, *Back in Love Canal, Neighborhood Spirit Isn't Going to Waste,* Wall St. J., Sept. 25, 1992, at A5G (sale of homes in reinhabited Love Canal at about 20% below market value).

24. B.A. Payne et al., *The Effects on Property Values of Proximity to a Site Contaminated With Radioactive Waste,* 27 Nat. Resources J. 579 (1987); Janet

E. Kohlhase, *The Impact of Toxic Waste Sites on Housing Values*, 30 J. Urb. Econ. 1 (1991).

25. Brian Baker, *Perception of Hazardous Waste Disposal Facilities and Residential Real Property Values*, 6 Impact Assessment Bull. 47 (1988) (publicity concerning Love Canal affected property values in other upstate New York communities near hazardous waste sites).

26. Smolen, *supra* ch. 5 note 23, at 4 (proposed LLRW facility); *but see* William C. Metz, *Perceived Risk and Nuclear Waste in Nevada: A Mixture Leading to Economic Doom?* 10 Impact Assessment Bull. No. 3, at 23 (1992) (proposed siting of HLW facility has not had discernable impact on gaming-related tourist industry in Nevada).

27. Greenberg and Anderson, *supra* ch. 4 note 3, at 145–46; In the Matter of the Application of CECOS Int'l Inc., slip op. at 127–28 (Report of Administrative Law Judge, Aug. 21, 1989), *rev'd on other grounds* (N.Y. St. Industrial Hazardous Waste Facility Siting Bd. Mar. 13, 1990); Michael Elliot-Jones, *Rents and Proximity to Toxic Sites* (Spectrum Economics, undated).

28. David L. Bezer and Beverley S. Phillips, *Contaminated Property Valuation Issues: An Overview* 675 (Industrial Development Section, June 1990); Bill Mundy, *The Impact of Hazardous Materials on Property Value*, Appraisal J., Apr. 1992, at 155.

29. *See also New York: Sludge Landfill Would Lower Values of Nearby Homes*, Solid Waste Digest (Northeast ed.), May 1993, at 5 (report says proposed landfill for paper mill sludge would cause bordering properties to lose 12% of their value, with effect diminishing over distance for a two-mile radius); Kenneth Greiner and Mary Greiner v. N.J. Dep't of Envt'l Protection and Energy, OAL No. ECA 5401-92 (April 26, 1993) (decision of administrative law judge for NJDEPE) ($6,500 awarded for loss in value of home approximately one mile from NPL site).

30. 42 U.S.C. §§ 10136(c)(2), 10138(b)(3), 10156(e), 10161(f), 10167, 10169, 10173, 10173a.

31. 42 U.S.C. § 9607(a)(4).

32. Daigle v. Shell Oil Co., 972 F.2d 1527 (10th Cir. 1992); Brewer v. Ravan, 680 F. Supp. 1176 (M.D. Tenn. 1988); Artesian Water Co. v. Government of New Castle County, 659 F. Supp. 1269 (D. Del. 1987), *aff'd and remanded*, 851 F.2d 643 (3d Cir. 1988).

33. *See* Duffy, *supra* ch. 4 note 46, at 755, 788 (discussing Utah Code Ann. § 26-14a-7).

34. *See* James F. McAvoy, *Hazardous Waste Management in Ohio: The Problem of Siting*, 9 Cap. U. L. Rev. 435, 448 (1980) (discussing Ohio law).

35. In one case, the remedy of rescission was attempted. Smith v. Clark, No. 28019 (N.Y. Sup. Ct. Cortland County Mar. 23, 1990) (parties who had contracted to buy land attempted to rescind when LLRW site was proposed nearby; court dismissed complaint).

36. 42 U.S.C. § 6972(f); *see* Environmental Defense Fund v. Lamphier, 714 F.2d 331 (4th Cir. 1983).

37. *See* Denis J. Brion, *An Essay on LULU, NIMBY, and the Problem of Distributive Justice,* 15 Envtl. Aff. 437 (1988) [hereinafter Brion 1988]; Richard A. Epstein, *Nuisance Law: Corrective Justice and Its Utilitarian Constraints,* 8 J. Legal Stud. 49 (1979); Boomer v. Atlantic Cement Co., 26 N.Y.2d 219, 257 N.E.2d 870, 309 N.Y.S.2d 312 (1970). Application of the strict liability standard would solve many of these problems, but few courts have used it in hazardous waste cases. An exception is T and E Industries, Inc. v. Safety Light Corp., 123 N.J. 371, 587 A.D.2d 1249 (1991). *See also* Daigle v. Shell Oil Co., 972 F.2d 1527 (10th Cir. 1992) (holding that Colorado law might apply strict liability standard to hazardous waste disposal); *contra,* Fox v. McCoy Elec. Co., 21 Env't Rep. Cas. (BNA) 1945 (D. Pa. 1984).

38. Village of Wilsonville v. SCA Servs. Inc., 86 Ill.2d 1, 426 N.E.2d 824 (1981). *See also* Warner v. Waste Management, Inc., 36 Ohio St.3d 91, 521 N.E.2d 1091 (1988) (certifying class in action against HW facility).

39. E.g., Twitty v. State, 85 N.C. App. 42, 354 S.E.2d 296, *review denied,* 320 N.C. 177, 358 S.E.2d 69 (1987); Ortega Cabrera v. Municipality of Bayamon, 562 F.2d 91 (1st Cir. 1977); Salter v. B.W.S. Corp., 290 So.2d 821 (La. 1974); *Waste-Site Residents Lack Commonality, California Court Says in Denying Class Status,* 7 Toxics L. Rep. (BNA) No. 34, at 974 (Jan. 27, 1993); *Radioactive Chemicals: Property Damage Suit by Residents Barred by Federal Court Based on De Minimis Defense,* 8 Toxics L. Rep. (BNA) No. 11 at 300 (Aug. 18, 1993).

40. E.g., Sterling v. Velsicol Chem. Corp., 855 F.2d 1188 (6th Cir. 1988); Smith v. City of Brenham, 865 F.2d 662 (5th Cir.), *cert. denied,* 493 U.S. 813, 110 S. Ct. 60, 107 L.Ed.2d 27 (1989); Ayers v. Township of Jackson, 106 N.J. 557, 525 A.2d 287 (1987); Green v. Castle Concrete Co., 181 Colo. 309, 509 P.2d 588 (1973). *See* Frank B. Cross, *Environmentally Induced Cancer and The Law* 184–94 (1989); Charles J. Doane, *Beyond Fear: Articulating a Modern Doctrine of Anticipatory Nuisance for Enjoining Improbable Threats of Catastrophic Harm,* 17 Envtl. Aff. 441 (1990); Donald F. Pierce, *Recovery for Increased Risk of Developing a Future Injury From Exposure to a Toxic Substance,* 19 Envtl. L. Rep. (Envt. L. Inst.) 10,256 (1989); Andrew H. Sharp, *Comment, An Ounce of Prevention: Rehabilitating the Anticipatory Nuisance Doctrine,* 15 Envtl. Aff. 627 (1988); Edward Felsenthal, *Risk-of-Illness Cases Are Getting Unsympathetic Ear From Courts,* Wall St. J., July 7, 1993, at B8. Two states do, however, have statutes that permit injunctive relief for anticipatory nuisances where the consequences are "to a reasonable degree certain." Ala. Code § 6–5–125; Ga. Code Ann. § 41-2-4.

41. E.g., Nevada v. Burford, 918 F.2d 854 (9th Cir. 1990), *cert. denied,* 500 U.S. 932 (1991); Richmond County v. North Carolina Low-Level Radioactive Waste Management Authority, 436 S.E.2d 113 (N.C. 1993); Granville County Bd. of Comm'rs v. North Carolina Hazardous Waste Management Comm'n, 329 N.C. 615, 407 S.E.2d 785 (1991). *See also* U.S. v. Taylor, 8 F.3d 1074 (6th Cir. 1993).

42. Kara M. Bruge-Holland, *Comment: Constitutional Law—Pre-Taking Activities Pursuant to the New Jersey Hazardous Waste Facility Siting Process Do Not Implicate Constitutional Just Compensation Privileges; Rather, Administrative Recognition of Certain Property Rights Resulting Therefrom Should Be Considered.* Littman *v.* Gimello, *115 N.J. 154, 557 A.2d 314,* cert. denied, *110 S.Ct. 324 (1989),* 22 Rutgers L.J. 485 (1991); Elizabeth Anne Barba, *Constitutional Law—Eminent Domain—Designation of Property as Potential Site for Hazardous Waste Facility Does Not Constitute a Compensable Taking,* 20 Seton Hall L. Rev. 335 (1989). But see Hendler v. United States, 952 F.2d 1364 (Fed. Cir. 1991) (compensation must be paid for sinking monitoring wells on property adjoining CERCLA site). *See also* Philip Weinberg, *Hendler v. United States: "I'll Let You Save Me—If You Pay Me for the Privilege,"* 17 Colum. J. Envtl. L. 401 (1992).

43. Rolf Lidskog and Ingemar Elander, *Reinterpreting Locational Conflicts: NIMBY and Nuclear Waste Management in Sweden,* 20 Policy and Politics 249, 258 (1992).

44. E.g., City of Santa Fe v. Komis, 114 N.M. 659, 845 P.2d 753 (1992) (landowner awarded damages because highway through his property would carry nuclear waste to Waste Isolation Pilot Plant, diminishing value of land due to fear of the waste); Texas Elec. Serv. Co. v. Nelon, 546 S.W.2d 864 (Tex. Civ. App. 1977) (allowed recovery for fear of owner of land adjacent to railroad through which nuclear wastes were transported); Heddin v. Delhi Gas Pipeline Co., 522 S.W.2d 886 (Tex. 1975) (damages awarded to landowner who feared that pipeline on adjoining land would explode); Lunda v. Matthews, 46 Or. App. 701, 613 P.2d 63 (1980) (emotional distress damages available for air emissions from cement plant). *See also* Day v. NLO, Inc., 3 F.3d 153 (6th Cir. 1993) (class action concerning nuclear fuel facility in Fernald, Ohio).

45. A more common outcome was that in Adkins v. Thomas Solvent Co., 440 Mich. 293, 487 N.W.2d 715 (1992), rejecting claims by homeowners for compensation of decreased property values they feared would result from soil and water contamination in area. *Similarly, see* Boughton v. Cotter Corp., 8 Toxics L. Rep. 458 (BNA) (D. Colo. No. 89-Z-1505, Sept. 9, 1993). Many of the cases on recovery for emotional distress are discussed in Martha A. Churchill, *Arguing Public Policy as a Defense to Environmental Toxic Tort Claims,* 8 Toxics L. Rep. (BNA) No. 17 at 505 (Sept. 29, 1993).

46. Ky. Rev. Stat. Ann. § 224.46-830(2)(a) (Michie/Bobbs-Merrill 1991).

47. 49 Fed. Reg. 47747 (Dec. 6, 1984).

48. Geo-Tech Reclamation Indus. Inc. v. Hamrick, 886 F.2d 662 (4th Cir. 1989). *See also* City of Cleburne v. Cleburne Living Center, Inc., 473 U.S. 432 (1985) (city could not zone out a home for the mentally retarded based on impermissible motives); In the Matter of the Application of Combined Air and Solid Waste Permit No. 2211-91-OT-1, 489 N.W.2d 811 (Minn. Ct. App. 1992) (generalized concern about possible adverse effects is insufficient to support permit denial).

Cases concerning the courts' shifting attitudes toward the ability of local governments to prohibit feared land uses are reviewed in David Bernstein, *From Pesthouses to AIDS Hospices: Neighbors' Irrational Fears of Treatment Facilities and Contagious Diseases,* 33 Colum. Human Rights L. Rev. 1 (1990), and Harold A. Ellis, *Neighborhood Opposition and the Permissible Purposes of Zoning,* 7 J. Land Use and Envtl. L. 275 (1992).

49. In the Matter of the Application of CECOS Int'l, Inc., slip op. at 3 (N.Y. St. Dep't of Envtl. Conservation, Decision of the Commissioner, Mar. 13, 1990). This hearing is discussed in detail in Michael R. Edelstein, *Psychosocial Impacts on Trial: The Case of Hazardous Waste Disposal, in Psychosocial Effects of Hazardous Toxic Waste Disposal on Communities* 153 (Dennis L. Peck ed., 1989) [hereinafter Edelstein 1989]. *See also* Metropolitan Edison Co. v. People Against Nuclear Energy, 460 U.S. 766, 103 S.Ct. 1556, 75 L.Ed.2d 534 (1983) (EIS on reopening of undamaged nuclear power plant at Three Mile Island need not discuss neighbors' fears).

50. Roger A. Bohrer, *Fear and Trembling in the Twentieth Century: Technological Risk, Uncertainty and Emotional Distress,* 1984 Wis. L. Rev. 83, 111. *Similarly,* Rodney Fort et al., *Perception Costs and NIMBY,* 38 J. Envtl. Management 185 (1993). Another review of the cases is found in William A. Barton, *Recovering for Psychological Injuries* (1985).

51. E.g., Anderson and Greenberg, *supra* ch. 4 note 49, at 207; Caskey, *supra* ch. 1 note 6, at 58; Letty G. Lutzker, *Making the World Safe for Chicken Little, or the Risks of Risk Aversion, in* Burns ed., *supra* ch. 3 note 11, at 175; Schmeidler and Sandman, *supra* ch. 5 note 2, at 29; Paul Slovic et al., *Perceived Risk, Trust, and the Politics of Nuclear Waste,* 254 Sci. 1603, 1603 (1991) [hereinafter Slovik 1991]; Tarlock 1984, *supra* ch. 4 note 43, at 433.

52. *See* Michael B. Gerrard, *The Dynamics of Secrecy in the Environmental Impact Statement Process,* 2 N.Y.U. Envt'l L.J. 279 (1993).

53. E.g., Michael P. Scott and Stephen T. Washburn, *The Role of Risk Assessment in the Siting of the OWMC Waste Management Facility* (Environ Corp., undated). *See* Luke W. Cole, *Remedies for Environmental Racism: A View from the Field,* 80 Mich. L. Rev. 1991, 1994 (1992) [hereinafter Cole 1992]. *See also* Suzanne Keller, *Ecology and Community,* 19 Envtl. Aff. 623, 624 (1992) (several reports prior to Three Mile Island accident predicted that risks of a serious accident at a nuclear power plant are tiny).

54. *See* Mary L. Lyndon, *Risk Assessment, Risk Communication and Legitimacy: An Introduction to the Symposium,* 14 Colum. J. Envtl. L. 289, 291 (1989).

55. Howard Kunreuther et al., *Decision-Process Perspective on Risk and Policy Analysis,* in Lake ed., *supra* ch. 1 note 11, at 260, 261.

56. National Research Council, *Environmental Epidemiology: Public Health and Hazardous Wastes* 1, 19 (1991); *see also* W. Norton Grubb et al., *The Ambiguities of Benefit-Cost Analysis: An Evaluation of Regulatory Impact*

Analysis Under Executive Order 12291, in Environmental Policy Under Reagan's Executive Order: The Role of Benefit-Cost Analysis (V. Kerry Smith ed., 1984) at 121, 149–51 (EPA study concluded that the benefits of CERCLA are "highly intangible" and that the health effects of cleaning up hazardous waste sites could not be estimated). *But see* Envtl. Health Network and National Toxics Campaign Fund, *Inconclusive By Design: Waste, Fraud and Abuse in Federal Environmental Health Research* (1992) (charges that federal studies of health risks at hazardous waste sites, especially those conducted by the Agency for Toxic Substances and Disease Registry, greatly understate risks).

57. Mary P. Harmon and Kathryn Coe, *Cancer Mortality in U.S. Counties With Hazardous Waste Sites*, 14 Population and Environment 463 (1993). *See also* Anthony C. Gatrell and Andrew A. Lovett, *Burning Questions: Incineration of Wastes and Implications for Human Health*, in *Waste Location: Spatial Aspects of Waste Management, Hazards and Disposal* (Michael Clark et al. eds., 1992) at 143 (found statistically significant correlation between laryngeal cancer and proximity to hazardous waste incinerator, but causation could not be established).

58. Raymond Neutra et al., *Hypotheses to Explain the Higher Symptom Rates Observed Around Hazardous Waste Sites*, 94 Envtl. Health Perspectives 31 (1991); Dennis Shusterman et al., *Symptom Prevalence and Odor-Worry Interaction Near Hazardous Waste Sites*, 94 Envtl. Health Perspectives 25 (1991).

59. Sandra A. Geschwind et al., *Risk of Congenital Malformations Associated With Proximity to Hazardous Waste Sites*, 135 Am. J. Epidemiology 1197 (1992). A similar, but not quite as strong, correlation was found in Mark C. Fulcomer et al., *Population-Based Surveillance and Etiological Research of Adverse Reproductive Outcomes and Toxic Wastes—Phase II: Correlational Analysis of Adverse Reproductive Outcomes and Environmental Pollution* (N.J. Dep't of Health, March 1992).

60. *Complex Cleanup, supra* ch. 3 note 8, at 8.

61. India Fleming et al., *Chronic Stress and Toxic Waste: The Role of Uncertainty and Helplessness*, 21 J. Applied Social Psychology 1889 (1991); Steven P. Schwartz et al., *Environmental Threats, Communities, and Hysteria*, 6 J. Public Health Policy 58 (1985).

62. *See* U.S. Environmental Protection Agency, Pub. No. 300-R92-008, *Enforcement Accomplishments Report FY91* (Apr. 1992). *See also Hazardous Waste: 'Cluster' Enforcement Action Brought Against Hazardous Waste Combustion Units*, 24 Env't Rep. (BNA) No. 22 at 995 (Oct. 1, 1993).

63. *RCRA Environmental Indicators, supra* ch. 2 note 12, at 5-5.

64. Clifford S. Russell, *Monitoring and Enforcement, in Public Policies for Environmental Protection* 243, 262 (Paul R. Portney ed., 1990).

65. Hunt v. Chemical Waste Management, Inc., 584 So.2d 1367 (Ala. 1991).

66. New York St. Dep't of Envtl. Conservation, *Report of On-Site Monitoring Activity at CWM Chemical Services, Inc. by NYS DEC Region 9* (quarterly reports).

67. In the Matter of the Application of CECOS Int'l, Inc., slip op. at 72–78 (Report of Administrative Law Judge Aug. 21, 1989) (discussing violations at BFI landfills in New York, Louisiana, and Ohio), *rev'd on other grounds* (N.Y. St. Industrial Hazardous Waste Facility Siting Bd. Mar. 13, 1990).

68. Jeff Bailey, *Concerns Mount Over Operating Methods of Plants That Incinerate Toxic Waste,* Wall St. J., Mar. 20, 1992, at B1; Julia Flynn, *The Ugly Mess at Waste Management,* Bus. Wk., Apr. 13, 1992, at 76; *Waste Management Unit's Plant is Curbed by EPA,* Wall St. J., Mar. 31, 1992; Harvey L. White, *Hazardous Waste Incineration and Minority Communities,* in Bryant and Mohai eds., *supra* ch. 3 note 120, at 126, 137.

69. U.S. Gen. Accounting Office, Pub. No. GAO/RCED-92–78, *Hazardous Waste: A North Carolina Incinerator's Noncompliance With EPA and OSHA Requirements* (June 1992).

70. *Hazardous Waste Facility Siting in New York State, supra* ch. 4 note 34, at 9.

71. *Feds Target Hazardous Waste Incinerator Safety,* Eng'g News-Record, June 17, 1991, at 5.

72. Robert Tomsho, *Small Town in Alberta Embraces What Most Reject: Toxic Waste,* Wall St. J., Dec. 27, 1991, at 1.

73. *Hazardous Incinerators?* 143 Science News 334 (May 22, 1993).

74. *Test Burn Data Show Dioxin Emissions Higher Than Expected From WTI Incinerator,* 24 Env't Rep. (BNA) No. 8 at 346, June 25, 1993.

75. Richard M. Sedman and John R. Esparza, *Evaluation of the Public Health Risks Associated with Semivolatile Metal and Dioxin Emissions from Hazardous Waste Incinerators,* 94 Envtl. Health Perspectives 181 (1991).

76. Memorandum from Sylvia K. Lowrance, director, Office of Solid Waste, EPA, to Waste Management Division directors, *Assuring Protective Operation of Incinerators Burning Dioxin-Listed Wastes,* Sept. 22, 1992. *See also New EPA Study Indicates WTI Dioxin Risk May Be 1,000 Times Above Previous Estimates,* 23 Env't Rep. (BNA) No. 43, at 2714 (Feb. 19, 1993).

77. For a theoretical analysis, *see* Venu G. Balagopal, *Total Probable Risk Analysis: A Technique for Quantitative Risk Evaluation of Hazardous Waste Disposal Options,* 6 Hazardous Waste and Hazardous Materials 315 (1989).

78. *Employee Health Risks at Toxic Waste Sites: What Don't We Know?: Hearings before Subcomm. on the Civil Service, House Comm. on Post Office and Civil Service,* 101st Cong., 2d Sess., 99, 100 (Sept. 18, 1990) (statement of David A. Lewis, Physicians for Social Responsibility); Mary H. Melville, *Temporary Workers in the Nuclear Power Industry: Implications for the Waste Management Program,* in Kasperson ed., *supra* ch. 3 note 4, at 229.

79. George Friedman-Jimenez, *Occupational Disease Among Minority Workers,* 37 Ass'n Occupational Health Nurses J. 64 (1989).

80. Ravinder Mamtani and Joseph A. Cimino, *Work Related Diseases Among Sanitation Workers of New York City,* 55 J. Envtl. Health 27 (1992).

81. Mary H. Melville, *Temporary Workers in the Nuclear Power Industry: Implications for the Waste Management Program, in* Kasperson ed., *supra* ch. 3 note 4, at 229.

82. The large degree of legislative and regulatory confusion in the setting of health-based and technology-based standards for hazardous waste facilities is discussed in John S. Applegate, *The Perils of Unreasonable Risk: Information, Regulatory Policy, and Toxic Substances Control,* 91 Colum. L. Rev. 261 (1991); Donald A. Brown, *EPA's Resolution of the Conflict Between Cleanup Costs and the Law in Setting Cleanup Standards Under Superfund,* 15 Colum. J. Envtl. L. 241 (1990); and Thomas A. Cinti, *The Regulator's Dilemma: Should Best Available Technology or Cost Benefit Analysis Be Used to Determine the Applicable Hazardous Waste Treatment, Storage, and Disposal Technology?* 16 Rutgers Computer and Tech. L.J. 145 (1990).

83. Benjamin A. Goldman et al., *Hazardous Waste Management: Reducing the Risk* 246 (1986); *Truck Transport of Wastes as Risky as Treatment, Disposal, Consultant Says,* 14 Env't Rep. (BNA) No. 18, at 733 (Sept. 2, 1983).

84. Robert D. Mutch, Jr., and W. Wesley Eckenfelder, Jr., *Out of the Dusty Archives: The History of Waste Management Becomes a Critical Issue in Insurance Litigation,* Hazmat World, Oct. 1993, at 59.

85. Green, *supra* ch. 2 note 24, at xviii (fear of illegal dumping resulting from capacity shortfalls was one of the reasons Congress enacted the capacity assurance provisions of CERCLA).

86. E.g., Village of Wilsonville v. SCA Servs., Inc., 86 Ill.2d 1, 426 N.E.2d 824, 838 (1981).

87. U.S. Environmental Protection Agency, *Interim Standards for Owners and Operators of New Hazardous Waste Land Disposal Facilities and EPA Administered Permit Programs: The Hazardous Waste Permit Program,* 46 Fed.Reg. 12414, 12416 (1981); Hazardous Waste Disposal Advisory Committee and Environmental Facilities Corp., *A Comprehensive Program for Hazardous Waste Disposal in New York State: Report to Governor Hugh L. Cary and the New York State Legislature* 13 (March 1980).

88. E.g., Morell and Magorian, *supra* ch. 1 note 8, at 69, 184; Landy et al., *supra* ch. 2 note 46, at 125; Andreen, *supra* ch. 1 note 8, at 847; Bacow and Milkey, *supra* ch. 1 note 6, at 267, 268; Colgazier and English, *supra* ch. 1 note 8; Delogu, *supra* ch. 4 note 159, at 200; Farkas, *supra* ch. 4 note 42, at 453; Gergits, *supra* ch. 5 note 6, at 101; Robert W. Craig and Terry R. Lash, *Siting Nonradioactive Hazardous Waste Facilities,* in Harthill, *supra* ch. 4 note 49, at 99, 104; Robert W. Hahn, *An Evaluation of Options for Reducing Hazardous Waste,* 12 Harv. Envtl. L. Rev. 201, 211 (1988); Jeffrey Kehne, *Encouraging Safety Through Insurance-Based Incentives: Financial Responsibility for Hazardous Wastes,* 96 Yale L.J. 403, 424 (1986); Matheny and Williams, *supra* ch. 4 note 46, at 78; Popper, *supra* ch. 1 note 11, at 10; Gerard M. McCabe, *The Validity of State Symmetry Requirements Banning Hazardous Waste from Environmentally Indolent States,* 10 Temp. Envtl. L. and Tech. J. 169, 173 (1991);

Robert Cameron Mitchell and Richard T. Carson, *Property Rights, Protest, and the Siting of Hazardous Waste Facilities*, 76 AEA Papers and Proc. 285, 285 (1986); Peter M. Sandman, *Getting to Maybe: Some Communications Aspects of Siting Hazardous Waste Facilities*, 9 Seton Hall Legis. J. 437, 442 (1985); Harlan T. Snider, *A New Approach to Pennsylvania's Hazardous Waste Siting Problem*, 5 Temp. Envtl. L. and Tech. J. 49, 52 (1986); Tarlock 1981, *supra* ch. 1 note 9, at 5; R. George Wright, *Hazardous Waste Disposal and the Problems of Stigmatic and Racial Injury*, 23 Ariz. St. L.J. 777, 792 n.86 (1992).

89. Richard F. Anders, *Public Participation in Hazardous Waste Facility Location Decisions*, 1 J. Plan. Literature 145, 147 (1986).

90. Douglas W. McNiel et al., *New Superfund Legislation: Major Provisions, Revenue Sources, and Economic Incentives for Environmental Protection*, 35 Oil and Gas Tax Q. 610, 617 n.18 (1987).

91. Mazmanian and Morell 1992, *supra* ch. 1 note 5, at 107.

92. This statement is meant to apply to the United States; it may not be accurate in certain other regions of the world. *See Asia/Pacific Moves to Address Mountainous Waste Problem; Lack of Infrastructure Leads to Indiscriminate Dumping*, Chemical Wk, Mar. 24, 1993, at 42.

93. Discrepancies in storage volume figures lead to the inference that some of this material is unlawfully entering nonqualified disposal facilities. *Partnerships Under Pressure*, *supra* ch. 3 note 53, at 3, 15.

94. Brown, *supra* ch. 5 note 82, at 262.

95. Id. at 257.

96. Id. at 250.

97. Letta Taylor, *Waste-Dumping Sting*, Newsday, May 13, 1992, at 38; Jerry Cassidy, *Toxic Avenger Busts 5 in Sting*, N.Y. Daily News (Long Island ed.), May 13, 1992, at QLI 1; Josh Barbanel, *Elaborate Sting Operation Brings Arrests in Illegal Dumping of Toxic Wastes by Businesses*, N.Y. Times, May 13, 1992, at B5.

98. O'Neil v. Picillo, 682 F. Supp. 706, 710 (D.R.I. 1988), *aff'd*, 883 F.2d 176 (1st Cir. 1989), *cert. denied* 493 U.S. 1071 (1990).

99. Seymour I. Schwartz and Wendy B. Pratt, *Hazardous Waste From Small Quantity Generators: Strategies and Solutions for Business and Government* 19 (1990).

100. McCarthy and Reisch, *supra* ch. 2 note 12, at 16. *Similarly, National Biennial Report*, *supra* ch. 2 note 8, at 1.

101. 40 C.F.R. § 260.10.

102. James K. Hammitt and Peter Reuter, *Measuring and Deterring Illegal Disposal of Hazardous Waste: A Preliminary Assessment* 11, 16 (RAND Corp. Pub. No. R-3657-EPA/JMO, 1988); Donald Rebovich, *Understanding Hazardous Waste Crime: A Multistate Examination of Offense and Offender Charac-*

teristics in the Northeast 22 (Northeast Hazardous Waste Project and N.J. Div. of Criminal Justice, June 1986); Schwartz and Pratt, *supra* ch. 5 note 99, at 13–30; Bruce A. Williams and Albert R. Matheny, *Testing Theories of Social Regulation: Hazardous Waste Regulation in the American States*, 46 J. Pol. 428, 452 (1984).

103. Hammitt and Reuter, *supra* ch. 5 note 102, at 18.

104. New York St. Assembly Envtl. Conservation Comm., *Organized Crime's Involvement in the Waste Hauling Industry* (1986); Joseph F. Sullivan, *12 Held in Trucking of Untaxed and Contaminated Oil*, N.Y. Times, May 28, 1993, at B5; Alan A. Block and Frank R. Scarpitti, *Poisoning for Profit: The Mafia and Toxic Waste in America* (1985); Andrew Szasz, *Corporations, Organized Crime, and the Disposal of Hazardous Waste: An Examination of the Making of a Criminogenic Regulatory Structure*, 24 Criminology 1, 2–3 (1986). Readers of the Block and Scarpitti work should be cautioned that its publisher, William Morrow and Co., was sued for libel after the book's publication; the suit was settled under terms that were to be kept confidential, the book was allowed to go out of print, and the publisher has no plans to reprint it. Telephone conversation with Robert Hawley, Law Department, William Morrow and Co., Oct. 28, 1992.

105. Hammitt and Reuter, *supra* ch. 5 note 102, at 22–25; Schwartz and Pratt, *supra* ch. 5 note 99, at 25–26; New York St. Assembly Envtl. Conservation Comm., *supra* ch. 5 note 104, at 137.

106. Donald J. Rebovich, *Policing Hazardous Waste Crime: The Importance of Regulatory/Law Enforcement Strategies and Cooperation in Offender Identification and Prosecution*, 9 Crim. Justice Q. 173 (1987); U.S. Gen. Accounting Office, Pub. No. GAO/RCED-85-2, *Illegal Disposal of Hazardous Waste: Difficult to Detect or Deter* (Feb. 22, 1985).

107. English 1992, *supra* ch. 1 note 8, at 117–60; Peter S. Wenz, *Environmental Justice* (1988); Been, *supra* ch. 4 note 99; Roger E. Kasperson et al., *Confronting Equity in Radioactive Waste Management: Modest Proposals for a Socially Just and Acceptable Program*, *in* Kasperson ed., *supra* ch. 3 note 4, at 331; Reg Lang, *Fair Siting in Waste Management* in Inst. for Soc. Impact Assessment, *supra* ch. 2 note 25; Ted F. Peters, *Ethical Considerations Surrounding Nuclear Waste Repository Siting and Mitigation, in Nuclear Waste: Socioeconomic Dimensions of Long-Term Storage* 41 (Steve H. Murdock et al. eds., 1983); Peter Timmerman, *The Ethical Sphere, in* Inst. for Soc. Impact Assessment *supra* ch. 2 note 25.

108. *An Introduction to the Principles of Morals and Legislation* (1789).

109. *Utilitarianism* (1863).

110. *Groundwork for the Metaphysics of Morals* (1785).

111. *A Theory of Justice* (1971).

112. Jacob, *supra* ch. 3 note 26, at 169–70.

113. 42 U.S.C. §§ 10161(g), 10165(g).

114. Contreras, *supra* ch. 1 note 2, at 517–19; English 1992, *supra* ch. 1 note 8, at 127.

115. E.g., Mich. Comp. Laws Ann. § 299.509(2)(f); N.Y. Envtl. Conserv. Law § 27-1102(2)(f) (McKinney Supp. 1992).

116. E.g., Panel on Social and Econ. Aspects of Radioactive Waste Management, National Research Council, *Social and Economic Aspects of Radioactive Waste Disposal: Considerations for Institutional Management* 73 (1984). This was the approach taken by New York State in choosing a site near a General Electric plant for disposal of PCBs dumped by that plant into the Hudson River. *See* Washington County Cease v. Persico, 99 A.D.2d 321, 473 N.Y.S.2d 610 (3d Dept. 1984), *aff'd mem.*, 64 N.Y.2d 923, 477 N.E.2d 1084, 488 N.Y.S.2d 630 (1985).

117. E.g., Edelstein 1988, *supra* ch. 4 note 61, at 186; Roger E. Kasperson, *Social Issues in Radioactive Waste Management: The National Experience, in* Kasperson ed., *supra* ch. 3 note 4, at 24, 50.

118. *See* Tom Anderson, *Residents Plead: Reject Sludge Plan,* Reporter-Dispatch (Westchester County, N.Y.), Jan. 22, 1993, at 5B ("Yonkers residents who said they were already burdened by a sewage treatment plant in their neighborhood beseeched county officials last night to reject a proposal for a sludge processing facility next to the sewage plant.").

119. This contrasts with the decision by the Canadian government to focus the search for a site for its high-level radioactive waste repository in the populous province of Ontario, because that is where most of Canada's nuclear power plants are located, rather than in more remote areas of the country. However, this repository has not yet been sited. Barry D. Solomon and Fred M. Shelley, *Siting Patterns for Nuclear Waste Repositories,* J. Geography, Mar.–Apr. 1988, at 59, 63.

120. *See* Olsen, *supra* ch. 4 note 3.

121. *See* Jonathan Kozol, *Savage Inequalities* 15–20 (1991); Scott McMurray, *Denying Paternity: Monsanto Case Shows How Hard It Is to Tie Pollution to a Source,* Wall. St. J., June 17, 1992, at 1.

122. Miles Corwin, *Vernon Redevelopment Plan is Sticky Business,* Los Angeles Times, Nov. 10, 1991, at B1.

123. *Hearings Set for Proposed Incinerator,* UPI, July 4, 1989, *available in* LEXIS, Nexis Library, UPI File; *Utah Board Grants RCRA Permit to USPC Facility,* Env't Wk., Nov. 7, 1991.

124. Conner Bailey and Charles E. Faupel, *Environmentalism and Civil Rights in Sumter County, Alabama, in* Bryant and Mohai eds., *supra* ch. 3 note 120, at 140, 150. *See also* Keith Schneider, *Plan for Toxic Dump Pits Blacks Against Blacks,* N.Y. Times, Dec. 13, 1993, at A12.

125. *See* Robert Suro, *Pollution-Weary Minorities Try Civil Rights Track,* N.Y. Times, Jan. 11, 1993, at 1.

126. E.g., Brian J.L. Berry ed., *The Social Burdens of Environmental Pollution: A Comparative Metropolitan Data Source* (1977); Jeffrey M. Zupan, *The Distribution of Air Quality in the New York Region* (1977); Texas Center for Policy Studies, *Toxics in Texas and Their Impact on Communities of Color* (Austin 1993). Most of these studies are assembled in Richard J. Lazarus, *Pursuing 'Environmental Justice': The Distributional Effects of Environmental Protection*, 87 Nw. U.L. Rev. 787 (1992), and Paul Mohai and Bunyan Bryant, *Environmental Justice: Weighing Race and Class as Factors in the Distribution of Environmental Hazards*, 63 U. Colo. L. Rev. 921 (1992). *See also* Maurice D. Hinchey et al., New York State Assembly Envtl. Conservation Comm., *Minorities and the Environment: An exploration into the effects of environmental policies, practices and conditions on minority and low-income communities* 5 (1991) (urban minority areas are especially prone to illegal dumping in vacant lots); Louisiana Advisory Committee to the U.S. Commission on Civil Rights, *The Battle for Environmental Justice in Louisiana . . . Government, Industry, and the People* (Sept. 1993).

127. U.S. Gen. Accounting Office, Pub. No. GAO/RCED-83–168, *Siting of Hazardous Waste Landfills and Their Correlation With Racial and Economic Status of Surrounding Communities* (1983). This study did not "determine why the sites were selected, the population-mix of the area when the site was established, the distribution of the population around the landfill, nor how the communities' racial and economic status compared to others in the State." Id. at 3.

128. This study has been favorably cited at the highest levels of government. *See* Al Gore, *Earth in the Balance* 149 (1992).

129. Comprehensive Environmental Response, Compensation and Liability Information System.

130. Commission for Racial Justice, United Church of Christ, *Toxic Wastes and Race in the United States* xiii (1987).

131. Id. at 13.

132. Id. at 53.

133. Robert D. Bullard, *Solid Waste Sites and the Black Houston Community*, 53 Soc. Inquiry 273 (1983).

134. Mohai and Bryant, *supra* ch. 5 note 126, at 927.

135. Pat Costner and Joe Thornton, *Playing With Fire: Hazardous Waste Incineration* (Greenpeace Publications ed., 1990).

136. Greenberg and Anderson, *supra* ch. 4 note 3, at 158–59.

137. Rae Zimmerman, *Risk and Public Controversy at Hazardous Waste Sites*, Executive Summary (Urban Research Center, N.Y. Univ., Feb. 1992); Rae Zimmerman, *Social Equity and Environmental Risk*, 13 Risk Analysis 649 (1993).

138. John A. Hird, *Environmental Policy and Equity: The Case of Superfund*, 12 J. Policy Analysis and Management 323 (1993).

139. Clean Sites, Inc., *Hazardous Waste Sites and the Rural Poor: A Preliminary Assessment* viii, ix (Mar. 1990).

140. Marianne Lavelle and Marcia Coyle, *Unequal Protection: The Racial Divide in Environmental Law,* Nat'l L.J., Sept. 21, 1992, at S1. *But see Georgia: State Report Looks at Waste Site Fines, Finds No Discrimination in Minority Areas,* 24 Env't Rep. (BNA) No. 6 at 284 (June 11, 1993). The Lavelle and Coyle study is attacked in Mary Bryant, *Unequal Justice? Lies, Damn Lies, and Statistics Revisited,* SONREEL News (American Bar Ass'n) 3 (Sept/Oct. 1993). A separate 1992 study also examined the correlation between the pace of cleanup of NPL sites and community racial composition. Rae Zimmerman, *Social Equity and Environmental Risk, supra* ch. 5 note 137. Anecdotal evidence is contained in Marc Cooper, *The Sickness on Evelina Street,* Village Voice, Sept. 7, 1993 at 33.

141. U.S. Environmental Protection Agency, *Environmental Equity: Reducing Risk for All Communities* 11 (June 1992); George A. Kaplan, et al., *Socioeconomic Status and Health,* in *Closing the Gap: The Burden of Unnecessary Illness* (Robert W. Amler and H. Bruce Dull eds., 1987) at 125; Austin and Schill, *supra* ch. 4 note 59, at 76–77.

142. Memorandum from Philip W. Johnston, secretary, Executive Office of Human Services, Commonwealth of Massachusetts, to John DeVillars, secretary, Executive Office of Environmental Affairs, Sept. 14, 1990. This finding was criticized by the facility's proponents as the result of a simple statistical error. *See* Michael O'Hare and Debra Sanderson, *Facility Siting and Compensation: Lessons From the Massachusetts Experience,* 12 J. Policy Analysis and Management 364, 374 (1993).

143. E.g., Kelly Michele Colquette and Elizabeth A. Henry Robertson, *Environmental Racism: The Causes, Consequences and Commendations,* 5 Tul. Envtl. L.J. 153, 170 (1991); Daniel Suman, *Robert Bullard: Dumping in Dixie: Race, Class and Environmental Quality,* 19 Ecology L.Q. 591, 596–97 (1992) (book review).

144. *See* Austin and Schill, *supra* ch. 4 note 59, at 69–70; Been, *supra* ch. 4 note 99; Bernstein, *supra* ch. 2 note 18, at 83–84; Robert W. Collin, *Environmental Equity: A Law and Planning Approach to Environmental Racism,* 11 Va. Envtl. L. J. 495 (1992); O'Hare 1977, *supra* ch. 4 note 159, at 454. *See also* Vicki Been, *Locally Undesirable Land Uses in Minority Neighborhoods: Disproportionate Siting or Market Dynamics?* 103 Yale L.J. 1383 (1994) (statistical analysis of demographic characteristics of communities at time HW facilities were sited).

145. *See* Greenberg and Anderson, *supra* ch. 4 note 3, at 158 (survey shows that towns with economic bases oriented toward industry are far more likely to have numerous dumpsites).

146. Robert D. Bullard, *Dumping in Dixie: Race, Class, and Environmental Quality* 7, 10 (1990).

147. Lawrence S. Bacow, *Waste and Fairness: No Easy Answers,* 8 Forum for Applied Research and Public Policy 43 (1993).

148. 42 U.S.C. § 1971.

149. Max Neiman and Ronald O. Loverridge, *Environmentalism and Local Growth Control: A Probe into the Class Bias Theories,* 6 Env't and Behavior 759, 760 (1981).

150. Id. at 768; Morell and Magorian, *supra* ch. 1 note 8, at 155; Kent E. Portney, *Siting Hazardous Waste Treatment Facilities: The NIMBY Syndrome* 71 (1991); Dorceta E. Taylor, *Blacks and the Environment: Toward an Explanation of the Concern and Action Gap Between Blacks and Whites,* 21 Env't and Behavior 175 (1989).

151. Two studies have examined this question. One found a correlation between voter turnout rates and HW facility expansion. James T. Hamilton, *Politics and Social Costs: Estimating the Impact of Collective Action on Hazardous Waste Facilities,* 24 RAND J. Econ. 101 (1993). The other, which looked at other surrogates for political power, did not find a correlation. Greenberg and Anderson, *supra* ch. 4 note 3, at 157.

152. *See* Foster Church, *Can Nevada Keep America's Sizzling Nuclear Waste Out of Its Backyard?* Governing, Apr. 1990, at 21; Easterling, *supra* ch. 4 note 47, at 465; Shapiro, *supra* ch. 3 note 24.

153. Kenneth M. Bachrach and Alex J. Zautra, *Assessing the Impact of Hazardous Waste Facilities: Psychology, Politics, and Environmental Impact Statements, in Exposure to Hazardous Substances: Psychological Parameters* 71, 74 (Allen Lebovitz et al. eds., 1986). *See also* Bullard 1983, *supra* ch. 5 note 133, at 275 (political explanation for siting of MSW facilities in Houston).

154. Luke W. Cole, *Empowerment as the Key to Equal Protection: The Need for Environmental Poverty Law,* 19 Ecology L.Q. 619 (1993) [hereinafter Cole 1993].

155. R.I.S.E., Inc. v. Kay, 1992 U.S. App. Lexis 26732 (4th Cir. Oct. 15, 1992), *aff'g* 768 F. Supp. 1144 (E.D. Va. 1991); East Bibb Twiggs Neighborhood Ass'n v. Macon-Bibb County Planning and Zoning Comm'n, 896 F.2d 1264 (11th Cir. 1989); Bean v. Southwestern Waste Management Corp., 482 F. Supp. 673 (S.D. Tex. 1979), *aff'd mem.,* 782 F.2d 1038 (5th Cir. 1986); Aiello v. Browning-Ferris, Inc., 1993 U.S. Dist. Lexis 16104 (N.D. Cal. Nov. 2, 1993). *But see* El Pueblo Para el Aire y Agua Limpio v. County of Kings, 22 Envtl. L. Rep. (Envtl. L. Inst.) 20357 (Cal. Super. Ct. Sacramento County, Dec. 30, 1991) (holding that environmental impact statement and other documents for proposed hazardous waste incinerator in Spanish-speaking area should have been translated into Spanish). *See also* Bowman v. City of Franklin, 980 F.2d 1104 (7th Cir. 1992), cert. denied, 113 S.Ct. 2417 (1993) (unsuccessful equal protection challenge to waste management facility; no allegations of racial discrimination).

156. *See* Arlington Heights v. Metropolitan Hous. Dev. Corp., 429 U.S. 252 (1977).

157. County of El Paso v. Texas Low-Level Radioactive Waste Disposal Authority, No. 2588-34 (Dist. Ct., Hudspeth County, Tex., Apr. 25, 1991).

158. 42 U.S.C. § 2000d. Lazarus, *supra* ch. 5 note 126.

159. Tsao, *supra* ch. 4 note 99, at 397.

160. Rachel D. Godsil, *Remedying Environmental Racism*, 90 Mich. L. Rev. 394, 421–22 (1991).

161. Edward Patrick Boyle, Note, *It's Not Easy Bein' Green: The Psychology of Racism, Environmental Discrimination, and the Argument for Modernizing Equal Protection Analysis*, 46 Vand. L. Rev. 937 (1993).

162. Cole 1992, *supra* ch. 5 note 53, at 1996–97.

163. U.S. Environmental Protection Agency, Environmental Appeals Board, In the Matter of Genesee Power Station Limited Partnership, PSD Appeal Nos. 93-1-93-7 (Oct. 22, 1993).

164. *See, generally,* Been, *supra* ch. 4 note 99.

165. *See* Bernstein, *supra* ch. 2 note 18, at 86, and Conner Bailey and Charles E. Faupel, *Environmentalism and Civil Rights in Sumter County, Alabama*, in Bryant and Mohai, *supra* ch. 3 note 120, at 150 (concerning Emelle); U.S. Gen. Accounting Office, Pub. No. GAO/RCED-83-168, *Siting of Hazardous Waste Landfills and Their Correlation With Racial and Economic Status of Surrounding Communities* 9 (1983) (concerning Warren County); Lawrence J. Straw, Jr., *Environmental Equity: A Controversial Catchphrase Confronts Environmentally Sensitive Projects,* 1992 Cal. Envtl. L. Rep. 508, 510 (concerning Kettleman Hills). *But see* Bullard 1990, *supra* ch. 5 note 146, at 35–38, 70.

166. *See* John Rawls, *A Theory of Justice* 284–93 (1971).

167. U.S. Const. pmbl.

168. Makhijani and Saleska, *supra* ch. 3 note 14, at 41.

169. Peters, *supra* ch. 5 note 107, at 41. For a contrarian view, *see* Robert L. Heilbroner, *What Has Posterity Ever Done for Me?* in *Responsibilities to Future Generations* 191 (Eugene Partridge ed., 1981).

170. *See* Harold P. Green, *Legal Aspects of Intergenerational Equity Issues, in* Kasperson ed., *supra* ch. 3 note 4, at 189.

171. Goble, *supra* ch. 3 note 119, at 139.

172. Richard A. Denison and John Ruston eds., *Recycling and Incineration: Evaluating the Choices* 8, 201 (1990).

173. Gore, *supra* ch. 5 note 128, at 157.

174. Alan Burdick, *The Last Cold-War Monument,* Harper's Mag., Aug. 1992, at 62.

175. Allen V. Kneese et al., *Economic Issues in the Legacy Problem, in* Kasperson, *supra* ch. 3 note 4, at 203.

176. In re Will of Getman, 30 A.D.2d 257, 291 N.Y.S.2d 395 (4th Dep't 1968).

177. E.g., N.Y. St. Dep't of Envtl. Conservation, *Perpetual Monitoring, Maintenance, and Care* (Section U of Module II of permit for SLF-12 hazardous waste

landfill, Model City, N.Y., 1989). *See also* Ray Pospisil, *Radical Change for Hazardous Waste Services,* Chemical Wk., Aug. 18, 1993, at 26, 28 ("perpetual post-closure monitoring" imposed on HW landfill in Ohio).

178. 42 U.S.C. § 7914(f)(2); 10 C.F.R. pt. 40 appx. A.

179. Jack A. Calwell, *Engineering Perspectives for Near-Surface Disposal, in* Reith and Thompson eds., *supra* ch. 3 note 49, at 161, 165.

180. U.S. General Accounting Office, Pub. No. GAO/RCED-93-124, *Nuclear Waste: Yucca Mountain Project Behind Schedule and Facing Major Scientific Uncertainties* 41 (May 1993).

181. Christopher Anderson, *Weapons Labs in a New World,* 262 Science 168, 170 (Oct. 8, 1993).

182. Earl R. Hoskins and James E. Russell, *Geologic and Engineering Dimensions of Nuclear Waste Storage, in* Murdock et al. eds., *supra* ch. 5 note 107, at 19, 28.

183. Resnikoff, *supra* ch. 3 note 9, at 50. (1987).

184. 46 Fed.Reg. 11,126, 11,128 (1981); 47 Fed.Reg. 32,373 (1982). *Similarly,* Hunt v. Chemical Waste Management, Inc., 584 So.2d 1367 (Ala. 1991), *rev'd on other grounds,* 112 S. Ct. 2009 (1992); U.S. Gen. Accounting Office, *Hazardous Waste: Funding of Postclosure Liabilities Remains Uncertain* 18 (1990); G. Fred Lee and R. Anne Jones, *Landfills and Ground-Water Quality,* 29 Ground Water 482 (1991); Peter Montague, *The Limitations of Landfilling, in Beyond Dumping* 3 (Bruce Piasecki ed., 1984).

185. *Mixed Waste: A Way to Solve the Quandary, supra* ch. 3 note 136, at 10711 nn. 48–49.

186. *See* Mazmanian and Morell, *supra* ch. 1 note 5, at 15, 127.

187. *See* Jeffrey Spear, *Remedy Selection Under CERCLA and Our Responsibilities to Future Generations,* 2 N.Y.U. Envtl. L.J. 117 (1993).

188. *Coming Clean: Superfund Problems Can Be Solved . . . , supra* ch. 2 note 51, at 140, 217; U.S. Gen. Accounting Office, Pub. No. GAO/RCED-92-138, *Superfund: Problems With the Completeness and Consistency of Site Cleanup Plans* (May 1992); U.S. Gen. Accounting Office, Pub. No. GAO/RCED-93-188, *Superfund: Cleanups Nearing Completion Indicate Future Challenges* (Sept. 1993) at 28, 49.

189. U.S. Gen. Accounting Office, Pub. No. GAO/NSIAD-92-117, *Hazardous Materials: Upgrading of Underground Storage Tanks Can Be Improved to Avoid Costly Cleanups* 18 (May 1992).

190. *Complex Cleanup, supra* ch. 3 note 8, at 7, 35–36.

191. U.S. General Accounting Office, Pub. No. GAO/RCED-93-156, *Nuclear Regulation: Cleanup Delays Continue at Two Radioactive Waste Sites in Ohio* (June 1993).

192. *See* Keith Schneider, *Wasting Away,* N.Y. Times Mag., Aug. 30, 1992, at 42, 56. After success in early rounds of litigation, Idaho agreed to accept addi-

tional shipments of spent naval fuel, but only upon winning numerous concessions from the Navy. Keith Schneider, *Pact Allows Navy to Send Nuclear Waste to Idaho*, N.Y. Times, Aug. 10, 1993, at A12; *Idaho Agrees to Allow Navy to Ship Spent Fuel to DOE Engineering Laboratory*, 24 Env't Rep. (BNA) No. 15 at 624 (Aug. 13, 1993).

193. Expansion of these storage facilities, necessitated by the delayed opening of Yucca Mountain, has been controversial in some locations. *See* Michigan v. NRC, 93CV67 (D.C. W.Mich. May 10, 1993); Prairie Island Mdewakanton Sioux Indian Community v. Northern States Power Co., No. C1-92-2314 (Minn. Sup. Ct. July 15, 1993).

194. Matthew L. Wald, *Uranium Rusting in Storage Pools is Troubling U.S.*, N.Y. Times, Dec. 8, 1993, at A1.

195. *Dismantling the Bomb and Managing the Nuclear Materials, supra* ch. 3 note 97, at 77.

196. E.g., Slovic 1991, *supra* ch. 5 note 51, at 1607; Makhijani and Saleska, *supra* ch. 3 note 14, at 108; Howard Hu et al., *Plutonium: Deadly Gold of the Nuclear Age* 152 (International Physicians for the Prevention of Nuclear War and Institute for Energy and Environmental Research 1992). Similarly, see Kai Erikson, *Out of Sight, Our of Our Minds*, N.Y. Times Magazine, March 6, 1994, at 34 (argues against permanent disposal of HLW, for that removes choice from future generations.)

197. Frans Berkhout, *Radioactive Waste: Politics and Technology* 177–78 (1991); Zorpette and Stix, *supra* ch. 3 note 40.

198. *Aging Nuclear Power Plants, supra* ch. 3 note 37, at 23.

199. Carol Hornibrook and Michael D. Naughton, *Lagging Regional LLRW Disposal Spotlights On-Site Storage*, Power, Mar. 1992, at 76.

200. Richard J. Bord, *The Low-Level Radioactive Waste Crisis: Is More Citizen Participation the Answer? in* Burns ed., *supra* ch. 3 note 11, at 193, 210 (citing Sierra Club publications); Makhijani and Saleska, *supra* ch. 3 note 14, at 93; Resnikoff, *supra* ch. 3 note 9, at 78; Radioactive Waste Campaign, *"Low-Level" Nuclear Waste: Options for Storage* (1985).

201. 42 U.S.C. § 6924(j). *See* Hazardous Waste Treatment Council v. EPA, 886 F.2d 355, 357 (D.C. Cir. 1989).

202. James M. Melius et al., *Facility Siting and Health Questions: The Burden of Health Risk Uncertainty*, 17 Nat. Resources Law. 467, 470 (1984); Barry G. Rabe, *When Siting Works, Canada-Style*, 17 J. Health Politics, Policy and L. 119, 138 (1992); Michael Rose, *Return to St-Basile*, Maclean's, Sept. 26, 1988, at 16A.

203. U.S. General Accounting Office, Pub. No. GAO/T-NSIAD-93-18, *Chemical Weapons Storage: Communities Are Not Prepared to Respond to Emergencies* (July 16, 1993).

204. Samuel S. Epstein et al., *Hazardous Waste in America* 337–78 (1982) (Sierra Club publication); Citizens Clearinghouse for Hazardous Waste, *How to Deal With a Proposed Facility* 5 (1986). *See also* Paul Kemezis, *Congress Kills Funds for Alabama Chemical Weapons Incinerator,* Chemical Wk., Oct. 14, 1992, at 14 (citizens opposed incineration of old chemical weapons at depot in Alabama, even though Centers for Disease Control said the health risk from continued storage of the weapons at the depot were greater than those from incineration).

205. Gary A. Davis, *Shifting the Burden Off the Land: The Role of Technical Innovation, in* Piasecki and Davis eds., *supra* ch. 2 note 1, at 43, 61.

206. *But see* Dan M. Berkovitz, *Pariahs and Prophets: Nuclear Energy, Global Warming, and Intergenerational Justice,* 17 Colum. J. Envtl. L. 245 (1992).

207. *See* Brion 1991, *supra* ch. 1 note 2, at 14; Morell and Magorian, *supra* ch. 1 note 8, at 188; Murray and Seneker, *supra* ch. 4 note 1, at 323.

208. Palumbo v. Waste Technologies Industries, 989 F.2d 156 (4th Cir. 1993). *See also* Greenpeace, Inc. v. Waste Technologies Industries, 9 F.3d 1174 (6th Cir. 1993); Margaret Kriz, *Slow Burn,* National J., Apr. 3, 1993, at 811; Liane Clorfene-Casten, *E.P.A. Fiddles While W.T.I. Burns,* Nation, Sept. 27, 1993, at 307; Marjorie Coeyman, *WTI Battles For Hearts and Minds of its Neighbors,* Chemical Wk., Dec. 8, 1993, at 50; *Reviewers Criticize WTI Risk Assessment on Accidental Releases, Upset Conditions,* 24 Env't Rep. (BNA) No. 33 at 1523 (Dec. 17, 1993).

209. *Hazardous Waste: Partial Trial Burn at WTI Incinerator Halted Because of Operational Problems, Ohio EPA Says,* 24 Env't Rep. (BNA) No. 34 at 1545 (Dec. 24, 1993).

210. *See* Popper, *supra* ch. 1 note 11, at 1, 5 (reporting on 1980 survey by Resources for the Future for the U.S. Council on Environmental Quality [CEQ]); Owen J. Furuseth, *Community Sensitivity to a Hazardous Waste Facility,* 17 Landscape and Urb. Plan. 357 (1989) (largely replicating CEQ study in Charlotte, North Carolina); Slovic 1991, *supra* ch. 5 note 51 (1989 survey; nuclear power plants were more acceptable than HW/RW facilities); Smith and Desvousges, *supra* ch. 5 note 23 (1984 survey in suburban Boston). *See also* Michael Dear, *Understanding and Overcoming the NIMBY Syndrome,* APA J., Summer 1992, 288, 292 (comparing public reactions to waste disposal facilities with reactions to social service facilities).

211. Popper, *supra* ch. 1 note 11; Slovic et al., *supra* ch. 5 note 51.

212. Smith and Desvousges, *supra* ch. 5 note 23; Lyons et al., *supra* ch. 1 note 1, at 89, 91–92. Other surveys are reported in Easterling, *supra* ch. 4 note 47; Christopher J. Smith and Robert Q. Hanham, *Any Place But Here! Mental Health Facilities as Noxious Neighbors,* 33 Prof. Geographer 326 (1981). For discussions of the inconsistent roles of proximity in shaping public opinion about facilities, *see* Panel on Social and Econ. Aspects of Radioactive Waste Management, *supra* ch. 5 note 116, at 101–02; William Hallman and Abraham Wan-

dersman, *Perception of Risk and Toxic Hazards, in* Peck ed., *supra* ch. 5 note 49, at 31, 45–47.

213. Michael E. Burns and William H. Briner, *Setting the Stage, in* Burns ed., *supra* ch. 3 note 11, at 1, 24.

214. George T. Mazuzan, *"Very Risk Business": A Power Reactor for New York City, in Technology and Choice* 179 (Marcel C. LaFollette and Jeffrey K. Stine eds., 1991).

215. Jacob, *supra* ch. 3 note 26, at 47.

216. Stanley M. Nealey and John A. Hebert, *Public Attitudes Toward Radioactive Wastes, in Too Hot to Handle? Social and Policy Issues in the Management of Radioactive Wastes* 94, 95 (Charles A. Walker et al. eds., 1983).

217. B.S. Forcade, *Public Participation in Siting, in* Harthill ed., *supra* ch. 4 note 49, at 111, 111.

218. Victor Gilinsky, *quoted in* Christian Joppke, *Mobilizing Against Nuclear Energy; A Comparison of Germany and the United States* 135 (1993).

219. *See* Panel on Social and Econ. Aspects of Radioactive Waste Management, *supra* ch. 5 note 116, at 9, 16–21 (1984); Mary Douglas and Aaron Wildavsky, *Risk and Culture* 10 (1982); Paul Slovic, *Perception of Risk and the Future of Nuclear Power,* 9 Ariz. J. Int'l and Comp. L. 191 (1992) [hereinafter Slovic 1992].

220. Portney, *supra* ch. 5 note 150, at 89, 95, 134; James L. Regens, *Siting Hazardous Waste Management Facilities, in Public Involvement and Social Impact Assessment* 121, 124–25 (Gregory A. Daneke et al. eds., 1983).

221. Edelstein 1988, *supra* ch. 4 note 61, at 64; John Sorensen et al., *Impacts of Hazardous Technology: The Psycho-Social Effects of Restarting TMI-1* 154 (1987); Holly L. Howe, *Public Concern About Chemicals in the Environment: Regional Differences Based on Threat Potential,* 105 Pub. Health Rep. 186 (1990); Chris Zeiss and James Atwater, *Waste Facilities in Residential Communities: Impacts and Acceptance,* 113 J. Urb. Plan. and Dev. 19 (1987); John M. Halstead et al., *An Examination of the NIMBY Syndrome: Why Not in My Backyard?* (unpublished manuscript, Dep't of Resource Economics and Development, U. of New Hampshire, Aug. 1992).

222. Anna Vari et al., *Public Concerns About LLRW Facility Siting: A Comparative Study,* 22 J. Cross-Cultural Psychology 83 (1991) (comparison of public opinion in United States, United Kingdom, and Hungary).

223. Richard J. Bord and Robert E. O'Connor, *Determinants of Risk Perceptions of a Hazardous Waste Site,* 12 Risk Analysis 411, 415 (1992).

224. Edelstein 1988, *supra* ch. 4 note 61, at 158–167; Piller, *supra* ch. 4 note 8, at 165; Heiman, *supra* ch. 4 note 34, at 359. *See also* Walter A. Rosenbaum, *The Politics of Public Participation in Hazardous Waste Management, in* Lester and Bowman eds., *supra* ch. 4 note 46, at 176, 191–92 (notes explosion of citizen activism on hazardous waste issues at both national and local levels in early

1980s); *Campaigning for Environmental Justice,* Everyone's Backyard, Feb. 1993, at 6, 10 ("Perhaps the most common denominator in grassroots fights for environmental justice is the issue of health effects. More people get involved in this movement because their children or other members of their family are ill from exposure to toxic chemicals than any other reason.").

225. William Greider, *Who Will Tell the People* 213–21 (1992).

226. Edelstein, *supra* ch. 4 note 61, at 167.

227. Numbers 12:10–15.

228. Seley 1988, *supra* ch. 3 note 45, at 5.

229. *See* Louise H. Feffer, *AIDSphobia, a New Entity,* N.Y. St. Bar J. 14 (Feb. 1993).

230. Edelstein 1988, *supra* ch. 4 note 61, at 49; Margaret S. Gibbs, *Psychopathological Consequences of Exposure to Toxins in the Water Supply, in* Lebovitz et al. eds., *supra* ch. 5 note 153, at 47. *See also* Michael R. Reich, *Toxic Politics: Responding to Chemical Disasters* (1991); Henry M. Vyner, *Invisible Trauma: The Psychosocial Effects of Invisible Environmental Contaminants* (Lexington Books, 1988).

231. *See* Richard Walker, *The Return of the Repressed: Freudian Theory, Hazardous Waste Siting, and Public Resistance, in* Peck ed., *supra* ch. 5 note 49, at 239.

232. *See* Gaston Bachelard, *The Psychoanalysis of Fire* (Alan C.M. Ross trans., 1964).

233. *See* William Chaloupka, *Knowing Nukes: The Politics and Culture of the Atom* (1992); Spencer R. Weart, *Nuclear Fear: A History of Images* (1988); Robert Jay Lifton, *The Broken Connection: On Death and the Continuity of Life* (1979).

234. Slovic 1992, *supra* ch. 5 note 219; Paul Slovic et al., *Perceived Risk, Trust, and the Politics of Nuclear Waste,* 254 Science 1603 (1991).

235. *But see* Commission on Physical Sciences, Mathematics, and Resources, National Research Council, *The Nuclear Weapons Complex: Management for Health, Safety and the Environment* 117 (1989) (eight "criticality accidents" have occurred in the nuclear weapons complex, when critical mass was achieved with plutonium or uranium solutions; several fatalities have resulted). *See also supra* ch. 3 notes 6, 7 concerning explosions at Soviet nuclear waste facilities in 1957 and 1993.

236. *See* Roger G. Noll and James E. Krier, *Some Implications of Cognitive Psychology for Risk Regulation,* 19 J. Legal Stud. 747 (1990).

237. Donald A. Redelmeier et al., *Understanding Patients' Decisions: Cognitive and Emotional Perspectives,* 270 J. Am. Med. Ass'n 72 (1993).

238. Reich, *supra* ch. 5 note 230, at 207 (citations omitted).

239. Kai Erikson, *Toxic Reckoning: Business Faces a New Kind of Fear,* Harv. Bus. Rev., Jan.–Feb. 1990, at 118, 121. Cf. Rhodes, *supra* ch. 4 note 4, at 594

(President Franklin Roosevelt refused to authorize use of poison gas over Iwo Jima before American invasion, even though it might have saved the lives of thousands of Allied soldiers, presumably because he remembered the world outcry that followed German use of poison gas in World War I.).

240. Erikson, *supra* ch. 5 note 239, at 121–22. *See also* Charles Perrow, *Normal Accidents* 326–28 (1984); William Hallman and Abraham Wandersman, *Perception of Risk and Toxic Hazards,* in Peck ed., *supra* ch. 5 note 49, at 31; Kasperson, *The Social Amplification of Risk: A Conceptual Framework,* 8 Risk Analysis 177 (1988).

241. Irving L. Janis, *Psychological Effects of Warning,* in *Man and Society in Disaster* 55 (G.W. Baker and D.W. Chapman eds., 1962).

242. O'Hare 1983, *supra* 1 note 9, at 58; Schmeidler and Sandman, *supra* ch. 5 note 2, at 54; Richard N.L. Andrews and Terrence K. Pierson, *Local Control or State Override: Experiences and Lessons to Date,* 14 Pol'y Stud. J. 90, 97 (1985); Daniel Burchard and Robert Hughes, *Beyond Capacity: Addressing the Concerns of Local Opposition in the Siting Process,* 6 Stan. Envtl. L.J. 145, 151 (1986–87).

243. Morell and Magorian, *supra* ch. 1 note 8, at 63.

244. *Cf.* Inhaber, *supra* ch. 1 note 2, at 55 (observes that many opposed to siting of monitored retrievable storage facility in Tennessee drove to hearing without wearing seat belts, and smoked during the hearing).

245. *See* Janet M. Fitchen, *When Toxic Chemicals Pollute Residential Environments: The Cultural Meanings of Home and Homeownership,* 48 Human Organization 313 (1989).

246. Edelstein 1988, *supra* ch. 4 note 61, at 64.

247. *Assurance of Hazardous Waste Capacity: Guidance to State Officials, supra* ch. 4 note 22, at 3. The same phenomenon has been observed abroad. *See* Stephen Tromans and Kathy Mylrea, *Siting Hazardous Waste Facilities in the United Kingdom,* 7 Natural Resources and Envt. No. 3 at 29 (Winter 1993).

248. *Review of EPA's Capacity Assurance Program, supra* ch. 2 note 11, at 56, 69 (statement of Doug MacMillan, National Solid Wastes Management Ass'n). *See also* U.S. EPA, Office of Toxic Substances, *The Toxics Release Inventory: A National Perspective, 1987* 17 (June 1989) (large amount of out-of-state transfers of toxic chemicals); Edward Repa, *Interstate Movement of Municipal Solid Waste—1992 Update,* Waste Age, Jan. 1994 at 37.

249. *National Biennial Report, supra* ch. 2 note 8, at 6.

250. C and A Carbone, Inc. v. Town of Clarkstown, 62 U.S.L.W. 4315 (May 16, 1994); Chemical Waste Management, Inc. v. Hunt, 112 S.Ct. 2009 (1992); Fort Gratiot Sanitary Landfill, Inc. v. Michigan Dep't of Natural Resources, 112 S.Ct. 2019 (1992); City of Philadelphia v. New Jersey, 437 U.S. 617 (1978).

251. O'Hare 1983, *supra* ch. 1 note 9, at 18–22 (operating facility in Wilsonville, Illinois was shut down after neighbors learned it would receive PCB-con-

taminated waste from Missouri); Michael E. Kraft and Ruth Kraut, *Citizen Participation and Hazardous Waste Policy Implementation, in* Davis and Lester eds., *supra* ch. 4 note 30, at 63, 69 (protests against landfill in Ohio that would accept waste from Kentucky and West Virginia); Charles J. McDermott, *Environmental Equity: A Waste Manager's Perspective,* 2 Land Use Forum No. 1 (Winter 1993) at 12, 14 (protests by neighbors of landfill in Louisiana against accepting lead-contaminated soil from a housing development in Texas).

252. Paul Kemezis, *Congress Kills Funds for Alabama Chemical Weapons Incinerator,* Chemical Wk., Oct. 14, 1992, at 14 (opposition to incineration of chemical weapons).

253. Crim, *supra* ch. 4 note 41, at 136 (incident in Texas).

254. Jonathan Walters, *Tempted by Trash,* Governing, July 1991, at 29 (proposed landfill in West Virginia); *Block Foreign Garbage* (editorial), Niagara Gazette, Mar. 3, 1993, at 11A (opposition to importation of MSW from Canada).

255. English 1992, *supra* ch. 1 note 8, at 127.

256. Sam Howe Verhovek, *New York Fears Burden and Leaves Waste Plan,* N.Y. Times, Oct. 18, 1989, at B1.

257. *See* Olsen, *supra* ch. 4 note 3 (hazardous waste facilities in western New York State that could take waste from remainder of state); Rae Zimmerman, *Public Acceptability of Alternative Hazardous Waste Management Services, in* Peck ed., *supra* ch. 5 note 49, at 197, 212 (concerning power project in Prattsville, New York, which would send electricity to New York City); BFI Medical Waste Systems v. Whatcom County, 983 F.2d 911 (9th Cir. 1992) (striking down county ordinance barring medical waste from outside county).

258. Linda Yglesias, *Dumpster: A Pastoral Upstate Town Decides It Isn't Taking Any More Crap,* N.Y. Daily News Mag., Oct. 29, 1989, at 17 (Town of Saugerties, New York opposed to taking entire county's MSW). There are many other examples of towns opposing efforts to site county MSW landfills in their borders.

259. Richard Briffault, *Voting Rights, Home Rule, and Metropolitan Governance: The Secession of Staten Island as a Case Study in the Dilemmas of Local Self-Determination,* 92 Colum. L. Rev. 775, 783 n.46 (1992) (secessionist movements in Staten Island have arisen partly from location of New York City's only MSW landfill there).

260. *See* Center for Investigative Reporting and Bill Moyers, *Global Dumping Ground: The International Traffic in Hazardous Waste* 17–30 (1990); Shirley E. Perlman, *In the Barge's Wake, in* Rush to Burn: Solving America's Garbage Crisis? 243 (Newsday 1989).

261. *See* William Bunch, *Another Stink Over Roving Trash Trains,* N.Y. Newsday, July 30, 1992, at 32; *Train Bearing Tainted Soil Rolls On,* N.Y. Times, Apr. 24, 1991 at A12.

262. J.S. Brown, *Putting the Lid on Out-of-State Garbage,* State Gov't News, Jan. 1990, at 22.

263. John Holusha, *In Some Parts the Battle Cry is "Don't Dump on Me,"* N.Y. Times, Sept. 8, 1991, at E5. *Similarly,* J.C. Barden, *Garbage is One Thing, But Garbage From New York? Forget It!* N.Y. Times, Feb. 12, 1989, at 26.

264. Alan R. Gold, *Threat to Garbage Traffic Upsets 2 States,* N.Y. Times, Sept. 19, 1990, at B1. *See also* John Krukowski, *The War Between the States,* Pollution Eng'g, June 15, 1992, at 37; Sue Darcy, *Pressured Congress Tackles Interstate Waste Legislation,* World Wastes, Oct. 1991, at 42.

265. 42 C.F.R. § 271.4(a).

266. Many of these attempts are recounted in W. Victoria Becker, *Legal Issues Affecting Interstate Disposal* (National Governors' Ass'n, 1989); Piller, *supra* ch. 4 note 8, at 73–74; *Review of EPA's Capacity Assurance Program, supra* ch. 2 note 11, at 153 (statement of Richard C. Fortuna, Hazardous Waste Treatment Council).

267. E.g., Northern States Power Co. v. Prairie Island Mdewakanton Sioux Indian Community, 991 F.2d 458 (8th Cir. 1993); In re Southeast Arkansas Landfill, Inc., 981 F.2d 372 (8th Cir. 1992); Chemical Waste Management, Inc. v. Templet, 967 F.2d 1058 (5th Cir. 1992); Diamond Waste, Inc. v. Monroe County, 939 F.2d 941 (11th Cir. 1991); Government Suppliers Consolidating Servs., Inc. v. Bayh, 975 F.2d 1267 (7th Cir. 1992); Hazardous Waste Treatment Council v. South Carolina, 945 F.2d 781 (4th Cir. 1991); Nat'l Solid Wastes Management Ass'n v. Alabama Dep't of Envtl. Management, 910 F.2d 713 (11th Cir. 1990), *modified,* 924 F.2d 1001 (11th Cir.), *cert. denied,* 501 U.S. 1206 (1991); Washington State Bldg. and Constr. Trades Council v. Spellman, 684 F.2d 627 (9th Cir. 1982), *cert. denied,* 461 U.S. 913 (1983); TENNSV Inc v. Illinois Envt'l Protection Agency, 1993 U.S. Dist. Lexis 10403, Index No. 92-503 (D.C. S. Ill., July 8, 1993); National Solid Wastes Management Ass'n v. Voinovich, 763 F. Supp. 244 (S.D. Ohio 1991); Stephen D. DeVito Jr. Trucking Inc. v. Rhode Island Solid Waste Management Corp., 770 F. Supp. 775 (D.C. R.I.), *aff'd,* 947 F.2d 1004 (1st Cir. 1991); Southern States Landfill v. Georgia Dep't of Nat. Resources, 801 F. Supp. 725 (M.D. Ga., 1992).

268. E.g., American Trucking Ass'ns Inc. v. Secretary of State, 595 A.2d 1014 (Maine Sup. Jud. Ct. 1991); Gilliam Co. v. Oregon Dep't of Envtl. Quality, 144 Or. App. 369, 837 P.2d 965 (1992).

269. Marshall I. Goldman, *Environmentalism and Ethnic Awakening,* 19 Envtl. Affairs 511, 513 (1992).

270. *See* Constance Mathiessen, *No Place Like Home,* California Lawyer, Aug. 1993, at 32.

271. Richard Walker, *The Return of the Repressed: Freudian Theory, Hazardous Waste Siting, and Public Resistance, supra* ch. 5 note 231, at 260.

272. O'Hare 1983, *supra* ch. 1 note 9, at 18–22.

273. E.g., Bartlett and Steele, *supra* ch. 3 note 10, at 269 (incident in Texas); Brion 1991, *supra* ch. 1 note 2, ch. 1 (experience in Massachusetts); Ryan, *supra* ch. 4 note 28, at A4, A18 (incidents in Florida and South Carolina); Arnold W. Reitze, Jr., and Andrew N. Davis, *Reconsidering Ocean Incineration as Part of*

a U.S. Hazardous Waste Management Program: Separating the Rhetoric From the Reality, 17 Envtl. Aff. 687, 708-09 (1990) (problems with Marine Shale Co. adding to its siting difficulties).

274. Matthijs Hisschemoller and Cees J.H. Midden, *Technological Risk, Policy Theories and Public Perception in Connection With the Siting of Hazardous Facilities,* in *Social Decision Methodology for Technological Projects* (Charles Vlek and George Cvetkovich eds., Kluwer 1989) at 173, 184–85. *Similarly, see* Conner Bailey and Charles E. Faupel, *supra* ch. 4 note 13, at 92 (considerable hostility to the hazardous waste landfill in Emelle, Alabama was caused by the secret manner in which the facility was sited).

275. Roger E. Kasperson et al., *Social Distrust as a Factor in Siting Hazardous Facilities and Communicating Risks,* 48 J. Social Issues 161 (1992); James Flynn et al., *Trust as a Determinant of Opposition to a High-Level Radioactive Waste Repository: Analysis of a Structural Model,* 12 Risk Analysis 417 (1992); *similarly,* Bord and O'Connor, *supra* ch. 5 note 223, at 415.

276. Allan G. Pulsipher, *Compensation: Will It Provide a Waste Site?* 8 Forum for Applied Research and Public Policy 108, 109 (1993).

277. Contreras, *supra* ch. 1 note 2, at 500–02; Roger E. Kasperson, *Social Issues in Radioactive Waste Management: The National Experience, in* Kasperson ed., *supra* ch. 3 note 4, at 24, 48–49, 57–58; Slovic 1992, *supra* ch. 5 note 219, at 196.

278. *Radioactive Waste: Distrust of DOE's Waste Management Activities Widespread, Long-Lasting, Advisory Board Says,* 24 Env't Rep. (BNA) No. 10 at 418 (July 8, 1993).

279. Erikson, *supra* ch. 5 note 239, at 125.

280. Sandman, *supra* ch. 5 note 88, at 444. *Similarly, see* Bachrach and Zautra, *supra* ch. 5 note 153, at 85 (Many neighbors of proposed HW facility in Arizona believed they were powerless to stop it, and "[a]s a consequence, the psychologically most fragile and vulnerable residents were underrepresented in community activities and thus less visible to government officials attempting to 'take the pulse' and assess the impact of the HWF on the local population.")

281. Ortwin Renn and Vincent Covello, *Medical Waste: Risk Perception and Communication, in Perspectives on Medical Waste* (Nelson A. Rockefeller Institute of Government, State University of New York, June 1989) at VII.1, VII.5.

282. Richard Walker, supra ch. 5 note 231, at 260.

283. E.g., Greenberg and Anderson, *supra* ch. 4 note 3, at 167 (facility in Logan Township, New Jersey); Gail Bingham and Daniel S. Miller, *Prospects for Resolving Hazardous Waste Siting Disputes Through Negotiation,* 17 Nat. Resources Law. 473, 485 (1984) (facility in Providence, Rhode Island); Walter and Getz, *supra* ch. 4 note 80, at 240–41 (facilities in Montana and Texas); Paul Slovic and Baruch Fischoff, *How Safe is Safe Enough? Determinants of Perceived and Acceptable Risk, in* Walker et al. eds., *supra* ch. 5 note 216, at 112, 116 (facility in Oregon). *See also* text accompanying ch. 4 notes 54–78 *supra* (con-

cerning relative ease in expanding existing facilities); Rae Zimmerman, *Public Acceptability of Alternative Hazardous Waste Management Services, in* Peck ed., *supra* ch. 5 note 49, at 197, 219, 225 (survey of newspaper coverage of siting controversies shows high degree of public acceptance of onsite disposal); Lyons et al., *supra* ch. 1 note 1, at 93 (public opinion poll shows strong preference for on-site disposal); Hugh F. Holman, *Siting New Hazardous Waste Facilities: Final Report* (Management Associates, Natick, Massachusetts, Nov. 1986) at 3 (statistical analysis shows much higher success rates for siting non-commercial [typically on-site] HW facilities than for commercial facilities).

284. *See* Morell and Magorian, *supra* ch. 1 note 8, at 38, 41, 55.

285. Ellen Goodbaum and David Rotman, *Hazardous Waste: Faced With Dwindling Choices, Companies Must Seek New Ways to Manage It,* Chemical Wk., Aug. 23, 1989, at 18.

286. Citizens Clearinghouse for Hazardous Waste, *How to Deal With a Proposed Facility* 3 (1986).

287. E.g., Peter Dorman, *Environmental Protection, Employment, and Profit: The Politics of Public Interest in the Tacoma/Asarco Arsenic Dispute,* 16 Rev. Radical Polit. Econ. 151 (1984). *See generally* Richard Kazis and Richard L. Grossman, *Fear at Work: Job Blackmail, Labor and the Environment* (1982).

288. *See* George Cvetkovich and Timothy C. Earle, *Hazard Images, Evaluations and Political Action: The Case of Toxic Waste Incineration, in Communicating Risks to the Public: International Perspectives* 327 (Roger E. Kasperson and Pieter Jan M. Stallen eds., 1991).

289. Walter and Getz, *supra* ch. 4 note 80, at 241. An exception is an HLW repository, which is expected to create 870 to 1,000 operation jobs. John K. Thomas et al., *The Socioeconomic Impacts of Repositories, in* Murdock et al. eds., *supra* ch. 5 note 107, at 103, 106.

290. Bailey and Faupel, *supra* ch. 4 note 13, at 143.

291. English 1992, *supra* ch. 1 note 8, at 85.

292. Robert W. Kates and Bonnie Braine, *Locus, Equity, and the West Valley Nuclear Wastes,* Kasperson ed., *supra* ch. 3 note 4, at 94, 104.

293. *Embattled Ohio Incinerator Wins Support; Hundreds Rally to Encourage Official Start,* 23 Env't Rep. (BNA) No. 35, at 2101, 2102 (Dec. 25, 1992).

294. Robert Tomsho, *Small Town in Alberta Embraces What Most Reject: Toxic Waste,* Wall St. J., Dec. 27, 1991, at 1.

295. U.S. Nuclear Regulatory Commission, Office of Nuclear Materials Safety and Safeguards, Pub. No. NUREG-1476, *Final Environmental Impact Statement to Construct and Operate a Facility to Receive, Store, and Dispose of 11e.(2) Byproduct Material Near Clive, Utah* 5–4 (Aug. 1993).

296. Carter, *supra* ch. 3 note 51, at 424.

297. Reith and Fischer, *supra* ch. 3 note 49, at 303, 308; Keith Schneider, *Wasting Away,* N.Y. Times, Aug. 30, 1992, § 6 (Magazine), at 42, 56.

298. Seley, *supra* ch. 3 note 45, at 93.

299. Portney, *supra* ch. 5 note 150, at 37; N.Y. St. Dep't of Envtl. Conservation, *Recommendations for Assistance to Localities Affected by Hazardous Waste Management Facilities* 23–24 (Apr. 1988); Bachrach and Zautra, *supra* ch. 5 note 153, at 71, 86; Richard J. Bord, *Judgments of Policies Designed to Elicit Local Cooperation in LLRW Disposal Siting: Comparing the Public and Decision Makers,* 7 Nuclear and Chemical Waste Mgmt. 99 (1987); Colglazier and English, *supra* ch. 1 note 8, at 645; Charles Davis, *Public Involvement in Hazardous Waste Siting Decisions,* 19 J. Northeastern Political Sci. Ass'n 296 (1986); Patricia K. Freeman et al., *Legislative Representation on a Technical Policy Issue: Hazardous Waste in Tennessee,* 26 Social Sci. J. 455, 460 (1989); Lyons et al., *supra* ch. 1 note 1, at 92; Mazmanian and Morell 1992, *supra* ch. 1 note 5, at 137; Lawrence E. Susskind, *The Siting Puzzle: Balancing Economic and Environmental Gains and Losses,* 5 Envtl. Impact Assess. Rev. 157 (1985). Local involvement in facility operations has also been found to reduce neighbor hostility to other kinds of installations. *See* Ann Marie Rizzo et al., *Strategies for Responding to Community Opposition in an Existing Group Home,* 15 Psychosocial Rehab. J. 85 (1992).

300. *See* Austin and Schill, *supra* ch. 4 note 59, at 76; Fred Setterberg and Lonny Shavelson, *Toxic Nation: The Fight to Save Our Communities from Chemical Contamination* 236 (1993). Similar attitudes toward chemical exposure in the workplace are often expressed by workers. Dorothy Nelkin and Michael S. Brown, *Workers at Risk: Voices From the Workplace* 181–82 (1984).

301. *Recommendations for Assistance to Localities Affected by Hazardous Waste Management Facilities, supra* ch. 5 note 299, at 35–39; Susan Saiter, *Local Opposition is Stalling Development of Waste Sites,* N.Y. Times, June 18, 1983, at 6; *Commissioners Approve Dump Site,* UPI, Aug. 15, 1983, *available in* LEXIS, Nexis Library, UPI file; *Colorado,* USA Today, July 22, 1991, *available in* LEXIS, Nexis Library.

302. Douglas and Wildavsky, *supra* ch. 5 note 219, at 14. *See also* Lee Clarke, *Acceptable Risk? Making Decisions in a Toxic Environment* (1989) (discusses role of organizational culture in reacting to risk, based on study of response to fire in state office building in Binghamton, New York, in 1981).

303. *Supra* at 194.

304. Reich, *supra* ch. 5 note 230, at 198. *Similarly,* Karl Dake, *Orienting Dispositions in the Perception of Risk: An Analysis of Contemporary Worldviews and Cultural Biases,* 22 J. Cross-Cultural Psychology 61 (1991).

305. Rabe, *supra* ch. 4 note 61, at 177; Oene Wiegman, *Verification of Information Through Direct Experiences With an Industrial Hazard,* 12 Basic and Applied Social Psych. 325 (1991). *Similarly,* Ray Weiss, *Jailhouse Blocks: Homeowners Feel Secure Near Prisons,* Gannett Suburban Newspapers (Westchester County, New York), Jan. 13, 1994, at C1 (concerning attitudes of neighbors of Sing-Sing Correctional Facility).

306. Carter, *supra* ch. 3 note 51, at 165; Bill Richards, *Nuclear Site Learns to Stop Worrying and Love the Boom,* Wall St. J., Aug. 28, 1992, at A1; Timothy Egan, *Richland Journal: Little Sentiment to Ban the Bomb,* N.Y. Times, Jan. 14, 1988, at A14.

307. English 1992, *supra* ch. 1 note 8, at 110–11.

308. E. Brent Sigmon, *Achieving a Negotiated Compensation Agreement in Siting: The MRS Case,* 6 J. Policy Analysis and Management 170 (1987); Dick Thompson, *Living Happily Near a Nuclear Trash Heap,* Time, May 11, 1992, at 53.

309. Suzanne Ruta, *Fear and Silence in Los Alamos,* Nation, Jan. 4, 1993, at 9.

310. Flynn et al., *supra* ch. 5 note 275.

311. Colglazier and English, *supra* ch. 1 note 8, at 647 n.27.

312. Paul Slovic and Baruch Fischoff, *How Safe is Safe Enough? Determinants of Perceived and Acceptable Risk, in* Walker, *supra* ch. 5 note 216, at 126. *See also* Ronald Smothers, *Plan to Destroy Toxic Weapons Polarizes a City,* N.Y. Times, Sept. 24, 1992, at A16 (some residents in Anniston, Alabama, support plan to destroy chemical weapons at local army depot, while others oppose plan).

313. Barbara Connell, *The Siting Approach for a Hazardous Waste Management Facility in Manitoba: A Guide to Nonconfrontational Siting Procedures, in* Resource Futures International ed., *supra* ch. 1 note 2, at 63, 72.

314. Armour 1991, *supra* ch. 5 note 2, at 40–41.

315. J. McQuaid-Cook, *Siting a Fully Integrated Hazardous Waste Management Facility With Incinerator and Landfill, Swan Hills, Alberta, Canada, in* Resources Futures International ed., *supra* ch. 1 note 2, at 123, 129.

316. Aaron Wildavsky and Karl Dake, *Theories of Risk Perception: Who Fears What and Why?* 119 Daedalus No. 4 at 41 (Fall 1990).

317. Benford et al., *supra* ch. 1 note 7, at 44; *see also* Bailey and Faupel, *supra* ch. 4 note 13.

318. Francoise Zonabend, *The Nuclear Peninsula* 104–07 (J.A. Underwood trans., 1993).

319. This takeover is described in James B. Stewart, *Den of Thieves* (New York: Simon and Schuster 1991), at 158–59.

320. N.Y. Environ. Conservation L. § 27-1102.

321. Teresa H. Sharp, *Citizens Group Stops Incinerator,* Niagara Gazette, June 10, 1993.

322. Timothy P. Henderson, *It's a Victory for Lewiston, Porter,* Niagara Gazette, June 30, 1993, at 11A.

323. Lisa Aug, *We're Selling Our Souls to Polluters,* Niagara Gazette, June 30, 1993, at 11A.

324. Corydon Ireland, *CWM Opposition is Quiet,* Niagara Gazette, July 27, 1993, at 3A.

325. The information presented here is drawn from U.S. Nuclear Regulatory Commission, Office of Nuclear Materials Safety and Safeguards, Pub. No. NUREG-1476, *Final Environmental Impact Statement to Construct and Operate a Facility to Receive, Store, and Dispose of 11e.(2) Byproduct Material Near Clive, Utah* (Aug. 1993); Donovan Webster, *Happiness is a Toxic Waste Zone,* Outside, Sept. 1993, at 58; Univ. of Utah, Grad. Sch. of Business, Bureau of Economic and Business Research, *Profile of Tooele County* (Aug. 1988); USPCI, *Grassy/Grayback Mountain Facility Audit Information* (Aug. 1992); Justice Martin, '*Imby, Please*', Fortune, Oct. 4, 1993, at 13; *Chem Weapons: Utah's Burning Desire for a Hospital,* Greenwire, Feb. 17, 1993; *Pollution News Shocking But Not Surprising,* Associated Press, June 20, 1989.

326. *See Transportation of Hazardous Materials: Hearings before the House Comm. on Public Works and Transportation,* 100th Cong., 2nd Sess. 760–61 (1988) [hereinafter *Transportation of Hazardous Materials Hearings*].

327. *See* Holman, *supra* ch. 5 note 283, at 14 (statistical analysis showing far greater success in siting storage, treatment, and recycling units than in siting landfills and incinerators).

Chapter 6

1. *See* Morell and Magorian, *supra* ch. 1 note 8, at 173; S.A. Carnes et al., *Incentives and Nuclear Waste Siting: Prospects and Constraints, in* Lake 1987, *supra* ch. 1 note 11, at 359.

2. Frank I. Michelman, *Property, Utility and Fairness: Comments on the Ethical Foundations of Just Compensation Law,* 80 Harv. L. Rev. 1165 (1967).

3. Michael O'Hare, "*Not On My Block You Don't*": Facility Siting and the *Strategic Importance of Compensation,* 25 Pub. Pol'y 407 (1977) [hereinafter O'Hare 1977]. *See also* O'Hare 1983, *supra* ch. 1 note 9. An earlier, less detailed exposition of the idea is set forth in Anthony J. Mumphrey et al., *A Decision Model for Locating Controversial Facilities,* 37 J. Am. Inst. Planners 397 (1971).

4. E.g., Bacow and Milkey, *supra* ch. 1 note 6, at 265; Bingham and Miller, *supra* ch. 5 note 283; Carnes et al., *supra* ch. 6 note 1, at 353; Robin Gregory et al., *Incentives Policies to Site Hazardous Waste Facilities,* 11 Risk Analysis 667 (1991); Herbert Inhaber, *Can We Find a Volunteer Nuclear Waste Community?* Pub. Utilities Fortnightly, July 15, 1991, at 19; C. Miller, *Efficiency, Equity and Pollution: The Case of Radioactive Waste,* 19 Envt. and Planning A. 913 (1987); Ronald Pushchak and Ian Burton, *Risk and Prior Compensation in Siting Low-Level Nuclear Waste Facilities: Dealing With the NIMBY Syndrome,* 23 Plan Canada No. 3, at 68 (1983).

5. Morell and Magorian, *supra* ch. 1 note 8, at 167. In many ways this is the inverse of state land use laws protecting environmentally fragile "critical areas" such as coastal zones, shorelands, and wetlands. "These are areas where the benefits of development, in terms of new jobs or expanded tax base, will be

enjoyed by local residents, while the environmental losses will be felt statewide." Richard Briffault, *Our Localism: Part I—The Structure of Local Government Law,* 90 Colum. L. Rev. 1, 65 (1990).

6. Green, *supra* ch. 2 note 24, at 93. Under some of these laws, the communities near existing facilities receive a gross receipts tax. *See* Teresa H. Sharp, *Tax Windfall Didn't Set Off a Spending Spree,* Niagara Gazette, Nov. 28, 1992, at 1 (the towns of Lewiston and Porter, New York, each received $1 million in 1992 from Chemical Waste Management for its operating hazardous waste landfill; the towns used a portion of those funds for legal and expert fees to oppose the company's plans to build HW incinerators on the site).

7. U.S. General Accounting Office, Pub. No. GAO/RCED-93-81, *Nuclear Waste: Connecticut's First Site Selection Process for a Disposal Facility* 28–31 (Apr. 1993).

8. 42 U.S.C. § 10173a(a)(1).

9. 42 U.S.C. §§ 10136(c), 10161(f).

10. 42 U.S.C. § 10173a(b).

11. Portney, *supra* ch. 5 note 150, at 27. Compensation appears to be successful in siting a temporary HLW storage facility in Japan. Nicholas Lenssen, *Confronting Nuclear Waste, in State of the World 1992* 46, 62 (Lester R. Brown ed., 1992).

12. This is confirmed by extensive polling data. *See* Portney, *supra* ch. 5 note 150, at 31; Richard J. Bord, *The NIMBY Syndrome: Why is Everyone So Upset? The Role of Risk Communication, Knowledge, Attitudes, and Power Sharing on Public Reactions, in* Inst. for Soc. Impact Assessment, *supra* ch. 2 note 25, at 30; Howard Kunreuther and Douglas Easterling, *Are Risk-Benefit Tradeoffs Possible in Siting Hazardous Facilities?* 80 AEA Papers and Proceedings 252 (1990); Kunreuther et al., *supra* ch. 1 note 7; Lyons et al., *supra* ch. 1 note 1, at 92. EPA acknowledges this finding. *See* U.S. Environmental Protection Agency, Pub. No. EPA/530-SW-90-019, *Sites for Our Solid Waste: A Guidebook for Effective Public Involvement* 84–85 (1990).

13. Some of the moral issues are discussed in Douglas and Wildavsky, *supra* ch. 5 note 219, at 67; Sagoff, *supra* ch. 5 note 1, at 68–69; Been, *supra* ch. 4 note 99, at 60–61; Peters, *supra* ch. 5 note 107, at 50.

14. Carnes et al., *supra* ch. 6 note 1, at 362; Bradford C. Mank, *The Two-Headed Dragon of Siting and Cleaning Up Hazardous Waste Dumps: Can Economic Incentives or Mediation Slay the Monster?* 19 Envtl. Aff. 239, 275 (1991).

15. Godsil, *supra* ch. 5 note 160, at 408; Lisa Aug, *We're Selling Our Souls to Polluters,* Niagara Gazette, June 10, 1993, at 11A.

16. *See* Bullard, *supra* ch. 5 note 146, at 91; Schmeidler and Sandman, *supra* ch. 5 note 2, at 111–12; Richard Walker, *supra* ch. 5 note 59, at 249–50. *See also* Austin and Schill, *supra* ch. 4 note 59, at 70, and Lynette Holloway, *28 Acres of Roof and a Place to Play in West Harlem,* N.Y. Times, Sept. 1, 1992,

at B1 (controversy concerning park on roof of new sewage treatment plant in minority community in New York City).

17. Elizabeth Royte, *Other People's Garbage: The New Politics of Trash,* Harper's, June 1992, at 54.

18. McKewen and Sloan, *supra* ch. 4 note 28.

19. Rabe, *supra* ch. 5 note 202. Respondents to a few public opinion surveys have said they would be more willing to accept HW facilities if they received compensation. These surveys are discussed by Vicki Been in an unpublished, untitled manuscript of 1993.

20. Bacow and Milkey, *supra* ch. 1 note 6, at 275 n.61; Jeff Bailey, *Economics of Trash: Some Big Waste Firms Pay Some Tiny Towns Little for Dump Sites,* Wall St. J., Dec. 3, 1991, at 1; Marvin G. Katz, *YIMBYism is Coming, But . . . ,* Waste Age, Jan. 1990, at 40; Roni Rabin, *LI's Trash is Town's Treasure,* Newsday, Oct. 21, 1991, at 5; Shuff, *Bribes Work in Wisconsin,* Waste Age, Mar. 1989, at 51; Jonathan Walters, *Tempted by Trash,* Governing, July 1991, at 29; Randy Woods, *Fighting NIMBY With Fire,* Waste Age, Sept. 1992, at 19; *Overcoming NIMBY: New Approaches to Resolving Siting Disputes,* The Public's Capital, Winter 1991, at 1.

21. Canter, *supra* ch. 4 note 44, at 453; Carnes et al., *supra* ch. 6 note 1, at 359; Duffy, *supra* ch. 4 note 46, at 786. *See also* "A Proposed 'Bill of Rights' for the Host Community," *in* DiMuzio, *supra* ch. 1 note 6, at 506.

22. Lyle S. Raymond, Jr., et al., *Winning When You Have Lost: Cutting Your Losses With Host Community Benefits,* 18 Law Studies (N.Y.S. Bar Assn.) No. 2 at 56 (1993).

23. *See* Garr, *supra* ch. 1 note 2, at 95, 104; William A. Ruskin, *Value Protection: An Alternative to Litigation,* 5 Toxics L. Rep. (BNA) No. 14, at 482 (Sept. 5, 1990); Daniel A. Spitzer, *Maybe in My Backyard: Strategies for Local Regulation of Private Solid Waste Facilities in New York,* 1 Buff. Envtl. L. J. 87, 135–38 (1993); Jack R. Lebowitz, *The Mechanics and Effectiveness of Property Value Guarantees in Solid Waste Management Facility Siting,* paper presented at the Seventh Annual Conference on Solid Waste Management and Materials Policy, New York, New York, Jan. 1991.

24. *See* Frederick R. Anderson et al., *Environmental Improvement Through Economic Incentives* 31 (1977).

25. E.g., Brion 1991, *supra* ch. 1 note 2, at 192–208; Austin et al., *The Implementation of Controversial Facility-Complex Programs,* 2 Geographical Analysis 315 (1970); Inhaber 1991, *supra* ch. 6 note 4, at 19; Howard Kunreuther et al., *A Compensation Mechanism for Siting Noxious Facilities: Theory and Experimental Design,* 14 J. Envtl. Econ. and Mgmt. 371 (1987); David Goetze, *A Decentralized Mechanism for Siting Hazardous Waste Disposal Facilities,* 39 Pub. Choice 361 (1982); O'Hare 1977, *supra* ch. 4 note 159, at 437–40; Arthur M. Sullivan, *Victim Compensation Revisited: Efficiency Versus Equity in the Siting of Noxious Facilities,* 41 J. Pub. Econ. 211 (1990).

26. See O'Hare 1977, supra ch. 4 note 159, at 447–48; Morell and Magorian, supra ch. 1 note 8, at 171.

27. Tom Arrandale, When the Poor Cry NIMBY, Governing, Sept. 1993, at 36, 39; English 1992, supra ch. 1 note 8, at 57.

28. 42 U.S.C. § 10173a.

29. See, e.g., Bailey and Faupel, supra ch. 4 note 13.

30. 42 U.S.C. § 10136(c)(3)(A).

31. Formation of County in Nevada Outback is Found to be Illegal, N.Y. Times, Feb. 13, 1988, at 7; Shapiro, supra ch. 3 note 24, at 64.

32. Lawrence E. Susskind, A Negotiation Credo for Controversial Siting Disputes, Negotiation J., Oct. 1990, at 309; Gail Bingham, Prospects for Negotiation of Hazardous Waste Siting Disputes, 15 Envtl. L. Rep. (Envt. L. Inst.) 10249; James E. McGuire, The Dilemma of Public Participation in Facility Siting Decisions and the Mediation Alternative, 9 Seton Hall Legis. J. 467 (1985); John Charles Sassaman, Jr., Siting Without Fighting: The Role of Mediation in Enhancing Public Participation in Siting Radioactive Waste Facilities, 2 Alb. L.J. Sci. and Tech. 207 (1992); Katherine R. Shanabrook, Low-Level Radioactive Waste Disposal Facility Sitings: Negotiating a Role for the Public, 3 J. Dispute Resol. 219 (1987); Snider, supra ch. 5 note 88.

33. E.g., Texas Water Commission, The Keystone Siting Process Handbook: A New Approach to Siting Hazardous and Nonhazardous Waste Management Facilities (Dec. 1987); U.S. EPA Office of Solid Waste and Emergency Response, Pub. No. SW-942, Using Compensation and Incentives When Siting Hazardous Waste Management Facilities (1982).

34. Hazardous Waste Facility Siting: A National Survey, supra ch. 1 note 1, at 16.

35. 42 U.S.C. § 10242(b)(2).

36. See Matthew L. Wald, Hired to Be Negotiator, But Treated Like Pariah, N.Y. Times, Feb. 13, 1991, at B5; see also David H. Leroy and T. Scott Nadler, Negotiate Way Out of Siting Dilemmas, 8 Forum for Applied Research and Public Policy 102 (1993).

37. E.g., Brion 1991, supra ch. 1 note 2, at 7–29; Schmeidler and Sandman, supra ch. 5 note 2, at 276–92; Bacow and Milkey, supra ch. 1 note 6; Bingham and Miller, supra ch. 5 note 283; Caskey, supra ch. 1 note 6; Bernd Holznagel, Negotiation and Mediation: The Newest Approach to Hazardous Waste Facility Siting, 13 Envtl. Aff. 329, 354–368 (1986).

38. Mary R. English, The Search for Political Authority in Massachusetts' Toxic Waste Management Law, 16 Envtl. Aff. 39, 41 (1988); Zinc Recycler Run Out of Mass., U.S. Water News, Oct. 1992, at 5.

39. Bingham and Miller, supra ch. 5 note 283, at 484. At least two storage, as opposed to disposal, facilities have been sited in Wisconsin. Another mediation process, known as the Keystone Process, has been tested in Texas, with no greater

success. Thomas O. McGarity, *Public Participation in Risk Regulation,* 1 Risk—Issues in Health and Safety 103, 125–28 (1990).

40. O'Hare and Sanderson, *supra* ch. 5 note 142, at 375.

41. Bingham and Miller, *supra* ch. 5 note 283, at 479.

42. *Overcoming NIMBY: New Approaches to Resolving Siting Disputes,* The Public's Capital, Winter 1991, at 1. *See also* Arnett, *supra* ch. 5 note 6; Harrington, *supra* ch. 4 note 159; Terry A. Trumball, *Using Citizens to Site Solid Waste Facilities,* Pub. Works, Aug. 1988, at 66.

43. Michael Heiman, *From 'Not in My Backyard!' to 'Not In Anybody's Backyard!',* American Planning Assn. J., Summer 1990, at 359, 361.

44. *See* Anthony J. Mumphrey and Julian Wolpert, *Equity Considerations and Concessions in the Siting of Public Facilities,* 49 Econ. Geography 109 (1973); Julian Wolpert, *Regressive Siting of Public Facilities,* 16 Nat. Resources J. 103 (1976).

45. *See* 40 C.F.R. pt. 124; U.S. Environmental Protection Agency, Pub. No. SW-865, *Hazardous Waste Facility Siting: A Critical Problem* 7 (1980) (calling on states to have full public participation in HW siting programs); 132 Cong.Rec. S14925 (daily ed., Oct. 3, 1986) (remarks of Sen. Chafee in floor debate on SARA, calling for full public participation).

46. Sherry Arnstein, *A Ladder of Citizen Participation, in The Politics of Technology* 243 (Godfrey Boyle et al. eds., 1977). *See also* Morell and Magorian, *supra* ch. 1 note 8, at 119. A different spectrum is presented in McGarity, *supra* ch. 6 note 39.

47. *Hazardous Waste Facility Siting: A National Survey, supra* ch. 1 note 1, at 16. For a description of a process in Ontario in which intervenors were given several million dollars to hire lawyers and experts for adjudicatory hearings, *see* Garr, *supra* ch. 1 note 2.

48. E.g., Bacow and Milkey, *supra* ch. 1 note 6, at 269 ("[F]inding sites for the safe disposal and processing of hazardous materials is largely a problem of managing local opposition."); *Hazardous Waste: Education Seen as Key to Overcoming Public Resistance to Incinerator Siting,* 23 Env't Rep. (BNA) No. 52 at 3196 (Apr. 23, 1993); Morell and Magorian, *supra* ch. 1 note 8, at 23, 65, 117, 120; Matheny and Williams, *supra* ch. 4 note 46, at 73.

49. O'Hare 1983, *supra* ch. 1 note 9, at 30; Portney, *supra* ch. 5 note 150, at 65; Cynthia-Lou Coleman, *What Policy Makers Can Learn From Public Relations Practitioners: The Siting of a Low-Level Radioactive Waste Facility in Cortland County, New York,* Pub. Relations Q., Winter 1989–90, at 26; James E. Lukaszewski and Terry L. Serie, *Public Consent Built on Credibility is the Goal,* Waste Age, Feb. 1993 at 45, 46. For projects not perceived as threatening, however, public participation has been found to be helpful in securing public acceptance. Richard A. Ellis and John F. Disinger, *Project Outcomes Correlate With Public Participation Variables,* 53 J. Water Pollution Control Fed. 1564 (1981).

50. Stanley M. Nealey and John A. Herbert, *Public Attitudes Toward Radioactive Wastes*, in Charles A. Walker et al. eds., *supra* ch. 5 note 216, at 94, 108; Bord and O'Connor, *supra* ch. 5 note 223, at 415; Bruce B. Clary and Michael E. Kraft, *Impact Assessment and Policy Failure: The Nuclear Waste Policy Act of 1982*, 8 Policy Studies Rev. 105 (1988); Wildavsky and Dake, *supra* ch. 5 note 316, at 41; Heiman, *supra* ch. 6 note 43, at 361.

51. Ed Brethour, *The Attempted Siting of a Physical-Chemical Hazardous Waste Treatment Facility in Hamiota, Manitoba*, in Resource Futures International ed., *supra* ch. 1 note 2, at 43, 53; Matheny and Williams, *supra* ch. 4 note 46, at 74.

52. *See* Brion 1988, *supra* ch. 5 note 37, at 449; Hugh Kaufman and Lynn Moorer, *The Nuke Dump NIMBY Game: Why Nebraska Was Targeted*, Pub. Utilities Fortnightly, July 15, 1991, at 16; Robert W. Muilenburg, *Balancing the Debate: A Case Study in Combatting Misinformation*, Solid Waste and Power, Feb. 1990, at 32; *Water Group Undertakes Education Program on Beneficial Uses of Clean Sewage Sludge*, 23 Env't Rep. (BNA) No. 32, at 1946 (Dec. 4, 1992).

53. James Flynn, *How Not to Sell a Nuclear Waste Dump*, Wall St. J., Apr. 15, 1992, at A20.

54. Roger E. Kasperson, *Social Issues in Radioactive Waste Management: The National Experience*, in Kasperson ed., *supra* ch. 3 note 4, at 58.

55. V. Kerry Smith and F. Reed Johnson, *How Do Risk Perceptions Respond to Information? The Case of Radon*, 70 Rev. Econ. and Stat. 1 (1988).

56. *See* Portney, *supra* ch. 5 note 150, at 39, 41, 46; Michael L. Poirier Elliott, *Improving Community Acceptance of Hazardous Waste Facilities Through Alternative Systems for Mitigating and Managing Risk*, 1 Hazardous Waste 397 (1984); Susan G. Hadden, *Public Perception of Hazardous Waste*, 11 Risk Analysis 47 (1991). *See also* R.W. Lake and L. Disch, *Structural Constraints and Pluralist Contradictions in Hazardous Waste Regulation*, 24 Env't and Plan. A 663, 665 (1992) (argues that the hazardous waste management system narrows the frame of debate so that public participation "can only be expressed in terms of self-interested local opposition to facility siting," rather than in a more meaningful discussion about whether facilities are needed).

57. *See* Charles E. Faupel et al., *Local Media Roles in Defining Hazardous Waste as a Social Problem: The Case of Sumter County, Alabama*, 11 Sociological Spectrum 293 (1991).

58. A voluntary approach to siting has been advocated by the World Health Organization. William M. Sloan, *Site Selection for New Hazardous Waste Management Facilities* 2 (World Health Organization Regional Publications, European Series, No. 46, 1993).

59. Audrey Armour, *Social Impact Assessment of New Hazardous Waste Facilities: Contrasting Processes in Alberta and Ontario Illustrate Practical Value of Social Considerations* (Maryland Hazardous Waste Facilities Siting Board, 1986);

Jennifer McQuaid-Cook, *Siting a Fully Integrated Hazardous Waste Management Facility With Incinerator and Landfill, Swan Hills, Alberta, Canada, in* Resource Futures International ed., *supra* ch. 1 note 2, at 123; Barry G. Rabe and John Martin Gillroy, *Intrinsic Value and Public Policy Choice: The Alberta Case, in Environmental Risk, Environmental Values, and Political Choices* 150 (John Martin Gillroy ed. 1993); Rabe, *supra* ch. 5 note 202; Schmeidler and Sandman, *supra* ch. 5 note 2, at 268–75; Robert Tomsho, *Small Town in Alberta Embraces What Most Reject: Toxic Waste,* Wall St. J., Dec. 27, 1991, at 1; Dave Wenger, *Siting the Alberta Special Waste Treatment Centre: A Public Consensus—An Alberta Success, in* Inst. for Soc. Impact Assessment, *supra* ch. 2 note 25, at 124.

60. Ed Brethour, *The Attempted Siting of a Physical-Chemical Hazardous Waste Treatment Facility in Hamiota, Manitoba, in* Resource Futures International ed., *supra* ch. 1 note 2, at 43; Barbara Connell, *The Siting Approach for a Hazardous Waste Management Facility in Manitoba: Comment and Critique,* paper submitted to International Workshop on Innovative Approaches to Siting of Waste Management Facilities, Montebello, Quebec (Apr. 27–30, 1991); R.J. Cooke and A.B. Richards, *A Co-Management Approach to Developing Environmentally Sensitive Projects,* paper submitted to Environment and the Economy—Partners in the Future conference, Winnipeg, Manitoba (May 18, 1989); Alun Richards, *Implementing a Voluntary and Responsive Siting Process in Rural and Urban Settings,* 10 Impact Assessment Bull. 89 (1992).

61. Audrey Armour et al., *New Institutional Approaches to Alleviating Facility Siting Problems,* Presentation to Workshop on Nuclear as a Large-Scale Global Energy Option, Oak Ridge National Laboratory (Sept. 28–29, 1993) [hereinafter Armour 1993].

62. Ristoratore, *supra* ch. 1 note 4; Mario Ristoratore, *Siting Toxic Waste Disposal Facilities: Best and Worst Cases in North America, in Land Rites and Wrongs: The Management, Regulation and Use of Land in Canada and the U.S.* 201 (Lincoln Institute of Land Policy, 1987).

63. Personal communication with Audrey Armour, Sept. 23, 1993; Audrey M. Armour, *Opting for Cooperation: Process for Siting a Low-Level Radioactive Waste Management Facility, in* Resource Futures International ed., *supra* ch. 1 note 2, at 31; Armour 1991, *supra* ch. 5 note 2, at 49–67.

64. Armour 1993, *supra* ch. 6 note 61.

65. English 1992, *supra* ch. 1 note 8, at 104.

66. *Nuclear Waste: On the Reservation,* Economist, Oct. 3, 1992, at 30.

67. Melinda Kassen, *Siting the MRS—A Lesson in How Even Bribes Don't Work,* 7 Natural Resources and Envt. No. 3, at 16, 19 (Winter 1993).

68. Sigmon, *supra* ch. 5 note 308.

69. Contreras, *supra* ch. 1 note 2, at 536 n.329.

70. Flynn et al., *supra* ch. 5 note 275.

71. English 1992, *supra* ch. 1 note 8, at 55–57; Gretchen D. Monti, *"All Politics is Local": Integrating Local Concerns Into Facility Site Selection, in* Inst. for Soc. Impact Assessment, *supra* ch. 2 note 25, at 36; *Martinsville Rejected for Central Midwest LLRW Facility,* Radioactive Exchange, Oct. 21, 1992, at 1.

72. *Towns Offer to Accept Low-Level Nuke Waste,* Eng'g News Record, Mar. 26, 1987, at 15.

73. Carter, *supra* ch. 3 note 51, at 151–53, 166, 175–76, 423–24; English 1992, *supra* ch. 1 note 8, at 87; Jacob, *supra* ch. 3 note 26, at 42; Shapiro, *supra* ch. 3 note 24, at 62–63; *Low-Level Waste: Utah Gov. Rules Out Site for Repository,* Greenwire, Jan. 15, 1993; Barnaby J. Feder, *The Saga of the Lonetree Landfill,* N.Y. Times, Dec. 22, 1992, at D1. *See also* Peter T. Kilborn, *Dying Town Considers Salvation in a Landfill,* N.Y. Times, Oct. 6, 1991, at 20, Royte, *supra* ch. 6 note 17, and John T. Aquino, *The Politics of Landfills,* Waste Age, March 1993, at 37 (West Virginia town agreed to accept MSW landfill in exchange for monetary compensation, but governor attempted to kill deal).

74. U.S. Gen. Accounting Office, Pub. No. GAO/RCED-91-149, *Nuclear Waste: Extensive Process to Site Low-Level Waste Disposal Facility in Nebraska* (July 1991); Richard Paton, *Issues Management in Radioactive Waste Disposal Decisions, in* Inst. for Soc. Impact Assessment, *supra* ch. 2 note 25, at 27; Kaufman and Moorer, *supra* ch. 6 note 52; *Nebraskans Vote No on Nuclear Dump Plan,* N.Y. Times, Dec. 10, 1992, at D19.

75. *Hazwaste Incinerator Fight,* Chemical Wk., Apr. 29, 1992, at 56.

76. Carter, *supra* ch. 3 note 51, at 177–78.

77. *See* Sassaman, *supra* ch. 6 note 32, at 211–12; *Talk of a Nuclear Dump Again Splits a Community,* N.Y. Times, Aug. 23, 1993, at B5; Sam Howe Verhovek, *Town Heatedly Debates Merits of a Nuclear Waste Dump,* N.Y. Times, June 28, 1991, at B1; Mayerat v. Town Bd. of Town of Ashford, 185 A.D.2d 699, 585 N.Y.S.2d 928 (4th Dep't 1992).

78. E.g., John A. Barnes, *Learning to Love the Dump Next Door,* Wall St. J., June 25, 1991 (Riverview, Mich.); Robert T. Nelson, *Dealing With Waste: Oregon, Here We Come,* Seattle Times, March 24, 1991, at B1 (Gilliam County, Oregon).

79. E.g., *Voters Accept Landfill Plan,* N.Y. Bus. Env't, Aug. 31, 1992, at 3; *In Brief,* N.Y. Waste Rep., Sept. 1992 at 6, and John Gayusky, *Organizing Toolbox: Polluters' Secret Plan Update,* Everyone's Backyard, Apr. 1993 at 20 (Eagle, N.Y.); Aisling A. Swift, *Green Island Says "Yes",* Record (Troy, N.Y.), June 3, 1992, at 1, and *Bull Market for Incinerators Stalled by Recession, Recycling, Local Opposition,* 23 Env't Rep. (BNA) No. 8, at 686 (June 19, 1992) (Green Island, New York); *Michigan Voters Approve Incinerator,* World Wastes, June 1992, at 48 (Oakland County, Michigan).

80. E.g., Eric Schmitt, *Town Fights With Neighbors Over Burning of Tons of Tires,* N.Y. Times, Oct. 15, 1989, at 1 (Sterling, Conn.).

81. *See* Sigmon, *supra* ch. 5 note 308.

82. Advertisement, Hazmat World, June 1992, at 27. *See also* Leslie Miller, *New Mexico County Courts Waste for Disposal Sites,* World Wastes, Sept. 1992, at 11.

83. Rabe 1991, *supra* ch. 5 note 202, at 131–32.

84. John Duncan Powell, *A Hazardous Waste Site: The Case of Nyanza,* in *Environmental Hazards: Communicating Risks as a Social Process* (Sheldon Krimsky and Alonzo Plough eds., Auburn House 1988) at 239.

85. *See* Jana L. Walker and Kevin Gover, *Commercial Solid and Hazardous Waste Disposal Projects on Indian Lands,* 10 Yale J. on Reg. 229 (1993); Pamela A. D'Angelo, *Waste Management Industry Turns to Indian Reservations as States Close Landfills,* 21 Env't Rep. (BNA) No. 35, at 1607 (Dec. 28, 1990); Kathleen Sheehan and John T. Aquino, *Waste Disposal on Indian Lands: A Boon or Bust Proposition?* Waste Age, Oct. 1991, at 58.

86. *See* Judith V. Royster, *Environmental Protection and Native American Rights: Controlling Land Use Through Environmental Regulation,* 1 Kan. J. L. and Pub. Pol'y 89 (1991); Douglas A. Brockman, *Note: Congressional Delegation of Environmental Regulatory Jurisdiction: Native American Control of the Reservation Environment,* 41 Wash. U.J. Urb. and Contemp. L. 133 (1992); Teresa A. Williams, *Note: Pollution and Hazardous Waste on Indian Lands: Do Federal Laws Apply and Who May Enforce Them,* 17 Am. Indian L. Rev. 269 (1992).

87. Keith Schneider, *Idaho Tribe Stops Nuclear Waste Truck,* N.Y. Times, Oct. 17, 1991, at A18.

88. Northern States Power Co. v. Prairie Island Mdewakanton Sioux Indian Community, 991 F.2d 458 (8th Cir. 1993); Keith Schneider, *Nominee Is a Veteran of Atomic-Waste Battle,* N.Y. Times, Jan. 9, 1993, at 8.

89. Ronald Smothers, *Future in Mind, Choctaws Reject Plan For Landfill,* N.Y. Times, Apr. 21, 1991, at 22.

90. Keith Schneider, *Grants Stir Interest in Nuclear Waste Site,* N.Y. Times, Jan. 9, 1992, at A14.

91. *See* Matthew L. Wald, *Tribe on Path to Nuclear Waste Site,* N.Y. Times, Aug. 6, 1993, at A12; *Nuclear Waste: On the Reservation,* Economist, Oct. 3, 1992, at 30; *Wyoming Governor Blocks Additional Research on Nuclear Waste Storage Facility Sought by DOE,* 23 Env't Rep. (BNA) No. 18, at 1290 (Aug. 28, 1992); *Even Without DOE Funds, MRS Study Will Continue, Mescaleros Pledge,* Radioactive Exchange, Oct. 21, 1992, at 18; *Utah Tribe Asks DOE for Grant to Conduct IIA MRS Study,* Radioactive Exchange, Nov. 2, 1992, at 17.

92. *See* Valerie Taliman, *Waste Merchants Intentionally Poison Natives,* 1 Voces Unidas (Fourth Quarter 1991); Cole 1993, *supra* ch. 5 note 154 n.6.

93. Keven Gover and Jana L. Walker, *Escaping Environmental Paternalism: One Tribe's Approach to Developing a Commercial Waste Disposal Project in Indian Country,* 63 U. Colo. L.R. 933, 942 (1992). *Similarly, see* Dirk Johnson, *Tribes' New Foe: Environmentalists,* N.Y. Times, Dec. 28, 1991, at 7.

94. Vine Deloria, Jr., and Clifford M. Lytle, *American Indians, American Justice* (University of Texas Press, 1983) at 108–09.

95. Tony Davis, *Apaches Split Over Nuclear Waste*, High Country News, Jan. 27, 1992, at 12; Sherry Robinson, *Mescalero Apaches Study Nuclear Waste Storage*, Enchanted Times (New Mexico Research Education and Enrichment Foundation); Matthew L. Wald, *Nuclear Storage Divides Apaches and Neighbors*, N.Y. Times, Nov. 11, 1993, at A18; Valerie Taliman, *"Chernobyl Chino's" MRS: Mescaleros Apaches in Classic Economic Blackmail and Environmental Racism Struggle*, 2 Voces Unidas 1 (First Quarter 1992).

96. Portney, *supra* ch. 5 note 150, at 137–59. Peter Huber had earlier discussed risk substitution in the products liability context. Huber, *supra* ch. 4 note 63, at 1073.

97. Mank, *supra* ch. 6 note 14. It was reported in 1990 that the New Jersey legislature was considering a bill with similar features. W.B. Clapham, Jr., *Some Approaches to Assessing Environmental Risk in Siting Hazardous Waste Facilities*, 12 Envtl. Prof. 32, 37 (1990). Note also that a citizen task force in Oak Ridge, Tennessee, proposed that the siting of the monitored retrievable storage facility for HLW in that area be linked to a schedule for DOE cleanup of existing contamination in the area. C.P. Wolf, *The NIMBY Syndrome: Its Cause and Cure*, 502 Annals N.Y. Acad. Sci. 216, 223 (1987).

98. *See* ch. 4 notes 60–61.

99. *Coming Clean: Superfund Problems Can Be Solved . . .* , *supra* ch. 2 note 51, at 179.

100. Washburn and Harris have raised two additional objections to Portney's proposal: it unfairly targets communities that are already beleaguered; and the sites most amenable to risk substitution may not be those most technically suitable for waste disposal facilities. Stephen T. Washburn and Robert H. Harris, *Necessary Evils*, Issues Sci. and Tech., Fall 1991, at 86.

101. James C. McKinley, Jr., *Plan on Garbage Backed by Council in New York City*, N.Y. Times, Aug. 28, 1992, at A1; James C. McKinley, Jr., *Civics Lesson in the Art of Persuasion: How Dinkins Turned City Council Opponents Into Friends of His Trash Plan*, N.Y. Times, Sept. 1, 1992, at B3.

102. *Mescaleros Would Help Clean Up Uranium Mines in Return for MRS*, Radioactive Exchange, Dec. 1, 1992, at 17.

Chapter 7

1. William M. Sloan, *What We Did Well, Good, and Wrong: A Critique of State Hazardous Waste Facility Siting in the 80s* 3, paper presented to Conference on State Policies on Siting Hazardous Waste Facilities, San Francisco (Sept. 2–3, 1992) (National Governors' Assn.).

2. *See* Neil R. Shortlidge and S. Mark White, *The Use of Zoning and Other Local Controls for Siting Solid and Hazardous Waste Facilities,* 7 Natural Resources and Envt. No. 3 at 3 (Winter 1993).

3. 42 U.S.C. §6929.

4. U.S. Environmental Protection Agency, Pub. No. SW-865, *Hazardous Waste Facility Siting: A Critical Problem* 7 (1980).

5. 40 C.F.R. § 271.4(b).

6. S. Rep. No. 99–11, 99th Cong., 1st Sess. (1985).

7. Ensco, Inc. v. Dumas, 807 F.2d 743 (8th Cir. 1986); *similarly,* Ogden Envtl. Servs., Inc. v. City of San Diego, 687 F. Supp. 1436 (S.D. Cal. 1988). These cases are discussed in detail in Patrick O'Hara, Note, *The N.I.M.B.Y. Syndrome Meets the Preemption Doctrine: Federal Preemption of State and Local Restrictions on the Siting of Hazardous Waste Disposal Facilities,* 53 La. L. Rev. 229 (1992). *But see* Lafarge Corp. v. Campbell, 813 F. Supp. 501 (W.D. Tex. 1993) (upholding state law that prevented siting of hazardous waste incinerator within a half-mile of a residence).

8. Green, *supra* ch. 2 note 24, at 6.

9. *See* Philip Shabecoff, *Waste-Plant Inquiry Taps Hot Water,* N.Y. Times, July 25, 1989, at A20; Bill Gifford, *Reilly's March to the Sea: How the EPA is Sowing Toxic Waste Through the Sea,* Village Voice, Feb. 27, 1990, at 25.

10. *North Carolina's Authority to Run Hazardous Waste Programs Upheld by EPA,* 21 Env't Rep. (BNA) No. 6, at 307 (June 8, 1990); Hazardous Waste Treatment Council v. Reilly, 938 F.2d 1390 (D.C. Cir. 1991).

11. *See* Pacific Gas and Elec. Co. v. State Energy Resources Conservation and Dev. Comm'n, 461 U.S. 190 (1983); Northern States Power Co. v. Minnesota, 447 F.2d 1143 (8th Cir. 1971), *aff'd mem.,* 405 U.S. 1035, 92 S.Ct. 1307, 31 L.Ed.2d 576 (1972); R.C. Kearney and R.B. Garvey, *American Federalism and the Management of Radioactive Wastes,* 42 Pub. Admin. Rev. 14 (1982).

12. *Hazardous Waste Facility Siting: A National Survey, supra* ch. 1 note 1, at 12. *See also Illinois Low-Level Waste Facility Law Eliminates Local Veto of Siting Decisions,* 23 Env't Rep. (BNA) No. 47 at 3028 (Mar. 19, 1993).

13. *See* Godsil, *supra* ch. 5 note 160, at 406; Tsao, *supra* ch. 4 note 99, at 371.

14. The cases are reviewed in McCabe, *supra* ch. 5 note 88; Melissa Thorme, *Local to Global: Citizen's Legal Rights and Remedies Relating to Toxic Waste Dumps,* 5 Tul. Envtl. L.J. 101 (1991); William B. Johnson, Annotation, *Validity of Local Regulation of Hazardous Waste,* 67 A.L.R.4th 822.

15. E.g., Andreen, *supra* ch. 1 note 8; Davidson, *supra* ch. 4 note 46; A. Dan Tarlock, *State Siting Laws, Local Land Use Laws, and Their Interplay,* 15 Envtl. L. Rep. (Envt. L. Inst.) 10236 (1985). *See also* Laurie Reynolds, *The Failure of Local Landfill Siting Control in Illinois,* 17 S. Ill. U. L.J. 1 (1992) (argues that excessive control has been given to municipalities in siting MSW landfills in Illinois).

16. E.g., O'Hare 1983, *supra* ch. 1 note 9, at 24; Andrews, *supra* ch. 4 note 30, at 117, 121–22; Colglazier and English, *supra* ch. 1 note 8, at 641. *See, generally,* Spitzer, *supra* ch. 6 note 23; Robert W. Lake and Rebecca A. Johns, *Legitimation Conflicts: The Politics of Hazardous Waste Siting Law,* 11 Urban Geography 488 (1990).

17. Bingham and Miller, *supra* ch. 5 note 283, at 477.

18. Delogu, *supra* ch. 4 note 159, at 209–10. A comparable approach is proposed in William David Bridgers, Note, *The Hazardous Waste Wars: An Examination of the Origins and Major Battles to Date, With Suggestions for Ending the Wars,* 17 Vermont L. Rev. 821 (1993).

19. 42 U.S.C. § 2021e(d)(2)(C).

20. New York v. United States, 112 S.Ct. 2408 (1992).

21. 42 U.S.C. § 9607(a)(1).

Chapter 8

1. *See* Cole 1992, *supra* ch. 5 note 53, at 1996; Heiman, *supra* ch. 6 note 43; Lillie Craig Trimble, *What Do Citizens Want in Siting of Waste Management Facilities?* 8 Risk Analysis 375 (1988).

2. *Cf.* Clifford S. Russell, *Economic Incentives in the Management of Hazardous Wastes,* 13 Colum. J. Envtl. L. 257, 262 (1988).

3. 42 U.S.C. § 6902(a)(6).

4. 42 U.S.C. § 6902(b). *See* Robert F. Blomquist, *Beyond the EPA and OTA Reports: Toward a Comprehensive Theory and Approach to Hazardous Waste Reduction in America,* 18 Envtl. L. 817 (1988).

5. 42 U.S.C. § 6922(b)(2).

6. 40 C.F.R. pt. 262 Appx.

7. *Id.*; *see also* 42 U.S.C. § 6921(d)(3) (relaxed manifest requirements for small-quantity generators).

8. 42 U.S.C. § 6922(a)(6)(C).

9. 58 Fed. Reg. 31114 (May 28, 1993).

10. Exec. Order No. 12856, 58 Fed. Reg. 41981 (Aug. 3, 1993). *See President Directs Federal Agencies to Take Lead in Pollution Prevention,* 24 Env't Rep. (BNA) No. 15 at 623 (Aug. 13, 1993).

11. S. Wolf, *supra* ch. 4 note 32, at 575.

12. Landy et al., *supra* ch. 2 note 46, at 125.

13. Hazardous Waste Treatment Council v. South Carolina, 945 F.2d 781 (4th Cir. 1991).

14. 42 U.S.C. § 13101.

15. *See* E. Lynn Grayson, *The Pollution Prevention Act of 1990: Emergence of a New Environmental Policy,* 22 Envtl. L. Rep. (Envtl. L. Inst.) 10392 (1992); Stephen M. Johnson, *From Reaction to Proaction: The 1990 Pollution Prevention Act,* 17 Colum. J. Envtl. L. 153 (1992).

16. 42 U.S.C. § 13109.

17. *See* Johnson, *supra* ch. 8 note 15. *See also* Stephen L. Kass and Michael B. Gerrard, *New York's Requirements for Reducing Hazardous Waste,* N.Y.L.J., Nov. 21, 1990, at 3. One exception is the New Jersey Pollution Prevention Act of 1991, which requires firms within the state to reduce their use of hazardous materials and reduce their production of HW by 50% over five years. N.J.S.A. §§ 13:ID-35 *et seq.* This statute was enacted in partial response to protests against proposals to site HW incinerators in the state. Lake, *supra,* ch. 1 note 8.

18. *But see* N.Y. Envtl. Conserv. L. § 8-0109(2)(i) (McKinney Supp. 1993) (EISs prepared under New York State Environmental Quality Review Act must discuss "effects of proposed action on solid waste management where applicable and significant.").

19. *See* text accompanying ch. 5 notes 7–9 *supra. See also The Toxics Release Inventory: A National Perspective, 1987, supra* ch. 2 note 9, at 266 (many reporting companies are attempting to reduce waste generation because of high treatment or disposal costs).

20. Goodbaum and Rotman, *supra* ch. 5 note 285.

21. *See* United States v. Monsanto Co., 858 F.2d 160, 171 (4th Cir. 1988), *cert. denied,* 490 U.S. 1106 (1989); United States v. Chem-Dyne Corp., 572 F. Supp. 802, 808–10 (S.D. Ohio 1983). *But see* United States v. Alcan Aluminium, 990 F.2d 711 (2d Cir. 1993).

22. David J. Sarokin et al., *Cutting Chemical Waste: What 29 Organic Chemical Plants Are Doing to Reduce Hazardous Wastes* 140, 142 (1985); Carol Dansereau, *Smokescreen: The Myth of Incinerator Need* 18 (Washington Toxics Coalition, 1992); Alex. Brown and Sons, *supra* ch. 2 note 40, at 9–10.

23. Holman, *supra* ch. 5 note 283, at 18.

24. Hunt v. Chemical Waste Management, Inc., 584 So.2d 1367, 1373 (Ala. 1991), *rev'd,* 112 S. Ct. 2009 (1992); Hammitt and Reuter, *supra* ch. 5 note 102, at 13, 36. For one example of regulatory uncertainty in the definition of HW, *see* Catherine L. LaCroix, *RCRA and Non-Traditional Hazardous Wastes,* 23 Env't Rep. (BNA) No. 26, at 1650 (Oct. 23, 1992).

25. N.Y. St. Dep't of Envtl. Conservation, *Revised Draft New York State Hazardous Waste Facility Siting Plan and Environmental Impact Statement* 4-22 (Aug. 1989). New York State attempted to prohibit nonhazardous waste from being sent to its hazardous waste landfills, but this effort was struck down on procedural grounds. CWM Chemical Services, Inc. v. Jorling, Index No. 70900 (Sup. Ct. Niagara County, July 23, 1991).

26. Ronald Begley, *TRI and Pollution Prevention Data Show Positive Trend, CMW Says,* Chemical Wk, Apr. 21, 1993; Keith Schneider, *Manufacturers Recycling Half of Chemical Wastes,* N.Y. Times, May 26, 1993, at A15.

27. John L. Warren, *The Potential for Waste Reduction, in Hazardous Waste Minimization* 15 (Harry Freeman ed., 1990).

28. U.S. Congress, Office of Technology Assessment, Pub. No. OTA-E-560, *Industrial Energy Efficiency* 3 (1993).

29. Ronald T. McHugh, *The Economics of Waste Minimization, in* Freeman ed., *supra* ch. 5 note 17, at 127, 128.

30. Johnson, *supra* ch. 8 note 15, at 157; Freeman, ed., *supra* ch. 5 note 17, at 5; Mark H. Dorfman et al., *Environmental Dividends: Cutting More Chemical Wastes* (1992); Joel S. Hirschhorn and Kirsten U. Oldenburg, *Prosperity Without Pollution: The Prevention Strategy for Industry and Consumers* (1991).

31. Clifford S. Russell and Walter O. Spofford, Jr., *A Quantitative Framework for Residuals Management Decisions, in* Kneese and Bower, *supra* ch. 5 note 22, at 115.

32. Sarokin et al., *supra* ch. 8 note 22; *similarly,* Rick Mullin, *New Direction on Hazwaste,* Chemical Wk., Jan. 20, 1993, at 26.

33. Sarokin et al., *supra* ch. 8 note 22, at 32, 138–39, 183–201.

34. Id. at 124–25.

35. Dansereau, *supra* ch. 8 note 22, at 109.

36. Jeff Bailey, *Costs of Toxic Waste Leave Landfills Unfilled,* Wall St. J., Apr. 30, 1993, at B1.

37. Alex. Brown and Sons, *supra* ch. 2 note 40, at 11.

38. Piasecki and Davis, *supra* ch. 2 note 1, at 6.

39. Robert G. Cochran and Nicholas Tsoulfanidis, *The Nuclear Fuel Cycle: Analysis and Management* 188–95 (American Nuclear Society 1990).

40. Aaron Wildavsky, *Searching for Safety* (New Brunswick: Transaction Books 1988) at 198–99; Terri Shaw, *Smoke Alarm Alert,* Washington Post, Nov. 28, 1991, at T11.

41. *E.g.,* Tom Arrandale, *The Most Powerful Incentive For Reducing Waste at the Source,* Governing, Jan. 1993, at 61; Britt Anne Bernheim, *Can We Cure Our Throwaway Habits by Imposing the True Social Cost on Disposable Products?,* 63 U. Colo. L. Rev. 953 (1992); *Federal Options for Reducing Waste Disposal, supra* ch. 5 note 17, at 5; Menell, *supra* ch. 5 note 1; Timothy E. Wirth and John Heinz, *Project 88—Round II—Incentives for Action: Designing Market-Based Environmental Strategies* 49 (1991).

42. *See* Bette K. Fishbein and Caroline Gelb, *Making Less Garbage: A Planning Guide for Communities* 101 (INFORM 1992); Katya Andresen, *Communities Weigh Merits of Variable Rates,* World Wastes, Nov. 1992, at 18; Reason Foundation, *Variable Rates For Municipal Solid Waste: Implementation Experience,*

96. *See* 42 U.S.C. §§ 7412(b)(1), 9601(14)(E); *see* Lynn L. Bergeson, *New CAA Chemicals Raise CERCLA Reporting Issues,* Pollution Eng'g, June 1991, at 23.

97. Loren D. Potter, *Desert Characteristics as Related to Waste Disposal, in* Reith and Thompson eds., *supra* ch. 3 note 49, at 21.

98. Letter, Deanna M. Wieman, U.S. E.P.A., to Bruce West, U.S. Bureau of Land Management (BLM), Oct. 15, 1992. The facility would be on private land, but the BLM's approval would be required for access roads. Telephone interview with Dick Forester, BLM, Jan. 26, 1993.

99. Matthew L. Wald, *Tribe on Path to Nuclear Waste Site,* N.Y. Times, Aug. 6, 1993, at A12.

100. Robert Reinhold, *States, Failing to Cooperate, Face a Nuclear-Waste Crisis,* N.Y. Times, Dec. 28, 1992, at A1.

101. Smolen *supra* ch. 5 note 23, at 4.

102. Jefferson, *supra* ch. 8 note 61, at 89.

103. Olsen, *supra* ch. 4 note 3.

104. Philip J. Landrigan et al., *Toxic Air Pollution Across a State Line: Implications for the Siting of Resource Recovery Facilities,* 10 J. Pub. Health Policy 309 (1989); Joseph F. Sullivan, *Debate Rages Over Site Proposed for Incinerator,* N.Y. Times, Mar. 28, 1992, at 28.

105. James J. Florio, *The Solid Waste Crisis,* 9 Seton Hall Legis. J. 399, 401 (1985).

106. Roberto Suro, *Texas Town and Fertilizer From That City,* N.Y. Times, Jan. 25, 1993, at B2.

107. Frances Frank Marcus, *Medical Waste Divides Mississippi Cities,* N.Y. Times, June 24, 1992, at A16.

108. Edward Cody, *Mexico Seeks Halt in U.S. Waste Plan; Texas Sites Raise Pollution Concerns,* Washington Post, Mar. 22, 1992, at A29. *See also* Phillip Elmer-DeWitt, *"Love Canals in the Making",* Time, May 20, 1991, at 51; Stephen P. Mumme, *Complex Interdependence and Hazardous Waste Management Along the U.S.-Mexico Border, in* Davis and Lester eds., *supra* ch. 4 note 46, at 224; Roberto A. Sanchez, *Health and Environmental Risks of the Maquiladora in Mexicali,* 30 Nat. Resources J. 163 (1990).

109. Gore, *supra* ch. 5 note 128, at 153, 157.

110. Michael Specter, *Pact on Garbage in New York City,* N.Y. Times, Aug. 18, 1992, at A1. *See also Massachusetts: MWRA Eyes Out-of-State Sites Over Walpole Landfill,* Solid Waste Digest (Northeast ed.), Feb. 1993, at 3.

111. In the Matter of Brooklyn Navy Yard Resource Recovery Facility, slip op. at 13 (N.Y. St. Dep't of Envtl. Conservation, Admin. L.J., Dec. 23, 1992).

112. *Massachusetts Pact With Utah Landfill Approved,* 24 Envt. Rep. (BNA) No. 22 at 1019 (Oct. 1, 1993).

113. *See Commercial Hazardous Waste Management: Recent Financial Performance and Outlook for the Future,* Hazardous Waste Consultant, Sept./Oct.

1990, at 4-1, 4-12, and Illinois v. Teledyne, Inc., 233 Ill.App.3d 495, 599 N.E.2d 472 (3d Dist. 1992) (incidents in Wilsonville, Illinois and Sheffield, Illinois).

114. Susan M. Brett et al., *Assessment of the Public Health Risks Associated With a Proposed Excavation at a Hazardous Waste Site, in Risk Assessment of Environmental Hazards: A Textbook of Case Studies* (Dennis J. Paustenbach ed., 1989), at 427.

115. Center for Investigative Reporting and Bill Moyers, *supra* ch. 5 note 260, at 5.

116. William L. Long, *Economic Aspects of Transport and Disposal of Hazardous Wastes,* 14 Marine Pol'y 198, 201 (1990).

117. 42 U.S.C. § 6938.

118. Asante-Duah, *supra* ch. 5 note 8; Ibrahim J. Wani, *Poverty, Governance, the Rule of Law, and International Environmentalism: A Critique of the Basel Convention on Hazardous Wastes,* 1 Kan. J. L. and Pub. Pol. 37 (1991).

119. *Sources Seek U.S. Role in Waste Export Treaty Talks, Despite Lack of Status,* Inside E.P.A., Nov. 20, 1992, at 13.

120. U.S. General Accounting Office, Pub. No. GAO/PEMD-93-24, *Hazardous Waste Exports: Data Quality and Collection Problems Weaken EPA Enforcement Activities* 14 (July 1993).

121. U.S. Army, Chemical Materiel Destruction Agency, Program Manager for Non-Stockpile Chemical Materiel, *Non-Stockpile Chemical Materiel Program: Survey and Analysis Report* 2-4 (Nov. 1993).

122. Waligory, *supra* ch. 4 note 107, at 685.

123. John W. Birks, *Weapons Forsworn: Chemical and Biological Weapons, in* Ehrlich and Birks eds., *supra* ch. 2 note 75, at 161, 172.

124. 33 U.S.C. §§ 1401 *et seq.*

125. Goldman, *supra* ch. 5 note 83, at 95.

126. 33 U.S.C. § 1414b.

127. *Use of Barges for Illegal Waste Dumping Said to Hamper Coast Guard Cleanup Efforts,* 23 Env't Rep. (BNA) No. 35, at 2095 (Dec. 25, 1992). *See also* Abandoned Barge Act of 1992, Pub. L. 102-587 § 5301.

128. Convention on the Prevention of Marine Pollution by Dumping Wastes and Other Matter, Dec. 29, 1972, 26 U.S.T. 2403, T.I.A.S. 8165, 1046 U.N.T.S. 120, entered into force Aug. 30, 1975, reprinted in 11 I.L.M. 1294 (1972).

129. *See* Clifton E. Curtis, *Legality of Seabed Disposal of High-Level Radioactive Wastes Under the London Dumping Convention,* 14 Ocean Dev. and Int'l L. 383 (1985); Daniel P. Finn, *Ocean Disposal of Radioactive Wastes: The Obligation of International Cooperation to Protect the Marine Environment,* 21 Va. J. Int'l L. 621 (1981); David G. Spak, *The Need for a Ban on All Radioactive Waste Disposal in the Ocean,* 7 Nw. J. Int'l L. and Bus. 803 (1986); Waligory, *supra* ch. 4 note 107.

130. David E. Sanger, *Nuclear Material Dumped Off Japan,* N.Y. Times, Oct. 19, 1993, at A1. *See also Radioactive and Other Environmental Threats to the United States and the Arctic Resulting From Past Soviet Activities: Hearings Before the Senate Select Comm. on Intelligence,* 102nd Cong., 2d Sess. (1992).

131. *Ocean Incineration: Its Role in Managing Hazardous Waste, supra* ch. 2 note 12, at 179.

132. Arnold W. Reitze and Andrew Davis, Reconsidering Ocean Incineration as Part of a U.S. Hazardous Waste Management Program: Separating the Rhetoric From the Reality, 17 Envtl. Aff. 687, 731 and n.336 (1990).

134. Seaburn, Inc. v. EPA, 712 F. Supp. 218 (D.D.C. 1989); Waste Management, Inc. v. EPA, 669 F. Supp. 536 (D.D.C. 1987). The Coast Guard has promulgated regulations for "vessels engaged in bulk hazardous waste incineration at sea." 46 C.F.R. § 150.200.

135. *See generally* Elaine L. Hughes, *Toxic Waste Incineration at Sea,* 24 U.B.C. L. Rev. 19 (1990); Christopher B. Kende, *Oceans and Coasts, in* Gerrard 1992, *supra* ch. 2 note 7, § 23.06; Christopher A. Walker, *The United States Environmental Protection Agency's Proposal for At-Sea Incineration of Hazardous Wastes—A Transnational Perspective,* 21 Vand. J. Transnat'l L. 157 (1988).

136. *Incineration of Hazardous Waste at Sea, supra* ch. 2 note 103, at 221.

137. *Ocean Incineration: Its Role in Managing Hazardous Waste, supra* ch. 2 note 12, at 517.

138. Bruce Piasecki and Hans Sutter, *Alternatives to Ocean Incineration in Europe, in* Piasecki and Davis eds., *supra* ch. 2 note 1, at 67.

139. *E.g.,* James Ehmann, *Chatty's Island* (1982).

140. 33 U.S.C. §§ 1501 *et seq.*

141. Outer Continental Shelf Lands Act, 43 U.S.C. §§ 1331 *et seq.*

142. *See* United States v. Ray, 423 F.2d 16 (5th Cir. 1970).

143. *E.g.,* Keith Schneider, *Scientists Suggest Dumping Sludge on Vast, Barren Deep Sea Floor,* N.Y. Times, Dec. 2, 1991, at A1; A. Aristides Yayanos, *Ocean Engineering: Sea-Burial of Toxic Wastes,* 25 Cal. Bus. No. 5, at 105 (May 1990). *But see WHOI Report Deep Sixes Ocean Dumping,* 261 Science 423 (1993).

144. Shapiro, *supra* ch. 3 note 24, at 64.

145. 42 U.S.C. § 10204.

146. P.J. Skerrett, *Nuclear Burial at Sea,* Tech. Rev., Feb./Mar. 1992, at 22.

147. Stansfield Turner, *Freeze-Dry the Bomb,* N.Y. Times, Apr. 21, 1992, at A23.

148. Jacob, *supra* ch. 3 note 26, at 36 (James Schlesinger advocated space disposal when he was chairman of Atomic Energy Commission); Tang and Saling, *supra* ch. 3 note 8, at 383; Charles D. Hollister and Harry W. Smedes, *Selecting Sites for Radioactive Waste Repositories, in* Harthill ed., *supra* ch. 4 note 49, at 63; William J. Broad, *Nuclear Accords Bring New Fears on Arms Disposal,* N.Y. Times, July 6, 1992, at 1.

149. Earl R. Hoskins and James E. Russell, *Geologic and Engineering Dimensions of Nuclear Waste Storage, in* Murdock et al. eds., *supra* ch. 5 note 107, at 19, 27.

150. Bernard K. Shafer, *Solid, Hazardous, and Radioactive Wastes in Outer Space: Present Controls and Suggested Changes,* 19 Cal. W. Int'l L.J. 1 (1988–89).

151. *Id.* at 11.

152. *Id.* at 11.

153. *See* Molly K. Macauley, *In Pursuit of a Sustainable Space Environment: Economic Issues in Regulating Space Debris,* Resources (Resources for the Future), Summer 1993, at 12.

154. Benjamin A. Goldman, *The Truth About Where You Live: An Atlas for Action on Toxins and Mortality* 128–29 (1991).

155. *See Workers Sue Rocket Fuel Ingredient Maker Seeking $75 Million for Renal Cell Cancer,* 7 Toxics L. Rep. (BNA) No. 12, at 344 (Aug. 19, 1992); *Morton Thiokol to Pay $4.65 Million, Complete Cleanup at Goose Farm Waste Site,* 3 Toxics L. Rep. (BNA) No. 11, at 338 (Aug. 10, 1988).

156. 42 U.S.C. § 7503(e).

157. Lenny Siegel, *No Free Launch,* Mother Jones, Sept./Oct. 1990, at 24; *see also* William J. Broad, *New Methods Sought to Dispose of Rockets, With No Harm to Earth,* N.Y. Times, Sept. 17, 1991, at C4.

158. *See* Keith Schneider, *Texas Calls Halt to Waste-Disposal Sites,* N.Y. Times, Feb. 19, 1991, at A12; *North Carolina Waste Commission to Close,* 23 Env't Rep. (BNA) No. 12, July 17, 1992, at 908; *Ontario Outlaws Future Incinerators,* Waste Age, Nov. 1992, at 11; *Rhode Island Bans Incineration, Sets 70% Recycling Rate,* Waste Age, Sept. 1992, at 9.

159. Heiman, *supra* ch. 6 note 43, at 361; *similarly, see* Blumberg and Gottlieb, *supra* ch. 2 note 129, at 77; Lois Marie Gibbs, *Celebrating Ten Years of Triumph,* Everyone's Backyard, Feb. 1993, at 2.

160. E.g., Barbara Dudley (executive director, Greenpeace U.S.), *A Burning Issue for Gore,* Wall St. J., Jan. 20, 1993, at A13.

161. Lake, *supra* ch. 1 note 8, at 88. *Similarly,* Lake and Disch, *supra* ch. 6 note 56, at 665.

162. Philip Shabecoff, *A Fierce Green Fire: The American Environmental Movement* 237 (1993).

Chapter 9

1. Professor Vicki Been, a former resident of one of the volunteer communities cited earlier (Naturita, Colorado), has written to the author, "I would argue that 'culture of risk' is just a euphemism for 'lack of alternatives'. The residents of

Naturita certainly fear HW/RW facilities, but they fear having their kids go hungry even more. . . . The 'cultural' factors . . . are inseparable from those towns' economic dependence upon risky activities." Letter to author, Feb. 18, 1993. This comment illuminates the reasons (and level of enthusiasm) behind Naturita's willingness to accept such a facility, but it does not refute the observation that many other equally needy communities have rejected such facilities, based at least in part on the depth of their fears.

2. *Coming Clean: Superfund Problems Can Be Solved . . .* , *supra* ch. 2 note 51, at 61.

3. *See* Gary Davis, *Shifting the Burden off the Land: The Role of Technical Innovation, in* Piasecki and Davis eds., *supra* ch. 2 note 1, at 43, 44.

4. *See* Hahn, *supra* ch. 5 note 88, at 208 (estimates of costs, per ton, of facilities of different sizes); John F. Williams and Daniel D. Costello, *Orphan Waste: Where Will It Go?* World Wastes, Dec. 1992, at 47 (economics of scale at materials recovery facilities); Kathi A. Mestayer and Paul Radford, *Regional Solid Waste Partnerships: Getting to Yes,* Waste Age, Dec. 1993, at 89 (economies of scale at landfills); Contreras, *supra* ch. 1 note 2, at 522–23 (economies of scale in LLRW facilities); *Partnerships Under Pressure, supra* ch. 3 note 53, at 14 (development of efficient treatment technologies for LLRW may stall because of small waste volumes at decentralized LLRW disposal facilities).

5. Leo Duffy, *Nevada Test Site Proposed As Solution to LLRW Disposal,* Radioactive Exchange, June 21, 1993, at 8.

6. Adrienne Redd, *Regionalization Brings Economies of Scale to East Coast,* World Wastes, July 1993, at 40. A nationwide system of very large regional MSW landfills was proposed in Judd H. Alexander, *In Defense of Garbage* (Praeger 1993) at 209–12.

7. Holman, *supra* ch. 5 note 283, at 22.

8. Gary Davis et al., *Government Ownership of Risk: Guaranteeing a Treatment Infrastructure, in* Piasecki and Davis eds., *supra* ch. 2 note 1, at 95.

9. Id. at 113.

10. John C. Buckley, *Reducing the Environmental Impact of CERCLA,* 41 S.C. L. Rev. 765, 808–10 (1990); Mank, *supra* ch. 6 note 14, at 239–40.

11. *See* McKewen and Sloan, *supra* ch. 4 note 28, at 251 (discusses difficulties "in actually developing a site [near existing contamination] under real-world regulatory conditions in which the greatest imperative seems to be to prove you didn't cause degradation"). *Cf.* 10 C.F.R. § 61.50(a)(2) (importance of ability to characterize, model, and monitor in selection of LLRW disposal sites).

12. Curtis C. Travis and Carolyn B. Doty, *Can Contaminated Aquifers at Superfund Sites Be Remediated?* 24 Envtl. Sci. and Tech. 1464 (1990); Randy M. Mott, *Aquifer Restoration Under CERCLA: New Realities and Old Myths,* 23 Env't Rep. (BNA) No. 18, at 1301 (Aug. 28, 1992); Roger L. Olsen and Michael C. Kavanaugh, *Can Groundwater Restoration Be Achieved?,* 5 Water Envt. and

Tech. No. 3 (March 1993); *Pump-and-Treat Remedy May Be Ineffective at Many Toxic Waste Sites, EPA Official Says,* 6 Toxics L. Rep. (BNA) No. 50, at 1549 (May 20, 1992); *Complex Cleanup, supra* ch. 3 note 8, at 6; Hazardous Waste Cleanup Project, *Technological Reality: The Limits of Technology in Dealing With Hazardous Waste Site Cleanups* (1993).

13. *Complex Cleanup, supra* ch. 3 note 8, at 6–7.

14. *Compare* V. Kerry Smith and William H. Desvousges, *The Valuation of Environmental Risks and Hazardous Waste Policy,* 64 Land Econ. 211 (Aug. 1988) *with* Robert L. Raucher, *The Benefits and Costs of Policies Related to Groundwater Contamination,* 62 Land Econ. 33 (Feb. 1986) (debate concerning wisdom of large expenditures to clean groundwater at Superfund sites).

15. *Clinton Proposes Expedited Cleanup of Military Bases Scheduled for Closure,* 24 Env't Rep. (BNA) No. 10, at 424 (July 9, 1993).

16. Congressional Budget Office, *Environmental Cleanup Issues Associated With Closing Military Bases,* Aug. 1992; Bill Turner and John McCormick, *The Military's Toxic Legacy,* Newsweek, Aug. 6, 1990, at 20.

17. Dan W. Reicher and Jason Salzman, *Cleanup or Buildup: Nuclear Weapons Production in the 21st Century, in Hidden Dangers: Environmental Consequences of Preparing for War* 144, *in* Ehrlich and Birks eds., *supra* ch. 2 note 75.

18. U.S. Gen. Accounting Office, Pub. No. EMD-79-77, *The Nation's Nuclear Waste: Proposals for Organization and Siting* iv–v (1979).

19. Robert Hanley, *Fort Dix May Become Federal Prison,* N.Y. Times, Aug. 30, 1992, at 33.

20. *Environmental Technology Seen as Possible New Focus for Labs That Developed Weapons,* 23 Env't Rep. (BNA) No. 23, at 1514 (Oct. 2, 1992). *See also* Kevin D. Murphy, *Making the Most of a Base Closing,* Governing, Sept. 1993, at 22 (community efforts to convert closed bases for economically productive purposes).

21. Community Environmental Response Facilitation Act, Pub. L. 102-426, 106 Stat. 2174 (1992) (amends CERCLA to require federal government, before termination of federal activities on any real property owned by the government, to identify land where no hazardous waste was stored, released, or disposed of). The issues in reuse of closed bases are discussed in Raymond Takashi Swenson et al., *Resolving the Environmental Complications of Base Closure,* Federal Facilities Envtl. J., Autumn 1992, at 279.

22. *Waste Management to Build Treatment Facility at Department of Energy's Hanford Reservation,* 22 Env't Rep. (BNA) No. 33, at 1971 (Dec. 13, 1991).

23. *Army Set to Burn Wastes at Arsenal,* 23 Env't Rep. (BNA) No. 31, at 1925 (Nov. 27, 1992). Incineration of obsolete chemical weapons at the depots where they are stored was discussed at text accompanying ch. 2 notes 75–87 *supra.*

24. *In the NRC,* Radioactive Exchange, Nov. 2, 1992, at 8.

25. *DOE Looks to Spent Fuel Storage at Federal Facilities,* Radioactive Exchange, Dec. 22, 1992, at 1.

26. Tera Corp., Part 361 Certificate of Environmental Safety and Public Necessity and Supplemental Draft Environmental Impact Statement for the Arc Pyrolysis Project, Model City, Niagara County, New York, 2–5 (July 10, 1985).

27. O'Hare 1983, *supra* ch. 1 note 9, at 144–46.

28. *See* Deborah Cooney et al., *Revival of Contaminated Industrial Sites: Case Studies* (Northeast-Midwest Institute 1992); Hazardous Waste Engineering Research Laboratory, U.S. EPA, Pub. No. EPA/600/2-86/066, *Reclamation and Redevelopment of Contaminated Land: Vol. 1, U.S. Case Studies* (Aug. 1986); Tom Arrandale, *Developing the Decontaminated City,* Governing, Dec. 1992, at 44; Kathleen M. Martin, *Siting on Contaminated Property: Development and Cleanup Through Public/Private Cooperation,* 7 Natural Resources and Envt. No. 3 at 20 (Winter 1993); Jim Ford et al., *Contaminated Sites: An Overlooked Site-Selection Opportunity?* Industrial Development Section, June 1991, at 643; Rodolfo N. Salcedo and Delbert H. Dettmann, *Cities Wrestle With Abandoned Properties,* Pollution Eng'g, June 1, 1993, at 76; *Regional Transportation Center Planned by EPA in Restoration of Massachusetts Superfund Site,* 23 Env't Rep. (BNA) No. 26, at 1641 (Oct. 23, 1992).

29. *See* Elizabeth S. Kiesche, *A Smaller Role for the Chemical Industry in New Jersey,* Chemical Wk., July 22, 1992, at 7 (New Jersey's Environmental Cleanup Responsibility Act, N.J.S.A. §§13:1K-6 *et seq.* (ECRA) has reportedly resulted in the abandonment of several industrial sites); Keith Schneider, *Rules Easing for Urban Toxics Cleanups,* N.Y. Times, Sept. 20, 1993, at A12; Matthew L. Wald, *Trenton Acts to Loosen Industrial Cleanup Law,* N.Y. Times, June 7, 1993, at B1 (New Jersey state legislature weakens ECRA because of concerns about its negative effect on economic development); Mank, *supra* ch. 6 note 14, at 255–56 (regulatory impediments to redevelopment of contaminated property).

30. Farkas, *supra* ch. 4 note 42, at 453 and n.8.

31. *E.g.,* Steven W. Setzer, *Army's Green is More Than Its Uniforms,* Eng'g News-Record, Nov. 30, 1992, at 30; David Evans, *Nuclear Cleanup Falling Into Gap,* Chicago Tribune, Sept. 12, 1991, at 29; Paul Hoversten, *Some Military Bases Will Never Be Cleaned Up,* USA Today, July 5, 1991, at 7A; Michael Satchell, *Uncle Sam's Toxic Folly,* U.S. News and World Rep., Mar. 27, 1989, at 20.

32. C.W. Thornthwaite, *Modification of Rural Microclimates, in* 2 *Man's Role in Changing the Face of the Earth* 567, 572 (William L. Thomas, Jr., ed., 1956).

33. *New Columbia Encyclopedia* 2910 (William H. Harris and Judith S. Levy eds., 1975).

34. Bartlett and Steele, *supra* ch. 3 note 10, at 331–33. A 1958 nuclear waste accident in Kyshtym, Soviet Union, seems to have had the same effect. Id. at 72.

35. *See Safety, Modernization, and Environmental Cleanup of the U.S. Nuclear Weapons Complex, supra* ch. 3 note 52, at 169 (remarks of Sen. Glenn); "some

sites may be irreversibly contaminated, and DOE may have to place them in long-term institutional care," *id.* at 345 (statement of Keith O. Fultz, General Accounting Office).

36. Hazardous Waste Cleanup Project, *Sticker Shock: Recognizing the Full Cost of Superfund Cleanups* (June 1993). *Similarly,* U.S. House of Representatives, Committee on Ways and Means, *1992 Green Book: Background Material and Data on Programs Within the Jurisdiction of the Committee on Ways and Means* (May 15, 1992).

37. Breyer 1993, *supra* ch. 2 note 123, at 55–81.

38. *Id.* at 23. *Similarly, see* Wildavsky, *supra* ch. 8 note 40, at 66.

39. William Glaberson, *Coping in the Age of "NIMBY,"* N.Y. Times, June 19, 1988, § 3, at 1.

40. Crim, *supra* ch. 4 note 41, at 132. *See also* Armour 1991, *supra* ch. 5 note 2, at 20–21 (reporting on another study's findings that neighborhoods that accepted mental health facilities "are those in which residents have few children, are well-educated, and predominantly English-speaking; where the population is relatively transient, the population density relatively high; and where there is a mixture of land uses with commercial development and public open space in addition to residential areas"); Lois Marie Gibbs and Brian Lipsett, *The Siting Game: A NIMBY Primer,* 8 Forum for Applied Research and Public Policy 36, 37 (1993) (report of similar siting effort in North Carolina targeted at politically weak communities).

41. Timothy Noah, *Gore Vows to Block Incinerator Start-Up, Suggesting He'll Play an Activist Role,* Wall St. J., Dec. 8, 1992, at B6.

42. Ann Markusen et al., *The Rise of the Gunbelt: The Military Remapping of Industrial America* 239–42 (1991).

43. Greenberg and Anderson, *supra* ch. 4 note 3, at 166–67; Morell and Magorian, *supra* ch. 1 note 8, at 57–58, 154.

44. Herbert Inhaber, *A Market-Based Solution to the Problem of Nuclear and Toxic Waste Disposal,* 41 J. Air Waste Mgmt. Ass'n 808 (1991); Herbert Inhaber, *Can We Find a Volunteer Nuclear Waste Community?* Pub. Utilities Fortnightly, July 15, 1991, at 19; Herbert Inhaber, *Of LULUs, NIMBYs, and NIMTOOs,* Pub. Interest, Spring 1992, at 52. Other auctionlike mechanisms are described in Mitchell and Carson, *supra* ch. 5 note 88; Swallow et al., *supra* ch. 4 note 49, at 294.

45. Armour 1993, *supra* ch. 6 note 61.

46. *Ibid.*

47. *Ibid.*

48. These studies are reviewed in Gary H. McClelland et al., *The Effect of Risk Beliefs on Property Values: A Case Study of a Hazardous Waste Site,* 10 Risk Analysis 485 (1990). *Similarly,* Benford et al., *supra* ch. 1 note 7.

49. *E.g.*, Keith Schneider, *Safety Fears Prompt Plants to Buy Out Neighbors*, N.Y. Times, Nov. 28, 1990, at 1; Jon Bowermaster, *A Town Called Morrisonville*, Audubon, July–Aug. 1993, at 42; Caleb Solomon, *How a Neighborhood Talked Fina Refinery Into Buying It Out*, Wall St. J., Dec. 10, 1991, at 1; *Neighbors of Texas Plant Offered Buy-Out by Defendant in $100 Million Injury Suit*, 7 Toxics L. Rep. (BNA) No. 16, at 461 (Sept. 16, 1992). Several agencies have developed guidelines for minimum buffer areas around hazardous facilities, especially those with risk of explosion or fire. *See* Greenberg and Anderson, *supra* ch. 4 note 3, at 226; N.Y. Comp. Codes R. and Regs. tit. 6, § 361.7(b)(9)(ii)(a) (New York State hazardous waste facility siting regulations mandate consideration of buffer zones established in the American Table of Distances for Storage of Explosives). The more extensive use of buffer zones is urged in Mank 1992, *supra* ch. 6 note 14. The practice is attacked in Alair MacLean, *Hush Money to Homeowners: Polluters Pay the Buying Game*, Nation, Oct. 25, 1993, at 456.

50. McKewen and Sloan, *supra* ch. 4 note 28.

51. Arthur C. Nelson, et al., *Price Effects of Landfills on House Values*, 68 Land Econ. 359, 364–65 (1992).

52. *See* Brion 1991, *supra* ch. 1 note 2, at 475, 481, 498; Edelstein 1988, *supra* ch. 4 note 61, at 62; John E. Seley and Julian Wolpert, *Equity and Location*, *in* Kasperson ed., *supra* ch. 3 note 4, at 69, 80.

53. *See* Richard Briffault, *Our Localism: Part II—Localism and Legal Theory*, 90 Colum. L. Rev. 346, 420 (1990).

54. Charles W. Johnson and Charles O. Jackson, *City Behind a Fence: Oak Ridge, Tennessee, 1942–1946* 8 (1981).

55. Michele Stenehjem Gerber, *On the Home Front: The Cold War Legacy of the Hanford Nuclear Site* 22–23 (1992).

56. Robert A. Caro, *The Power Broker: Robert Moses and the Fall of New York* 344 (1974).

57. U.S. Tennessee Valley Authority, Technical Report No. 13, *A Comprehensive Report on the Planning, Design, Construction, and Initial Operation of the Kentucky Project* 545–46 (1951). *See also* Poletown Neighborhood Council v. Detroit, 410 Mich. 616, 304 N.W.2d 455 (1981) (upholds condemnation of community in Detroit to make way for new production facility for General Motors).

58. Lynn L. Bergeson, *The SRRP: Making Pollution Prevention Work*, Pollution Eng'g, July 1993, at 73.

59. U.S. Environmental Protection Agency, Office of Solid Waste and Emergency Response, OSWER Directive No. 9010.02, *Guidelines for Capacity Assurance Planning: Capacity Planning Pursuant to CERCLA § 104(c)(9)* (May 1993).

60. Mazmanian and Morell 1992, *supra* ch. 1 note 5, at 196–97.

61. *Cf.* Mason Willrich and Richard K. Lester, *Radioactive Waste Management and Regulation* 119 (1977) (proposed creation of national Radioactive Waste Authority for HLW and TRU).

62. Pub. L. 101-510, 104 Stat. 1808. *See* Dalton v. Specter, 62 U.S.L.W. 4340 (May 23, 1994); County of Seneca v. Cheney, 806 F. Supp. 387 (W.D.N.Y. 1992). *Similarly,* Eric Schmitt, *A Mission Accomplished: In Deciding Which Military Bases to Close, Commission Was a Fortress Against Politics,* N.Y. Times, June 29, 1993, at A10.

63. *See* Natalie Hanlon, *Military Base Closings: A Study of Government by Commission,* 62 U. Colo. L. Rev. 331 (1991); Schmitt, *id.*

64. *See* Green, *supra* ch. 2 note 24, at 91–93 (suggests some possible mechanisms for allocating among the states the obligation to create various amounts of disposal capacity for RCRA HW, largely proportionate to the amounts generated).

65. MacMillan, *supra* ch. 2 note 16, at 29.

66. *See generally* Rabe 1990, *supra* ch. 4 note 61.

67. Frank J. Popper has suggested allowing communities to trade all kinds of locally undesirable land uses; e.g., a neighborhood that agreed to accept a HW facility could decline the next three halfway houses. Frank J. Popper, *LULUs and Their Blockage: The Nature of the Problem, The Outline of the Solutions, in* DiMento and Graymer, *supra* ch. 4 note 41, at 13, 24.

68. There is some evidence that, in a few instances, states have informally allocated burdens among themselves. *See* Colglazier and English, *supra* ch. 1 note 8, at 647 n.28 (Washington and Oregon have a tacit agreement that Washington will continue to host an LLRW facility for the region, and Oregon will continue to host an HW facility). On the other hand, in meetings of the Southeast Compact (Alabama, Florida, Georgia, Mississippi, North Carolina, South Carolina, Tennessee, and Virginia) to pick a host state for an LLRW facility, Alabama contended that, because it had the Emelle HW landfill, it should not have to receive LLRW; the other states rejected this argument, partly because if Emelle were factored in, then existing or proposed large-scale waste facilities in the other states would need to be considered as well. *See* English 1992, *supra* ch. 1 note 8, at 121.

69. *See* 42 U.S.C. § 7410(c)(1).

70. E.g., 40 C.F.R. § 264.18 (location standards for RCRA hazardous waste landfills); 40 C.F.R. § 761.75 (EPA standards for PCB landfills under TSCA); 10 C.F.R. § 61.50 (NRC standards for LLRW facilities). However, many of these regulatory standards may not be sufficiently detailed for this purpose.

71. Seymour Simon et al., *Martinsville: Report of the Illinois Low-Level Radioactive Waste Disposal Facility Siting Commission on its Inquiry into the Martinsville Alternative Site* 449 (Dec. 18, 1992).

72. New York v. United States, 112 S.Ct. 2408 (1992).

73. 42 U.S.C. §§ 7503(c), 7511a(a)(4), 7511a(b)(5), 7511a(c)(10), 7511a(d)(2), 7511a(e)(1).

Chapter 10

1. *Facing America's Trash: What's Next for Municipal Solid Waste? supra* ch. 2 note 130, at 342.

2. *See* Margaret A. Walls and Barbra L. Marcus, *Should Congress Allow States to Restrict Waste Imports?* Resources (Resources for the Future), Winter 1993, at 7 (argues that host community fees internalize the external costs of siting waste facilities).

3. *See* 42 U.S.C. § 9607(a)(4).

4. *See* National Solid Wastes Management Ass'n v. Alabama Dep't of Envtl. Management, 910 F.2d 713, 720 (11th Cir. 1990), *modified*, 924 F.2d 1001 (11th Cir.), *cert. denied*, 111 S.Ct. 2800 (1991) (complaints from Alabama about truck traffic to Emelle facility); City of New York v. United States Dep't of Transp., 715 F.2d 732 (2d Cir. 1983) (New York City's unsuccessful attempt to prevent transportation of nuclear waste from Brookhaven National Laboratory through city highways), *cert. denied*, 465 U.S. 1055 (1984). *See generally* Stan Millan and Andrew J. Harrison, Jr., *A Primer on Hazardous Materials Transportation Law of the 1990s: The Awakening*, 22 Envtl. L. Rep. (Envtl. L. Inst.) 10583 (1992); Edward A. Nolfi, Annotation, *State or Local Regulation of Transportation of Hazardous Materials as Pre-Empted by Hazardous Materials Transportation Act (49 USCS §§ 1801 et seq.)*, 78 A.L.R. Fed. 289.

5. Rae Zimmerman, *Public Acceptability of Alternative Hazardous Waste Management Services, in* Peck ed., *supra* ch. 5 note 49, at 207; Cong. Rec. H11081 (daily ed. Dec. 5, 1985) (statement of Rep. Lagomarsino concerning trucking of hazardous waste to facility in his district); Gerald Jacob and Andrew Kirby, *On the Road to Ruin: The Transport of Military Cargoes* 71 *in* Ehrlich and Birks eds., *supra* ch. 2 note 75; Matthew L. Wald, *L.I. Agency Drops Plan to Ship Shoreham Fuel Via New York,* N.Y. Times, Sept. 15, 1993, at B6.

6. Office of Tech. Assessment, U.S. Cong., Pub. No. OTA-SET-304, *Transportation of Hazardous Materials* 103 (1986). Similarly, NRC has said that there are 1.3×10^{-6} accidents per kilometer of truck traffic. *Final Environmental Impact Statement . . . Clive, Utah, supra* ch. 5 note 325, at 5–16. EPA has devised formulas for the fraction of the annual quantity of liquids expected to be released in transportation mishaps. For bulk liquids carried in tanker trucks, the fraction released is predicted to be $(9.5 \times 10^{-8} \times D) + (7.6 \times 10^{-6})$; for liquids contained in drums on flatbed trucks, the fraction is $(2.4 \times 10^{-6} \times D) + (2.9 \times 10^{-4})$ (where D is the distance, in miles, to the treatment or disposal facility.). U.S. Environmental Protection Agency, Office of Policy, Planning and Evaluation and Office of Solid Waste, *Pollution Prevention Benefits Manual* C-2 (Oct. 1989).

7. *Transportation of Hazardous Materials Hearings, supra* ch. 5 note 326, at 273. *See also id.* at 760, 780 (six accidents involving spent fuel rods).

8. *Id.* at 274.

9. *Transportation of Hazardous Materials, supra* ch. 10 note 6, at 35, 41.

10. U.S. Environmental Protection Agency, Office of Policy Analysis, *1986–1987 Survey of Selected Firms in the Commercial Hazardous Waste Management Industry: Final Report* 3–19 (Mar. 31, 1988). *Cf.* Hammitt and Reuter, *supra* ch. 5 note 102 (uses figure of $0.20 per ton per mile).

11. *Hazardous Waste Industry Grows With Public Concern,* Chemical Marketing Rep., Jan. 26, 1991, at 9. This report said that consulting and engineering services were the second largest market segment, with revenues of $1.2 billion, followed by remediation services, at $920 million. A much different breakdown (but still showing transportation as a major market segment) is presented in *No Quick Recovery for Hazardous Waste,* Chemical Wk., Jan. 6/13, 1993, at 26.

12. E.g., Sewall, *supra* ch. 5 note 17, at 78–87; R. Batta and S.S. Chiu, *Optimal Obnoxious Paths on a Network: Transportation of Hazardous Materials,* 36 Operations Research 34 (1988); Cerry M. Klein, *A Model for the Transportation of Hazardous Waste,* 22 Decision Sci. 1091 (1991); Ram Gopalan et al., *Modeling Equity of Risk in the Transportation of Hazardous Materials,* 38 Operations Research 961 (1990).

13. E.g., List et al., *supra* ch. 1 note 9; George List and Pitu Mirchandani, *An Integrated Network/Planar Multiobjective Model for Routing and Siting for Hazardous Materials and Wastes,* 25 Transp. Sci. 146 (1991); Charles ReVelle et al., *Simultaneous Siting and Routing in the Disposal of Hazardous Wastes,* 25 Transp. Sci. 138 (1991).

14. Marcus V. Voth and Warren F. Witzig, *Determination of Optimum Alternative Low-Level Radioactive Waste Disposal Site/Disposal Technology Combinations,* 78 Nuclear Tech. 312 (1987).

15. MacMillan, *supra,* ch. 2 note 16, at 34.

16. Sagoff, *supra* ch. 5 note 1, at 46.

17. *See* Margaret Kriz, *Slow Burn,* National J., April 3, 1993, at 811, 813 (reporting on study by incinerator company showing that commercial hazardous waste incinerators would have much lower emissions than would steel mills, chemical plants, and coal-fired power plants).

18. *See* Markusen et al. *supra* ch. 9 note 42.

19. The demographics of these beneficiaries of hazardous waste generation are unknown. I am unaware of any studies concerning the income distribution impacts of higher hazardous waste disposal prices, for example. Such work has been done for the costs of air pollution and water pollution control. *See* A. Myrick Freeman III, *The Incidence of the Costs of Controlling Automotive Air Pollution, in The Distribution of Economic Well-Being* 163 (F. Thomas Juster ed., 1977); Elizabeth E. Lake et al., *Who Pays for Clean Water? The Distribution of Water Pollution Control Costs* (1979).

20. E.g., National Solid Wastes Management Ass'n v. Alabama Dep't of Envtl. Management, 910 F.2d 713, 717 n.6 (11th Cir. 1990), *modified,* 924 F.2d 1001 (11th Cir.), *cert. denied,* 111 S.Ct. 2800 (1991) (Alabama); State of Arizona v.

Motorola, Inc., 774 F. Supp. 566, 575 (D. Ariz. 1991) (Arizona); William Poole, *Gambling With Tomorrow,* Sierra, Sept./Oct. 1992, at 50, 52 (Nevada); Polsgrove, *supra* ch. 5 note 11, at 22 (Texas); Government Suppliers Consol. Servs. v. Bayh, 753 F. Supp. 739, 745 (S.D. Ind. 1990) (Indiana); Maritza Peck, *How to Save Your Neighborhood, City, or Town: The Sierra Club Guide to Community Organizing* 67 (1993) (North Carolina). In 1990, thirteen importing states formed a group, States for Responsible and Equitable Waste Management, because they were "tired of being the country's hazardous waste dumping grounds." Mazmanian and Morell 1992, *supra* ch. 1 note 5, at 139.

21. Matthijs Hisschemoller and Cees J.H. Midden, *Technological Risk, Policy Theories and Public Perception in Connection With the Siting of Hazardous Facilities,* in Vlek and Cvetkovich eds., *supra* ch. 5 note 274, at 173.

Chapter 11

1. Melinda Merriam, *Waste Project Lures Hard-Luck Areas,* High Country News, Jan. 27, 1992, at 15.

2. Armour 1993, *supra* ch. 6 note 61; Rabe 1992, *supra* ch. 5 note 202, at 131.

3. *See* Elaine Vaughan and Brenda Nordenstam, *The Perception of Environmental Risks Among Ethnically Diverse Groups,* 22 J. Cross-Cultural Psychology 29 (1991).

4. Richard W. Stoffle et al., *Risk Perception Mapping: Using Ethnography to Define the Locally Affected Population for a Low-Level Radioactive Waste Storage Facility in Michigan,* 93 American Anthropologist 611 (1991).

5. *Aging Nuclear Power Plants, supra* ch. 3 note 37, at 99.

Index

About the Author

Michael B. Gerrard is a partner in the New York City law firm of Berle, Kass & Case, where he has practiced environmental law since 1978. He is also a member of the adjunct faculty of Columbia Law School, and co-chair of New York University's annual Summer Institute in Environmental Law. He is General Editor of the six-volume *Environmental Law Practice Guide* (Matthew Bender 1992), which the Association of American Publishers named the best law book of 1992. He is co-author of *Environmental Impact Review in New York* (Matthew Bender 1990); co-author, since 1986, of the monthly environmental law column in the *New York Law Journal*; and editor, since 1989, of a monthly newsletter, *Environmental Law in New York*. He is a member of the Executive Committee of the Association of the Bar of the City of New York, and former chair of its Committee on Land Use Planning and Zoning; and Vice Chair of the Environmental Law Section of the New York State Bar Association. He is a member of the boards of the Council on the Environment of New York City; the Environmental Planning Lobby; and the Westchester Land Trust. He received his B.A. from Columbia University and his J.D. from New York University.